Textile Science and Clothing Technology

Series Editor

Subramanian Senthilkannan Muthu, Chief Sustainability Officer, Green Story, Canada, Tung Chung, Lantau, Hong Kong

This series aims to broadly cover all the aspects related to textiles science and technology and clothing science and technology. Below are the areas fall under the aims and scope of this series, but not limited to: Production and properties of various natural and synthetic fibres; Production and properties of different yarns, fabrics and apparels; Manufacturing aspects of textiles and clothing; Modelling and Simulation aspects related to textiles and clothing; Production and properties of Nonwovens; Evaluation/testing of various properties of textiles and clothing products; Supply chain management of textiles and clothing; Aspects related to Clothing Science such as comfort; Functional aspects and evaluation of textiles; Textile biomaterials and bioengineering; Nano, micro, smart, sport and intelligent textiles; Various aspects of industrial and technical applications of textiles and clothing; Apparel manufacturing and engineering; New developments and applications pertaining to textiles and clothing materials and their manufacturing methods; Textile design aspects; Sustainable fashion and textiles; Green Textiles and Eco-Fashion; Sustainability aspects of textiles and clothing; Environmental assessments of textiles and clothing supply chain; Green Composites; Sustainable Luxury and Sustainable Consumption; Waste Management in Textiles; Sustainability Standards and Green labels; Social and Economic Sustainability of Textiles and Clothing.

Mohammad Neaz Morshed · Mohammad Shahid
Editors

Advances in Surface Modification and Functionalization in Textiles

Springer

Editors
Mohammad Neaz Morshed
Department of Textile Technology
Faculty of Textiles, Engineering, and Business
The Swedish School of Textiles
University of Borås
Borås, Sweden

Mohammad Shahid
Department of Chemistry
Koneru Lakshmaiah Education Foundation
(Deemed to be University)
Guntur, Andhra Pradesh
India

ISSN 2197-9863 ISSN 2197-9871 (electronic)
Textile Science and Clothing Technology
ISBN 978-3-032-23376-9 ISBN 978-3-032-23377-6 (eBook)
https://doi.org/10.1007/978-3-032-23377-6

This Springer imprint is published by the registered company Springer Nature Switzerland AG
The registered company address is: Gewerbestrasse 11, 6330 Cham, Switzerland

Preface

Textile functionalization and surface modification technologies have fundamentally transformed the textile industry, enabling the creation of advanced materials with tailored properties that extend far beyond traditional uses. At the core of this transformation lies the integration of chemical, physical, biotechnological, and nanotechnological approaches, which allow precise manipulation of textile surfaces to impart functionalities such as antimicrobial activity, water repellency, UV protection, flame retardancy, self-cleaning behavior, and even electronic responsiveness while preserving the comfort, flexibility, and processability that make textiles indispensable in everyday life. Among these advances, nanotechnology has played a particularly pivotal role. The incorporation of nanomaterials, such as titanium dioxide, silver, zinc oxide, and copper nanoparticles, has revolutionized textile functionalization by enabling durable, high-performance surface effects at low material loadings. Titanium dioxide, for instance, is widely valued for its photocatalytic self-cleaning, antimicrobial, and UV-protective properties, combined with its nontoxicity and cost-effectiveness. Recent developments have focused on enhancing its visible-light activity and long-term durability through multiphase heterojunctions, ion and metal doping, and coupling with complementary semiconductors. In parallel, functional coating technologies including dip-coating, spray-coating, sol–gel processing, and layer-by-layer assembly have enabled the controlled integration of flame retardants, antimicrobial agents, and self-cleaning materials, increasingly drawing on bioactive substances such as chitosan and plant-derived antimicrobials to improve sustainability alongside performance. Grafting modification has emerged as another powerful strategy, allowing the covalent attachment of functional polymers to textile substrates to impart hydrophobicity, antimicrobial activity, UV resistance, flame retardancy, and improved dyeability. These chemical approaches are complemented by physical and plasma treatments, electrospinning, and microencapsulation techniques, which expand the surface modification toolkit while minimizing changes to bulk textile properties. Collectively, these technologies have enabled the emergence of multifunctional and smart textiles with applications spanning healthcare, protective clothing, environmental monitoring, wearable electronics, and beyond.

In healthcare, surface-functionalized textiles are increasingly used in wound dressings, drug delivery systems, bioactive face masks, and biosensors, where tailored surface chemistry enhances biocompatibility, antimicrobial efficacy, and controlled release behavior. In protective and performance textiles, integrating flame-retardant, antimicrobial, and self-cleaning capabilities enhances safety and durability in military, industrial, and sports applications. Phosphorus-based flame retardants and photocatalytic agents such as titanium dioxide and silica nanoparticles exemplify how high performance can be achieved alongside improved environmental compatibility. At the same time, the rise of wearable electronics and smart textiles enabled by conductive materials, nanostructures, and embedded sensors has opened new pathways for real-time monitoring, adaptive responses, energy harvesting, and interactive textile systems. Sustainability remains a central concern in the development of advanced functional textiles. Increasing emphasis is being placed on green chemistry, biodegradable materials, and bio-based functionalization strategies to reduce the environmental footprint of textile processing. Enzymatic treatments, natural colorants, and biosynthesized nanoparticles are being actively explored as alternatives to conventional chemical routes. Despite significant progress, challenges persist in terms of scalability, cost-effectiveness, durability of functional finishes, compatibility with existing manufacturing processes, and the potential environmental and health impacts of certain nanomaterials. Regulatory compliance and standardized testing protocols are therefore becoming increasingly important, particularly as functionalized textiles enter sensitive application areas such as healthcare and personal protection.

This book, *Advances in Surface Modification and Functionalization in Textiles*, brings together contributions from leading researchers and practitioners to address these developments in a coherent and comprehensive manner, with each chapter focusing on a specific aspect of theory, technology, or application in textile surface engineering. Chapter 1 introduces the fundamental concepts, theories, and principles of textile functionalization, establishing the physicochemical basis for surface- and bulk-modification strategies. Chapter 2 explores the role of surface chemistry and surface physics in textile functionalization, highlighting interfacial interactions, surface energy, wettability, and structure–property relationships. Chapter 3 presents key physicochemical techniques for textile surface modification, detailing methods such as sol–gel processing, grafting, and plasma treatments, along with their advantages and limitations. Chapter 4 focuses on oxidative surface modification approaches, including chemical, plasma, ozone, photocatalytic, electrochemical, and enzymatic oxidation routes, with emphasis on performance enhancement and sustainability. Chapter 5 discusses resource-efficient and environmentally conscious processes for textile functionalization, evaluating emerging technologies such as plasma treatment, inkjet printing, electrospinning, and biofunctionalization in terms of efficiency and scalability. Chapter 6 is dedicated to supercritical CO_2 processing for textile surface modification and functionalization, reviewing its principles, applications, and potential for water- and solvent-free textile processing. Chapter 7 examines the growing role of 2D and 3D printing technologies in textile surface functionalization, demonstrating how printed structures and coatings enable multifunctionality and smart textile applications. Chapter 8 reviews emerging nanomaterials used in textile surface modification, covering their integration methods, functional performance, sustainability considerations, and future industrial potential. Chapter 9 addresses instrumental analysis and characterization techniques essential for understanding surface-specific changes in modified textiles, with emphasis on surface chemistry, morphology, wettability, and functional performance. Chapter 10 explores the application of machine learning and artificial intelligence in textile surface modification, highlighting data-driven process optimization, property prediction, and defect detection.

By integrating perspectives from materials science, chemistry, physics, engineering, and data-driven modeling, this volume addresses the key challenges and opportunities outlined above, including advanced surface functionalization strategies, nanomaterial-enabled performance enhancement, sustainability and resource efficiency, durability and scalability of treatments, rigorous surface characterization, and the growing role of digital and intelligent tools in textile processing. The chapters collectively demonstrate how fundamental theory, innovative surface modification techniques, environmentally responsible processes, and data-driven approaches can be combined to design next-generation functional textiles. We hope this book will serve as a comprehensive reference and a practical guide for graduate students, researchers, engineers, and professionals seeking to develop high-performance, sustainable, and intelligent textile materials.

Borås, Sweden — Mohammad Neaz Morshed
Guntur, India — Mohammad Shahid

Contents

Editors and Contributors

About the Editors

Dr. Mohammad Neaz Morshed, Ph.D. is a researcher at the Department of Textile Technology under the Faculty of Textiles, Engineering, and Business (including The Swedish School of Textiles) at the University of Borås, Sweden. He holds Ph.D. degree in Mechanics of Solids, Materials, Structures and Surfaces from University of Lille (France), Ph.D. in Textile Materials Technology from University of Borås (Sweden) and Ph.D. in Textile Engineering from Soochow University (China) under the framework of Erasmus Mundus joint doctorate fellowship program about sustainable management and design for textiles (SMDTex). Dr. Morshed is a recognized expert in material chemistry, focusing on surface modification and functionalization of textiles, with an emphasis on eco-friendly, resource-efficient preparation methods. He teaches textile chemistry, textile biotechnology, textile functionalization, and textile finishing courses in various capacities at the University of Boras. Dr. Morshed has published over 25 scientific publications (~700 citations with h-index 16) encompassing peer-reviewed journal articles, reviews, and book chapters. e-mail: mohammad_neaz.morshed@hb.se

Dr. Mohammad Shahid, Ph.D. is an assistant professor in the Department of Chemistry, Koneru Lakshmaiah Education Foundation, India. Before joining KLEF, he worked at various esteemed institutions, including Jamia Millia Islamia, New Delhi; Soochow University, China; Henan University, China; University of Glasgow, UK; Institute of Chemical Technology, Mumbai; Dr. K. N. Modi University, Tonk; and Mewar University, Chittorgarh. He earned his Master's degree in Chemistry from Shibli National College, Azamgarh, in 2006 and his Ph.D. in Chemistry from Jamia Millia Islamia, New Delhi, in 2014. Dr. Shahid has received several prestigious fellowships to support his research pursuits, including grants from the UGC, India, the China Postdoctoral Fellowship, the EU Marie Curie Fellowship, and the Dr. D. S. Kothari Fellowship. He has contributed to over 50 scientific publications with ~4600 citations and an h-index of 27. In addition, has also edited and authored several books with renowned international publishing houses.

Contributors

Molla Tadesse Abate Engineering Department, Division Sustainable Textiles, Hof University of Applied Sciences, Hof, Germany;
Ethiopian Institute of Textile and Fashion Technology, Bahir Dar University, Bahir Dar, Ethiopia

Faisal Ahmed Department of Textile Engineering, BGMEA University of Fashion and Technology (BUFT), Dhaka, Bangladesh

Mahmuda Akter Department of Apparel Engineering, Bangladesh University of Textiles, Dhaka, Bangladesh

Nahida Akter Department of Islamic History and Culture, Government Debendra College Manikganj, National University, Gazipur, Bangladesh

Abdulrhman Hamdy Mustafa Aly Ethiopian Institute of Textile and Fashion Technology, Bahir Dar University, Bahir Dar, Ethiopia

A. M. Ehsanul Haque Department of Textile Machinery Design and Maintenance, Bangladesh University of Textiles, Dhaka, Bangladesh

S. M. Fijul Kabir Engineering (Fibers), Procter and Gamble, Iowa City, IA, USA

Md. Motakabbir Hasan Department of Wet Process Engineering, Bangladesh University of Textiles, Dhaka, Bangladesh

Md. Hasibul Hasan Department of Apparel Engineering, Bangladesh University of Textiles, Dhaka, Bangladesh

Md. Himel Mahmud Department of Textile Engineering, Northern University Bangladesh, Ashkona, Dhaka, Dakshinkhan, Bangladesh

Ibrahim Hossain Department of Apparel Engineering, Bangladesh University of Textiles, Dhaka, Bangladesh

Md. Rashedul Islam Department of Wet Process Engineering, Bangladesh University of Textiles, Dhaka, Bangladesh

Saiful Islam Department of Textile Engineering, Atish Dipankar University of Science and Technology, Dhaka, Bangladesh

Tariful Islam Department of Textile Engineering, University of Scholars, Dhaka, Bangladesh

Md. Kamrul Hassan Chowdhury Department of Apparel Engineering, Bangladesh University of Textiles, Dhaka, Bangladesh

Ayub Nabi Khan Department of Textile Engineering, BGMEA University of Fashion and Technology (BUFT), Dhaka, Bangladesh

Jörn Felix Lübben Sustainable Engineering (STE), Albstadt-Sigmaringen University, Albstadt, Germany

Sakil Mahmud Department of Chemistry and Physics, Ivory V. Nelson Center for the Sciences, Lincoln University, Oxford, PA, USA;
Department of Textile Engineering, Faculty of Engineering, Daffodil International University, Dhaka, Bangladesh

Shah Md. Maruf Hasan Department of Apparel Engineering, Bangladesh University of Textiles, Dhaka, Bangladesh

Badhon C. Mazumder Department of Textile Engineering, Northern University Bangladesh, Ashkona, Dhaka, Dakshinkhan, Bangladesh

Mohammad Raza Miah Department of Textile Engineering, BGMEA University of Fashion and Technology (BUFT), Dhaka, Bangladesh;
Key Laboratory of Bio-Based Polymeric Materials Technology and Application of Zhejiang Province, Ningbo Institute of Materials Technology and Engineering, Chinese Academy of Sciences, Ningbo, People's Republic of China

Mohammad Neaz Morshed Department of Textile Technology, Faculty of Textiles, Engineering and Business (Including the Swedish School of Textiles), University of Borås, Borås, Sweden

Md. Nahid-Ull-Islam Department of Apparel Engineering, Bangladesh University of Textiles, Dhaka, Bangladesh

Michael Rauch Engineering Department, Division Sustainable Textiles, Hof University of Applied Sciences, Hof, Germany

Md. Reazuddin Repon Department of Bioproducts and Biosystems, School of Chemical Engineering, Aalto University, Espoo, Finland

Mustakimul Hossain Saeef Department of Textile Materials Engineering, Bangladesh University of Textiles, Dhaka, Bangladesh

Abu Saifullah Portsmouth Centre for Advanced Materials and Manufacturing, Walthamstow, London, UK

Siam Sarker Department of Apparel Engineering, Bangladesh University of Textiles, Dhaka, Bangladesh

Sina Seipel Department of Textile Technology, Faculty of Textiles, Engineering and Business (Including the Swedish School of Textiles), University of Borås, Borås, Sweden

Md. Shohave Sarkar Department of Textile Engineering, Northern University Bangladesh, Dhaka, Bangladesh

Melkie Getnet Tadesse Sustainable Engineering (STE), Albstadt-Sigmaringen University, Albstadt, Germany;
Ethiopian Institute of Textile and Fashion Technology, Bahir Dar University, Bahir Dar, Ethiopia

Nowrin Ahmed Urmi Department of Textile Engineering, Northern University Bangladesh, Dhaka, Bangladesh

Jin Zhu Key Laboratory of Bio-Based Polymeric Materials Technology and Application of Zhejiang Province, Ningbo Institute of Materials Technology and Engineering, Chinese Academy of Sciences, Ningbo, People's Republic of China

Chapter 1
Theory and Principle of Textile Functionalization

Mohammad Neaz Morshed and Sakil Mahmud

Abstract Textile functionalization represents a transformative advancement in chemistry and textile engineering that surpasses the inherent limitations of conventional textiles. By incorporating advanced physical, chemical, and biological treatments along with smart material systems, textiles are designed to actively respond to environmental stimuli, enhance user performance, and provide multifunctional protection. This functional evolution of textiles enables the development of textiles with integrated properties such as flame retardancy, radiation protection, microbial protection, antistatic properties and electrical conductivity. The advancement of textile functionalization relies on interdisciplinary approaches at the crossroads of theory and principles of surface and bulk modification, as well as hybrid techniques. This chapter begins by defining textile functionalization, its importance, and the physicochemical basis of functionalization, and then proceeds to a discussion of the theoretical foundations and key principles. It presents an overview of the theory and principles of textile functionalization. It concludes by describing the critical role of these theoretical foundations in guiding the design and development of functional textiles and highlighting the potential of textile functionalization to revolutionize their applications across diverse industries.

Keywords Functional textiles · Surface modification · Sustainability · Textile engineering

M. N. Morshed (✉)
Department of Textile Technology, Faculty of Textiles, Engineering and Business (Including the Swedish School of Textiles), University of Borås, Borås, Sweden
e-mail: mohammad_neaz.morshed@hb.se

S. Mahmud
Ivory V. Nelson Center for the Sciences, Department of Chemistry and Physics, Lincoln University, Oxford, PA, USA

Department of Textile Engineering, Faculty of Engineering, Daffodil International University, Dhaka, People's Republic of Bangladesh

S. Mahmud
e-mail: smahmud@lincoln.edu

M. N. Morshed and M. Shahid (eds.), *Advances in Surface Modification and Functionalization in Textiles*, Textile Science and Clothing Technology,
https://doi.org/10.1007/978-3-032-23377-6_1

1.1 Introduction

The primary purpose of textiles has historically been to cover the human body, but as science has advanced, clothing has taken on new purposes beyond simple covering. It is now feasible for textiles to respond to environmental factors such as humidity, microbes, pH, temperature, radiation, light, wind, and more, thanks to ongoing advances in chemistry and textile engineering. Durability used to be the primary determinant of textile choice, with comfort and aesthetic appeal coming in second and third. Users today, however, are interested in textiles with additional functions, including but not limited to microbial protection; fire retardance; UV blocking; radiation protection; water-, oil-, and soil-repellent; sensing; antistatic and electrical conductivity, etc., along with their conventional determinants. Textiles with any of the additional functional properties, along with their inherent characteristics, are referred to as *Functional Textiles*. According to Dominique and Crégo [1], "functional textile material is a material to which a specific function is added via the material, its composition, its construction, and/or its finish".

Functional textiles are user-governed, engineered products designed to meet the user's performance requirements. According to a 2025 growth market report, the global functional textiles market in 2024 was valued at USD 219.4 billion. The market is expected to grow at a compound annual growth rate (CAGR) of 8.2% from 2025 to 2033, potentially reaching USD 438.7 billion by the end of the forecast period, driven by ongoing technological innovation and rising demand across key end-user industries. This upward trajectory is supported by increasing investment in smart and high-performance textile solutions, expanding applications in healthcare, sports, and defense, as well as in clothing and coverings, and growing consumer interest in multifunctional, advanced fabrics, as highlighted in Fig. 1.1 [2].

A significant factor contributing to this growth is the rapid adoption of cutting-edge technologies, including nanotechnology, biotechnology, and the integration of electronics into textile production. These advancements are enabling the creation of functional textiles that can sense and respond to environmental changes, thereby improving wearer safety, comfort, and performance [3]. In particular, the rise of

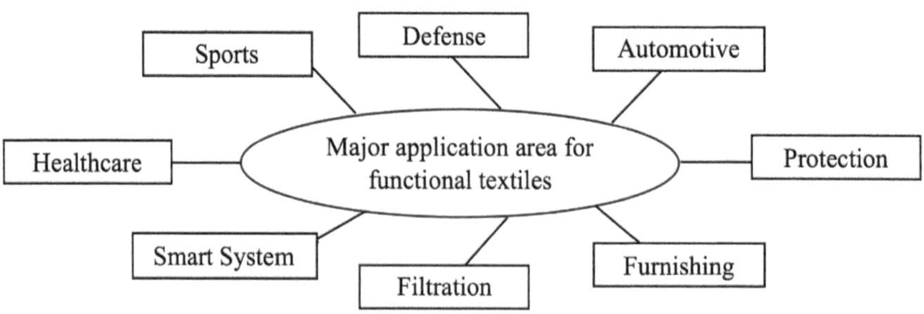

Fig. 1.1 Overview of major application areas for functional textile materials along with clothing and covering

medical textiles with antimicrobial, antiviral, and healing capabilities, spurred in part by recent global health challenges, has driven widespread adoption in healthcare settings. Simultaneously, the sports and fitness segment remains a vital driver of market growth, as athletes and active consumers seek textiles with advanced features such as moisture-wicking, UV protection, and thermal regulation. The growing global fitness trend and participation in outdoor activities have boosted demand for textiles that enhance physical performance and prevent injury [4]. The other segments of the functional textiles market span a broad spectrum of applications, including automotive interiors, filtration systems, protection and defense, intelligent systems, and home furnishings. These areas are experiencing growing demand driven by heightened focus on energy efficiency, environmental sustainability, and improved user comfort. In the automotive industry, for instance, textiles designed for better insulation and noise reduction are becoming increasingly important, particularly in electric vehicles [5]. Similarly, filtration textiles play a vital role in environmental management and various industrial operations [6]. The continued expansion of functional textile applications across these diverse sectors highlights their flexibility, performance benefits, and capacity to meet evolving industry needs. While functional textiles are defined by the advanced properties they offer, understanding how these functionalities are introduced reveals the critical processes involved in textile functionalization.

Textile functionalization is the process of modifying the properties of textiles to impart specific, desired functionalities beyond their basic mechanical properties, transforming them into materials that react with environmental elements. [3, 7, 8]. At its core, textile functionalization involves integrating functional agents/materials, such as nanoparticles, polymers, enzymes, or chemical compounds, into the textile structure [9, 10]. These agents can be incorporated through surface modification, bulk modification, or a combination of both (Fig. 1.2).

Surface modification is the process of altering the outermost layer of textile fibers or fabrics to introduce specific functional properties while maintaining the textile's inherent structure [11]. The goal of surface modification is to impart new characteristics without significantly altering the textile's bulk properties. This method is particularly valuable because it enables targeted performance improvements without compromising the fabric's feel, strength, or appearance. This can be done using a range of physical, chemical, biotechnological, or hybrid techniques. The most commonly used strategy for surface modification is coating and lamination. Coatings are typically applied as thin layers of functional agents, such as polymers, nanoparticles, or biomaterials, onto the textile surface [12]. Lamination, on the other hand, involves bonding a protective film to the textile surface, as seen in applications such as protective clothing or industrial textiles that require enhanced durability against environmental factors [13, 14]. Surface modification offers several advantages, including the ability to precisely control specific properties without altering the overall textile structure [15]. It is also a relatively cost-effective method that can be applied to finished textiles, allowing for on-demand functionalization. However, a key challenge is that surface treatments can wear off over time, particularly from repeated washing, abrasion, or exposure to environmental elements. As a result, the longevity of surface-modified textiles may be shorter than that of bulk-modified

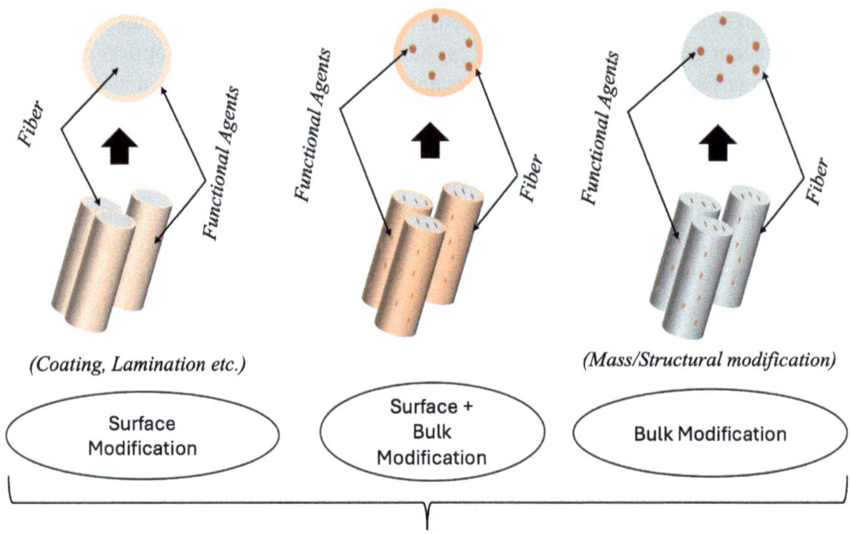

Fig. 1.2 Schematic illustration of textile functionalization involving surface modification, bulk modification, or a combination of both to integrate functional agents into the fabric structure

fabrics. Bulk modification, in contrast to surface modification, involves altering the entire structure of the textile fiber by incorporating functional agents directly into the fiber during the manufacturing process. This approach ensures that the functional agents are evenly distributed throughout the fiber matrix, providing a more durable, long-lasting effect. Bulk modification is advantageous when the textile must maintain its functionality over time, even under harsh conditions such as frequent washing, mechanical stress, or exposure to UV radiation. One of the primary methods of bulk modification is the incorporation of functional agents during fiber spinning. In this technique, functional agents such as nanoparticles, flame retardants, conductive materials, or antimicrobial agents are mixed into the polymer solution before the fiber is spun [16]. This results in a fiber that has inherent functional properties distributed uniformly throughout its structure. For example, adding silver nanoparticles during the spinning of polyester fibers can produce fabrics with permanent antimicrobial properties, ideal for medical or sportswear applications [17]. The choice of modification method depends on the desired functionality, end-use application, durability requirements, and cost considerations. Effective functionalization requires a deep understanding of theoretical foundations in materials science, surface chemistry, polymer physics, molecular interactions, and textile engineering, making it a unique multidisciplinary field central to the development of next-generation textiles.

1.2 Theoretical Foundations of Textile Functionalization

The theoretical foundation of textile functionalization is rooted in the principles of materials science, surface chemistry, polymer physics, and textile engineering. Key concepts include surface energy, molecular adhesion, diffusion mechanisms, inter- and intramolecular forces, colloid and interface science, fiber morphology, and polymer chain dynamics.

Surface energy refers to the excess energy present at the surface of a material compared to its bulk. Manipulating surface energy is essential for textile functionalization because it influences how well a functional agent spreads, adheres, or binds to the fabric. High-energy surfaces tend to be more wettable and allow better adhesion of coatings. In contrast, low-surface-energy surfaces may require surface activation to improve binding, as illustrated in Fig. 1.3a. *Molecular adhesion* describes the forces that enable functional molecules or particles to stick to textile fibers (Fig. 1.3b). This adhesion can occur through intermolecular and intramolecular interactions, ensuring that functional agents adhere effectively to fibers and maintain their performance over time [18]. Intermolecular forces act between molecules and include weak interactions, such as van der Waals forces, as well as stronger interactions, such as hydrogen bonds.

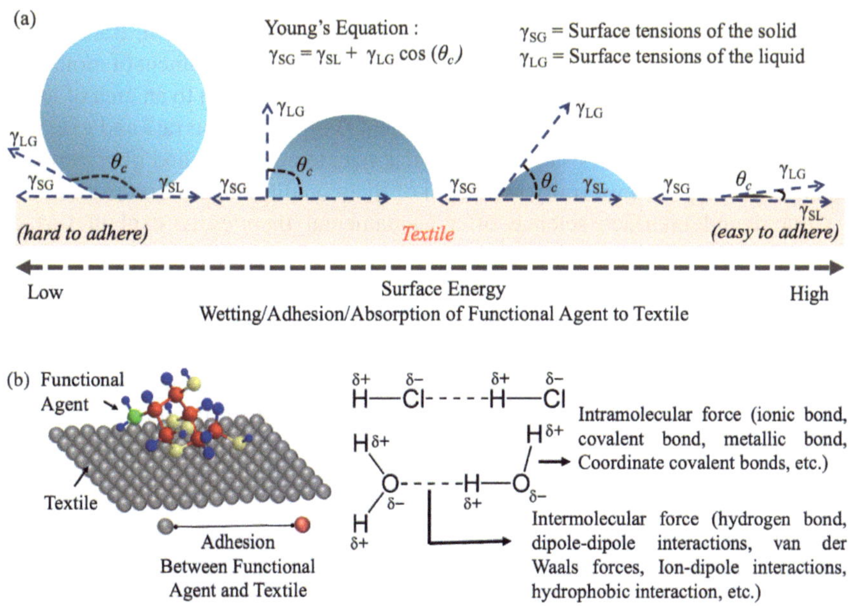

Fig. 1.3 Schematic illustration of **a** a basic relationship between contact angle, wetting, adhesion, absorption of functional agent to textile, and surface energy; **b** molecular adhesion between functional agents and textile

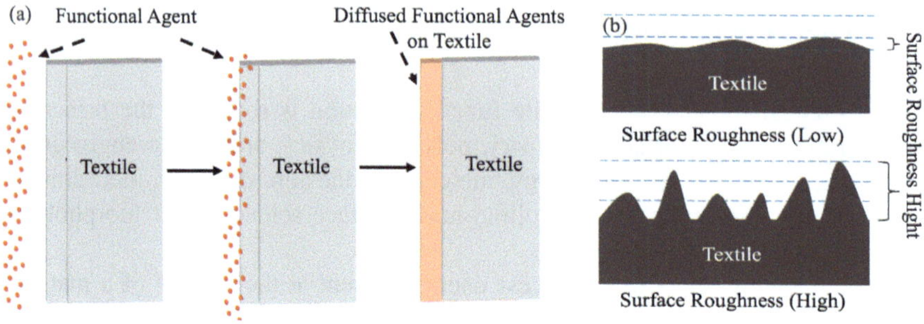

Fig. 1.4 Schematic illustration of **a** diffusion of functional agent on textile, and **b** surface roughness of textile

On the other hand, intramolecular forces such as covalent bonds maintain the structural integrity of the functional agents themselves (Fig. 1.3b). These forces ensure that active compounds retain their intended shape, stability, and chemical reactivity during application and under real-world use conditions. Covalent bonding can also occur between the functional agents and the textile, creating strong, permanent attachments that enhance the treatment's durability and wash resistance. Such covalent linkages are significant in high-performance applications, where long-term effectiveness is required despite repeated laundering or environmental exposure.

Diffusion mechanisms play a key role in textile functionalization, transporting functional agents into and throughout textiles. It refers to the movement of molecules from an area of higher concentration (e.g., on the textile surface) to an area of lower concentration driven by a concentration gradient (Fig. 1.4a). The rate and extent of diffusion depend on factors such as molecular size, temperature, and fiber porosity, thereby affecting the uniformity and depth of functionality imparted.

Colloid and interface science offer fundamental theories to explain surface phenomena critical to textile functionality. Wettability is described by Young's equation (relating the contact angle to surface energies) and modulated by surface roughness (Fig. 1.4b) via the Wenzel and Cassie-Baxter models. Self-cleaning effects (e.g., the lotus effect) emerge from these principles, in which micro- and nanoscale roughness and low surface energy minimize the adhesion of contaminants. Antimicrobial properties often exploit electrostatic interactions (DLVO theory) or controlled release kinetics (Fickian diffusion) at fiber interfaces. Meanwhile, capillary action in textiles follows the Lucas-Washburn equation, and adsorption thermodynamics (e.g., Langmuir isotherm) governs molecular binding to fiber surfaces.

Understanding fiber morphology is also essential, as the internal structure and surface characteristics of fibers influence their interactions with functional agents. Natural fibers like cotton (cellulose-based) [19], silk (fibroin) [20], or wool (protein-based) [21, 22] contain reactive groups, such as hydroxyl or amino groups, that readily participate in chemical treatments. In contrast, synthetic fibers like polyester or nylon have smoother [9], less reactive surfaces, often requiring surface activation

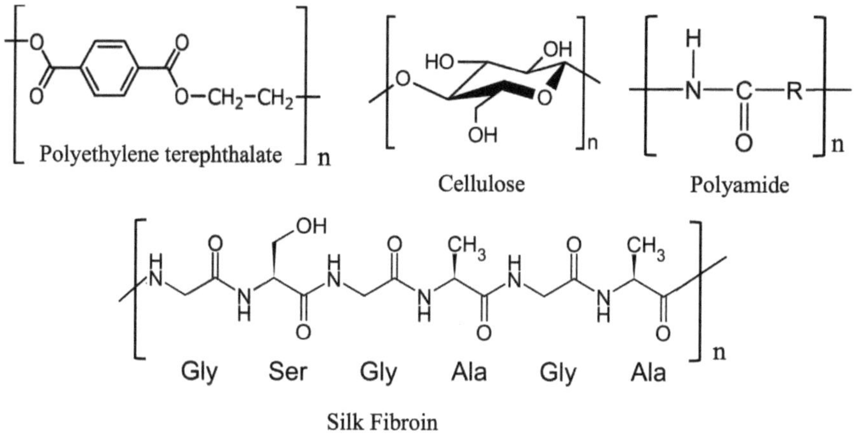

Fig. 1.5 Chemical structure of common textile polymers

(e.g., plasma or chemical etching) before functionalization. The chemical structure of common textile polymers is presented in Fig. 1.5.

The dynamics of polymer chains, including chain entanglement and segmental mobility, play a crucial role in determining how functional molecules penetrate and bind to textile fibers. These dynamics describe the movement and flexibility of polymer chains within the fiber structure. Chain entanglements can restrict diffusion, while increased segmental mobility facilitates deeper penetration of functional molecules, improving their interaction with the fiber matrix. Enhanced mobility typically leads to more uniform distribution and stronger binding, boosting the durability and effectiveness of functional treatments.

1.3 Functional Agents in Textile Functionalization

Functional agents are active substances that interact with textiles to modify or enhance their properties. These agents can be small molecules, polymers, nanoparticles, or biological molecules, such as enzymes and proteins (Fig. 1.6). The goal of using functional agents is to introduce specific characteristics to textiles separate from their inherent properties such as (a) antibacterial or antimicrobial activity, (b) water or oil repellency, (c) UV protection, (d) flame retardancy, (e) anti-odor and anti-static properties, (f) moisture management, (g) thermal regulation, (h) electrical conductivity, (i) self-cleaning or stain resistance, etc.

Functional agents can be broadly classified by origin, mode of interaction, chemical nature, and function. Understanding these classifications helps in selecting the right agents for textile functionalization. Based on origin, these agents are either natural, synthetic, or bio-derived. Natural agents are derived from biological sources and include substances such as chitosan, alginate, plant extracts, essential oils, and

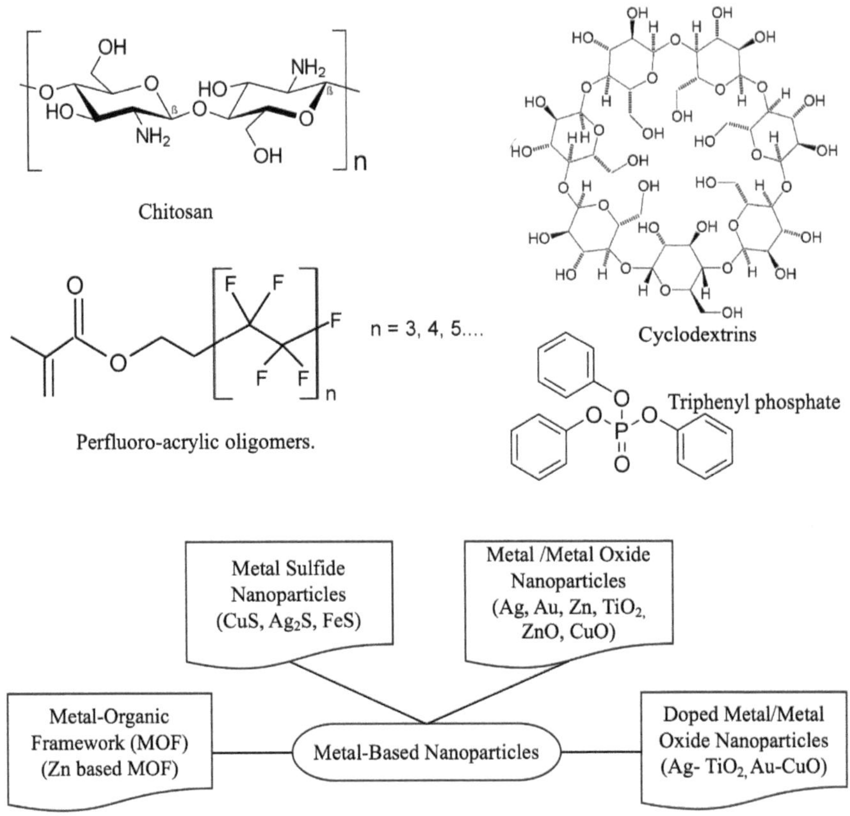

Fig. 1.6 Chemical structure of common functional agents in textile functionalization

enzymes [18, 19, 23]. Synthetic agents are chemically manufactured and include fluorocarbons, metallic salts, silicone softeners, and crosslinking polymers. Bio-derived agents are derived from natural substances, such as enzymatically tailored polysaccharides or green-synthesized nanoparticles [24].

In terms of their mode of interaction with textiles, functional agents may act in physical, chemical, or hybrid modes, such as biochemical finishes. Physical finishes involve the superficial application of agents, such as wax coatings or softeners, that do not form strong bonds with the fibers and are usually less durable. Chemical finishes, such as crosslinking agents or flame retardants, form covalent bonds with textile fibers, thereby improving durability and wash resistance. Biochemical finishes involve the use of biological catalysts, such as enzymes, which modify fiber surfaces through specific, efficient reactions under mild conditions, for example, enzymatic desizing or biopolishing. From a chemical nature perspective, functional agents are categorized as inorganic, organic, or biological. Inorganic agents include metal salts and oxides, such as silver nitrate, titanium dioxide, and zinc oxide, which are commonly used for antimicrobial or UV-protective properties. Organic agents are carbon-based compounds, such as fluoropolymers, silicones, and formaldehyde-based resins. Biological agents are derived from living systems and include chitosan, enzymes, and microbial extracts.

Classification by function refers to the specific performance imparted to textiles. These functions include antimicrobial activity (e.g., silver ions, chitosan), UV protection (e.g., titanium dioxide, zinc oxide), flame retardancy (e.g., triphenyl phosphate), water and oil repellency (e.g., fluorocarbons, silicones), wrinkle resistance (e.g., formaldehyde-free resins), conductivity for smart textiles (e.g., carbon nanotubes, conductive polymers), anti-odor effects (e.g., zinc compounds, cyclodextrins, probiotics), thermal regulation (e.g., phase-change materials), and fragrance release (e.g., microencapsulated essential oils).

The growing demand for conductive textiles in wearable electronics has driven the use of metals as functional materials [25]. Early approaches involved co-weaving metal fibers with polymeric fibers to impart conductivity while enhancing mechanical properties. However, the inherent rigidity and weight of metal fibers limited wearability and posed processing challenges. Advances in chemical and physical deposition have enabled the application of ultrathin metallic coatings, such as aluminum, silver, copper, and nickel, via electroless plating or sputtering [26]. These nanoscale coatings provide conductivity while preserving the textile's flexibility and softness. Silver-coated textiles, in particular, have become widely adopted for antibacterial applications and commercial wearable devices. For instance, electroless deposition of Ni, Co, Cu, and Ag on cotton achieved high conductivity (4–200 kS m^{-1}) without compromising softness, suitable for wearable batteries [27]. Atomic layer deposition (ALD) of Pt on textiles, including cotton, silk, nylon, wool, Kevlar, and Nomex, produced conductive fabrics with excellent wear resistance, bending stability, and long-term durability.

Inorganic materials have long been used in textiles for their excellent optical and electrical properties. With the advent of nanotechnology, nanoengineering has enabled these materials to exhibit distinct properties from their bulk counterparts, offering new opportunities for textile functionalization. Various nanomaterials (including nanoparticles, nanofibers, and 2D materials) have been employed to tune the physical, mechanical, optical, electrical, chemical, and biological properties of textiles. Among these, nanoparticles are particularly prevalent, either incorporated into synthetic fibers or coated onto fabrics, due to their small size and versatility in imparting functional characteristics [28].

The eco-friendliness of functional agents is a critical consideration in modern textile functionalization, as the industry increasingly seeks sustainable, environmentally responsible solutions. Agents vary widely in their environmental impact depending on their chemical composition, origin, and biodegradability. Generally, natural and bio-derived agents tend to be more eco-friendly because they are derived from renewable sources, are less toxic, and biodegrade more readily than many synthetic counterparts. For example, chitosan and essential oils offer antimicrobial properties without the environmental burden associated with heavy metals or fluorinated compounds. However, natural agents can sometimes pose challenges in terms of durability or cost-effectiveness. Synthetic agents, such as fluoropolymers, often deliver excellent performance but can persist in the environment and bioaccumulate, raising concerns about their long-term ecological footprint. Bio-derived agents synthesized through green chemistry approaches offer a promising middle ground, combining functionality with reduced environmental impact. Ultimately, selecting functional agents involves balancing performance needs with sustainability goals, making eco-friendliness ratings an essential guide for responsible textile innovation. Table 1.1 shows how different functional agents used in textiles are grouped, with examples and their eco-friendliness.

1.4 Strategies for Textile Functionalization

As functional materials evolve rapidly, their integration into textiles has become a priority. Since processing techniques play a decisive role in determining the final product's properties, the development and refinement of functionalization methods have attracted significant attention. In conventional textile manufacturing, raw fibers, either staple or filament, are first spun, twisted into yarns, and then assembled into fabrics through processes such as weaving, knitting, embroidery, or bonding with adhesives, before undergoing finishing to achieve the desired performance [3]. To introduce functionality into textiles, two main strategies have been pursued: (a) creating functional fibers directly, either by incorporating advanced materials during spinning or by employing emerging technologies such as electrospinning and additive manufacturing of textile structures. (b) Post-fabrication approaches or surface modification approach, where functional materials are deposited onto the surface of existing textiles through chemical, physical, physico-chemical, or biotechnological approaches.

1.4.1 Functionalizing Textiles at the Spinning

The functionalization of textiles during spinning is based on the direct incorporation of functional agents into fibers at the stage of fiber formation [3]. In this approach, polymers such as polyester, nylon, or cellulose derivatives are blended with additives like nanoparticles, functional agents, conductive fillers, or dyes, before or during spinning. The modified polymer solution is then processed using spinning techniques such as melt spinning, wet spinning, electrospinning, or solution extrusion, where the additives are uniformly dispersed and embedded within the fiber matrix (Fig. 1.7). As the fibers solidify, the functional materials become an integral part of the structure, ensuring stable and durable performance. This method enables the production of fibers with intrinsic properties, such as conductivity, UV resistance, antibacterial activity, or energy storage, offering superior wash resistance and long-term stability compared to surface-finishing methods.

Table 1.1 Classification of functional agents with examples and perceived eco-friendliness rating

Function	Examples of functional agents	Eco-friendliness rating
Antimicrobial	Silver nanoparticles, triclosan, chitosan, copper salts, zinc pyrithione, QACs, essential oils, and neem extract	Chitosan ★★★★, Neem/EOs ★★★★, Silver ★★☆☆, Triclosan ★☆☆☆, QACs ★☆☆☆
UV protection	Titanium dioxide (TiO_2), zinc oxide (ZnO), benzotriazole, stilbenes, cinnamates, benzophenones, lignin-based agents	ZnO/TiO_2 ★★★☆, Lignin ★★★★, Benzophenones ★★☆☆, Stilbenes ★★☆☆
Flame retardant	Phosphorus-based compounds, boron salts, nitrogen systems, melamine, aluminum hydroxide, expandable graphite	Boron/Phosphorus ★★★☆, Expandable Graphite ★★★★, Halogenated agents ★☆☆☆
Water/oil repellent	Fluorocarbons (C6/C8), silicones, paraffin wax, dendrimers, PFPEs, polyurethane dispersions	PFPEs/C8 FCs ★☆☆☆, Silicones ★★☆☆, Paraffin ★★☆☆, Bio-based repellents ★★★☆
Conductivity	CNTs, polyaniline, graphene, PEDOT: PSS, silver-coated yarns, copper nanowires, conductive polymers	Graphene ★★★☆, Silver/CNTs ★★☆☆, PEDOT/PANI ★★☆☆
Odor control	Cyclodextrins, zeolites, enzymes, activated carbon, bamboo charcoal, triclosan, chitosan	Enzymes ★★★★, Cyclodextrins ★★★☆, Triclosan ★☆☆☆, Bamboo charcoal ★★★★
Thermoregulation	PCMs (paraffin), microencapsulated PCMs, PEG, hydrated salts, thermochromic pigments	Hydrated salts ★★★☆, PEG ★★☆☆, Paraffin ★★☆☆, Bio-PCMs ★★★★
Anti-wrinkle	DMDHEU, polycarboxylic acids (citric acid, BTCA), epoxy agents, maleic anhydride copolymers	Citric/BTCA ★★★★, DMDHEU ★☆☆☆, Epoxy agents ★★☆☆
Moisture management	Surfactants, hydrophilic polymers (PEG), wicking agents, amphiphilic copolymers, PVA, bio-based polyesters	PEG/PVA ★★☆☆, Bio-polyesters ★★★★, Surfactants (nonionic) ★★☆☆

Perceived eco-friendliness rating (on a qualitative scale: ★☆☆☆ = low, ★★☆☆ = moderate, ★★★☆ = good, ★★★★ = excellent), based on general information related to biodegradability, toxicity, renewability, and environmental persistence of the agents

Some examples of functionalization of textiles during spinning include (a) ***Melt Spinning of Graphene–Polyamide Fibers***: Graphene nanoparticles were blended into a polyamide-6 (PA6) matrix and processed by melt spinning to produce conductive multifilament yarns with enhanced mechanical strength and electrical conductivity [29]. (b) ***Wet Spinning of Graphene Oxide Fibers***: Researchers prepared graphene oxide (GO) fibers using wet spinning, achieving high mechanical strength and flexibility, making them suitable for electronic textiles and sensors [30]. (c) ***Microfluidic Spinning of Protein Fibers***: Using a methanol–water coagulation bath,

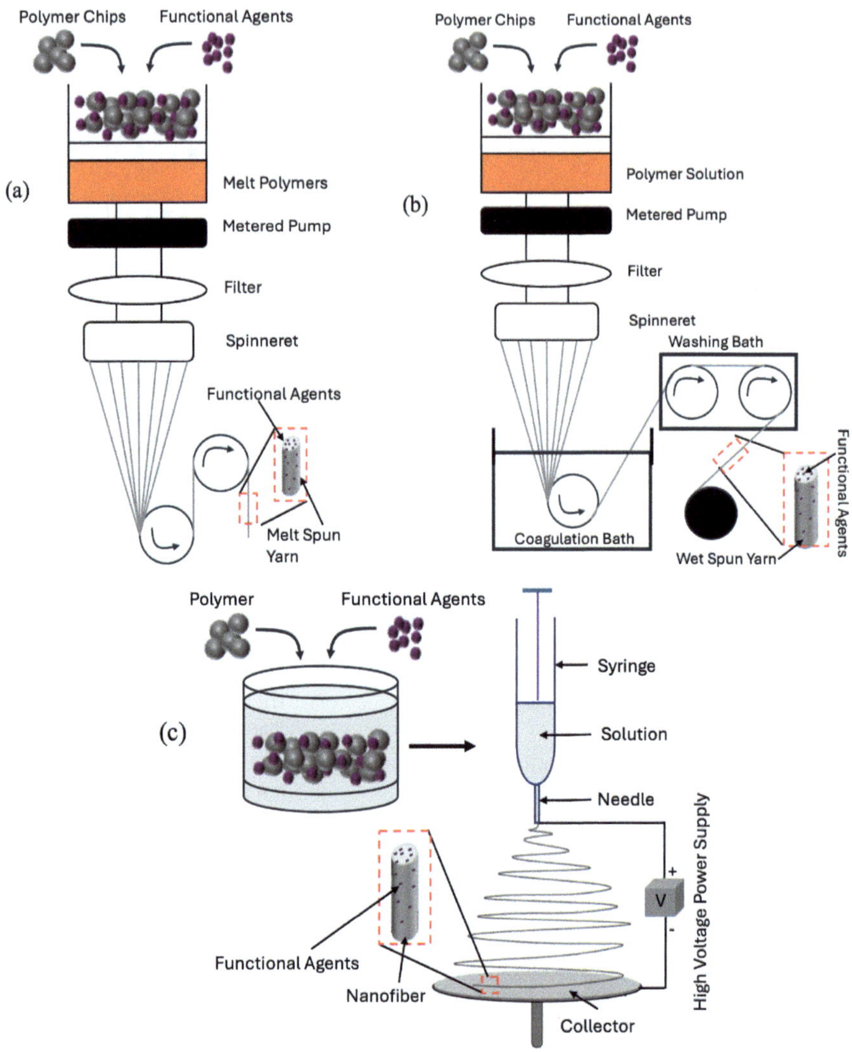

Fig. 1.7 Schematic illustration of incorporation of functional agents into fibers at the stage of fiber formation: **a** melt spinning, **b** wet spinning, and **c** electrospinning

bulk protein fibers were fabricated through microfluidic spinning, demonstrating potential for biomedical and smart textile applications [31]. (d) ***Thermal Drawing of Patterned Fibers***: Thermal drawing techniques enabled the creation of fibers with micro-patterned internal structures, offering controlled optical and electronic functionalities [32]. (e) ***Solution Blow Spinning of Functional Nonwovens***: Nonwoven fabrics made from poly(vinylidene fluoride-co-hexafluoropropylene) [P(VDF-HFP)] and poly(4-vinylpyridine) exhibited excellent porosity and chemical resistance, making them useful in protective textiles [33]. (f) ***Industrial-Scale Solution Extrusion for Fiber Batteries***: A large-scale fiber battery production process demonstrated how electrochemically active components can be integrated directly into fibers during spinning, opening pathways for energy-storing textiles [34].

Examples of electrospinning functionalization of textiles include [35, 36]: (a) Nitrogen-doped carbon nanofibers with ultrahigh conductivity and flexibility were fabricated via electrospinning, offering applications in flexible electronics and energy storage, (b) modified electrospinning with rotating collectors enabled patterned deposition of poly(L-lactic acid) fibers into multidimensional architectures, functional for biomedical scaffolds and wearable devices, (c) a coaxial electrospinning process was used to create liquid–metal/polymer fibers that were woven into large-area textiles with excellent electrical conductivity and stretchability, (d) a bath-assisted system produced continuous MXene/nylon hybrid nanoyarns, up to 50 cm in length, demonstrating scalable fabrication for smart electronic textiles, (e) continuous electrospinning enabled the large-scale production of nano-micro hybrid fibers, which were directly integrated into high-performance triboelectric nanogenerators, (f) incorporation of phase-change materials into electrospun fibers provided thermal regulation properties for smart textiles, demonstrating potential for protective and sportswear applications, (g) silver-nanoparticle-loaded electrospun fibers exhibited durable antibacterial activity, making them useful for medical and hygienic textiles [3]. Additive manufacturing, commonly known as 3D printing, is transforming the textile industry by allowing the construction of intricate textile structures layer by layer, often incorporating specialized functional properties. This approach provides significant benefits, including high levels of customization, minimized material waste, streamlined supply chains, and the ability to produce three-dimensional textile forms such as weaves and fur-like textures. Additionally, it enables the integration of advanced features, such as wearable health sensors or temperature-regulating elements.

1.4.2 Post-fabrication/Surface Modification for Textile Functionalization

Beyond direct incorporation of functional agents into fibers or yarns during their fabrication, considerable attention has also been directed toward fiber surface modification for textile functionalization. This strategy initially emerged as a means to overcome limitations in fiber functionalization, particularly issues of efficiency, process complexity, and the inherent difficulties of incorporating functional agents during fiber formation. Owing to its simplicity, versatility, and compatibility with existing processes, surface modification has emerged as a fundamental approach for textile functionalization. Several methods have been developed for the surface modification of textile fibers, including chemical, physical, physicochemical, and biotechnological methods, each offering distinct advantages depending on the desired functionality.

1.4.2.1 Chemical Approach for Textile Functionalization

Chemical functionalization involves deliberately introducing new chemical moieties onto textile fibers to impart specific functional properties. This is achieved through covalent or non-covalent interactions, often utilizing grafting reactions, cross-linking, esterification, amidation, or radical polymerization to chemically modify the fiber surface or internal structure. The introduced functionalities can significantly alter the physicochemical behavior of the textile material, including surface energy, hydrophilicity/hydrophobicity, and reactivity [8]. For instance, grafting quaternary ammonium compounds can confer antimicrobial properties by disrupting microbial membranes, while phosphorus- or nitrogen-based grafts can improve flame retardancy by promoting char formation and reducing flammable volatiles. Similarly, fluorinated alkyl chains or silicone-based compounds can be covalently bonded to create water- and oil-repellent finishes, essential for high-performance outdoor fabrics [37].

Depending on the fiber type (such as cotton [rich in hydroxyl groups], polyester [containing ester linkages], or nylon [with amide groups]), different chemical pathways are used to impart functional properties. For instance, hydroxyl-rich cellulose in cotton can be modified through esterification with acyl chlorides or anhydrides to impart hydrophobicity, or by carbodiimide coupling to attach carboxylic acid-containing agents such as antimicrobial molecules [38]. Additionally, periodate oxidation can be employed to generate dialdehyde groups for subsequent Schiff base formation with amines, enabling enzyme immobilization or antibacterial functionalization. Epoxide-based grafting using glycidyl methacrylate (GMA) is another approach to introduce hydrophilic or reactive groups.

In the case of polyester, which has a relatively inert ester backbone, surface activation is often required. This can be achieved through alkaline hydrolysis, which generates carboxyl and hydroxyl groups that serve as reactive sites for grafting silanes, polycarboxylic acids, or epoxy compounds. [39, 40]. Plasma treatment and ozonation are also effective for introducing surface reactivity, while aminolysis with ethylenediamine introduces amine groups that can be further modified with isocyanates or acid chlorides. Nylon fibers, being polyamides, can undergo acid or base hydrolysis to expose terminal amine or carboxyl groups [41]. These can be functionalized via reactions with aldehydes or epoxides, or by grafting vinyl monomers using free-radical initiators such as benzoyl peroxide or ceric ammonium nitrate. These tailored chemical strategies enable the incorporation of diverse functionalities, including antimicrobial activity, flame retardancy, UV resistance, hydrophobicity, and improved dyeability, depending on the specific fiber chemistry and the targeted application.

1.4.3 Physical Approach for Textile Functionalization

Physical approaches for textile functionalization employ advanced, machine-driven, energy-based techniques to modify fabric surfaces and structures while minimizing chemical use. These methods offer precise control over textile properties through physical interactions at micro- and nano-scales. Plasma treatment stands out as a versatile dry process that employs ionized gas to either activate fiber surfaces for better wettability and adhesion or deposit nanoscale functional coatings for water repellency, antimicrobial effects, or flame retardancy [42, 43]. Laser processing provides exceptional precision in surface patterning, enabling the creation of hierarchical microstructures that impart superhydrophobicity, enhance light absorption in solar textiles, or generate conductive tracks for wearable electronics. Electrospinning has emerged as a powerful technique for producing nanofibrous membranes with tunable porosity and functionality, allowing incorporation of nanoparticles for applications ranging from air filtration to drug-eluting wound dressings. Ultrasonic bonding offers a mechanical approach to embed functional microcapsules containing phase-change materials, fragrances, or therapeutic agents directly into textile fibers while maintaining fabric flexibility. Complementary techniques, such as magnetron sputtering and physical vapor deposition, enable the application of ultra-thin metallic or ceramic coatings that confer properties such as electrical conductivity, electromagnetic shielding, and infrared reflectance.

Additive manufacturing methods, including 3D printing of textile structures, enable the direct integration of functional elements such as sensors or energy storage components [3, 44]. High-pressure mechanical finishing processes, such as nanoimprinting and micro-embossing, physically alter surface topography to achieve directional liquid transport, enhanced thermal regulation, or improved tactile properties. These physical approaches collectively provide sustainable alternatives to conventional chemical treatments, offering advantages in durability, precision, and environmental compatibility while enabling advanced functionalities for smart textiles, protective gear, and biomedical applications. Mechanical techniques such as calendering and embossing represent critical physical approaches for textile functionalization, in which controlled pressure, temperature, and mechanical action modify fabric surfaces without chemical treatments. These methods can be combined with other physical treatments; for instance, calendering after plasma treatment can further enhance coating adhesion, while embossing can complement laser processing to create hierarchical surface structures for advanced liquid repellency. The scalability and adaptability of these methods make them increasingly popular for developing next-generation functional textiles with tailored performance characteristics.

1.4.4 Physico-chemical Approach for Textile Functionalization

The physico-chemical approach for textile functionalization involves the combined use of physical and chemical techniques to modify textile surfaces in a controlled and efficient manner. This method typically begins with a physical treatment such as plasma irradiation, UV exposure, or corona discharge that activates the fiber surface by introducing functional groups (like hydroxyl or carboxyl groups) or increasing surface energy [39, 45, 46]. These activated sites then serve as anchors for subsequent chemical reactions, such as grafting or coating with functional agents. By combining these two approaches, the physico-chemical method ensures stronger adhesion of functional materials to the textile substrate, thereby enhancing the durability and performance of the treated fabrics. This strategy is especially beneficial for chemically inert fibers like polyester or polypropylene, which are challenging to modify using traditional chemical methods alone. Techniques such as plasma-assisted grafting, UV-induced polymerization, sol–gel processing, and layer-by-layer deposition allow the creation of textiles with advanced properties, including antimicrobial activity, UV protection, self-cleaning surfaces, and improved wettability. Layer-by-layer assembly and sol–gel processes are also examples of methods for depositing chemical coatings using physical deposition methods. This hybrid approach can yield more uniform, durable, and multifunctional coatings, enhancing properties such as antimicrobial activity, self-cleaning, and conductivity. A physico-chemical approach is advantageous for environmental impact, as it often requires fewer chemicals and milder processing conditions.

1.4.5 Biotechnological Approach for Textile Functionalization

Biotechnological functionalization focuses on imparting specific functional properties to textiles using biological agents, such as enzymes, microorganisms, and bio-based polymers. Enzymatic surface modification of textiles entails treating fibers or biopolymers to enhance surface hydrophilicity, remove specific surface components via hydrolysis, oxidation, or eliminative cleavage, and selectively introduce functional groups onto the fiber surface. Enzymatic surface modification of textiles utilizes a variety of enzymes, each offering unique mechanisms to tailor fiber properties and introduce specific functionalities.

For example, cellulases are widely used to selectively hydrolyze cellulose on the fiber surface, improving softness, enhancing hydrophilicity, and removing microfibrillar fuzz from cotton and other cellulosic fabrics. Pectinases break down pectic substances in plant-based fibers, aiding degumming and improving fiber wettability and dye uptake. Xyloglucan endotransglycosylases (XETs) facilitate the remodeling of hemicellulosic networks on the fiber surface, allowing controlled modification of surface architecture without compromising bulk strength. Tyrosinases catalyze the oxidation of phenolic compounds, enabling the in situ formation of antimicrobial, antioxidant, or UV-protective coatings on textile surfaces. Cutinases, commonly applied to polyester or other synthetic fibers, hydrolyze surface ester bonds, increasing surface roughness and introducing reactive functional groups, thereby enhancing the adhesion of dyes, coatings, or bio-based polymers. Collectively, these enzymes demonstrate the versatility of biocatalytic approaches for textile functionalization, offering eco-friendly, selective, and often mild processing routes to achieve improved performance, multifunctionality, and sustainability in modern textile materials [47].

Enzymes such as glucose oxidase and horseradish peroxidase can be immobilized onto textile fibers (see Fig. 1.8) to catalyze the production of reactive oxygen species, which impart self-sterilizing, antiviral, and antifouling properties to the fabrics [45]. This enzymatic functionalization not only enhances hygiene but also extends the service life of textiles in healthcare, sports, and protective applications. Beyond enzymes, engineered microorganisms can be employed to deposit functional coatings directly onto textile surfaces, enabling the creation of bioluminescent, UV-responsive, or pH-sensitive materials. Such biologically derived coatings pave the way for smart textile applications, where the fabric itself can respond dynamically to environmental stimuli. A particularly significant application of these strategies is the development of textile-based biosensors, in which immobilized enzymes or other biorecognition elements selectively detect specific chemical or biological analytes. These biosensors provide real-time monitoring, interactive functionality, and the potential for integration into wearable health-monitoring systems, environmental sensors, and point-of-care diagnostic platforms [48].

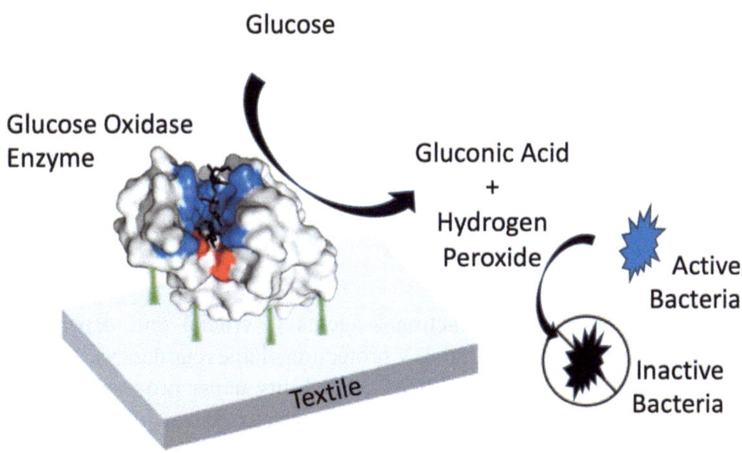

Fig. 1.8 Schematic illustration of antibacterial application of *glucose oxidase* enzyme immobilized textile

In general, biotechnologically driven functionalizations offer sustainable, selective, and often reversible or biodegradable alternatives to conventional synthetic chemical additives. By combining enzyme immobilization, microbial coatings, and functional polymer grafting, it is possible to fabricate intelligent, eco-friendly textiles that not only meet performance and aesthetic demands but also adhere to principles of sustainability and circularity, paving the way for next-generation smart and functional fabrics.

1.5 Influencing Factors in Textile Functionalization

The successful functionalization of textiles is governed by a complex interplay of factors ranging from fiber chemistry and fabric structure to the choice of functional agents, strategies, and processing conditions. By understanding and optimizing the interrelated factors, it becomes possible to design and optimize the textile functionalization. The following are the key factors that majorly influence textile functionalization, shaping the effectiveness, durability, and performance of functional textiles.

1.5.1 Fiber Type and Fabric Structure

The composition and morphology of the fiber significantly influence the outcomes of functionalization. Natural fibers like cotton, wool, and silk have abundant reactive groups (e.g., hydroxyl and amino groups), making them more receptive to both chemical and enzymatic treatments. In contrast, synthetic fibers [49–51] such as polyester, polyimide, or polypropylene often require pretreatment (e.g., plasma or hydrolysis) due to their inert and hydrophobic nature. Additionally, the fabric structure (woven, knit, or nonwoven) affects surface area, porosity, and the accessibility of functional agents. Denser or multilayered structures may hinder penetration and uniformity of treatment.

1.5.2 Choice Functional Agents

The selection of appropriate functional agents is crucial and depends on the desired property (e.g., antimicrobial, UV protection, flame retardancy). The chemical compatibility of the agent with the fiber, its stability under processing conditions, toxicity, biodegradability, and regulatory status must also be considered. For instance, bio-based agents such as chitosan are preferred for sustainable applications, whereas metal nanoparticles may require stabilization to prevent agglomeration.

1.5.3 Choice of Strategies

Functionalization strategies, whether applied during spinning or through post-fabrication treatments, largely determine how functional properties are imparted to textiles. In spinning-based approaches, functional agents are incorporated directly into fibers or yarns, ensuring uniform distribution and often providing durable performance. By contrast, post-fabrication functionalization (such as dyeing, coating, or surface modification) offers greater flexibility, allowing for the introduction of multiple functionalities onto finished fabrics. The choice between these strategies depends on factors such as the required functionality, compatibility with fiber chemistry, the durability of the intended effect, and the cost-effectiveness of the process. In some cases, hybrid strategies (e.g., are employed to maximize efficiency.

1.5.4 Choice of Techniques and Process

Specific techniques such as melt spinning, wet spinning, padding, exhaustion, spraying, sol–gel processing, or layer-by-layer assembly determine how the agent is applied and fixed to the textile surface.

1.5.5 Processing Conditions and Intended Use

Parameters such as time, temperature, pH, drying/curing conditions, and the use of crosslinkers or additives significantly influence the effectiveness and durability of functionalization. For instance, enzyme treatments require tightly controlled pH and temperature conditions to preserve activity. In contrast, some chemical agents may need high-temperature curing to ensure fixation. The intended application, e.g., medical textiles, protective clothing, or activewear, also defines the performance criteria, wash durability, biocompatibility, or resistance to environmental stressors. Special use-case scenarios may necessitate multi-functionalization, such as combining antimicrobial and moisture-wicking properties in a single fabric layer.

1.6 Environmental Issues in Textile Functionalization

Textile functionalization, while offering advanced performance and added value to fabrics, also raises several environmental concerns that must be addressed to ensure sustainable development. The most critical issues include: (a) Many conventional functionalization methods rely on toxic chemicals, solvents, and auxiliaries that can pose risks to human health and the environment. For instance, fluorocarbon-based finishes used for water- and oil-repellency can release persistent pollutants, while specific antimicrobial agents may leach into water systems and disrupt ecosystems. (b) Processes such as wet-chemical finishing, exhaustion, and padding often require large volumes of water and high energy inputs, contributing to resource depletion and an increased carbon footprint. Wastewater generated from these processes frequently contains dyes, salts, and finishing agents that are difficult to treat. (c) Post-functionalization processes can generate solid and liquid waste, including unused chemicals, off-specification fabrics, and wastewater treatment sludge. Improper handling of these wastes can lead to environmental contamination. (d) While functionalization is often intended to improve the durability of textiles, specific treatments may reduce a fabric's recyclability or biodegradability at the end of its life cycle. Coatings, laminates, or nanoparticles may interfere with fiber recovery or contribute to microplastic pollution. (e) The growing use of nanoparticles and nanostructured coatings in functional textiles raises new concerns about their environmental release. Nanoparticles can enter air, soil, or water during production, washing, or disposal, and their long-term ecological effects remain insufficiently understood. (f) Increasing awareness and tightening regulations, particularly in the EU and North America, are pushing the textile industry to move away from hazardous substances and adopt greener processes. At the same time, consumers are demanding eco-friendly products, putting additional pressure on manufacturers to adopt sustainable solutions. (g) To address these challenges, research is focusing on environmentally benign alternatives such as bio-based finishing agents, enzyme-assisted processes, supercritical CO_2 technologies, and waterless plasma treatments. These methods not only minimize chemical and water use but also improve energy efficiency and reduce waste, aligning textile functionalization with the principles of a circular economy.

While textile functionalization poses environmental challenges, it also offers several key contributions when applied responsibly. By extending the lifespan of fabrics through finishes such as antimicrobial, stain-resistant, or UV-protective coatings, functionalization reduces the need for frequent replacement and lowers overall resource demand. Functional textiles that incorporate self-cleaning or odor-control properties can also minimize washing frequency, thereby conserving water, detergents, and energy during use. Moreover, advances in eco-friendly processes, such as enzyme-assisted treatments, bio-based agents, plasma technologies, and supercritical CO_2 finishing, offer cleaner, more sustainable alternatives to conventional chemical methods. In addition, functionalization can support broader sustainability goals by enabling recyclable or biodegradable textiles, improving user safety through protective features, and even contributing to energy savings via thermoregulating or energy-harvesting fabrics. In summary, textile functionalization presents both environmental challenges and opportunities; its future lies in balancing performance with sustainability by adopting cleaner technologies, safer materials, and circular design principles.

Acknowledgements The authors would like to acknowledge the institutional support from the Department of Textile Technology, Faculty of Textiles, Engineering and Business (including the

Swedish School of Textiles), University of Borås, Sweden, and the Ivory V. Nelson Center for the Sciences, Department of Chemistry and Physics, Lincoln University, Oxford, USA, during the preparation of this chapter. Authors would also like to thank Dr. Sina Seipel from the University of Borås for her support during the preparation of the chapter manuscript.

AI-Usage Declaration During the preparation of this chapter, author(s) used artificial intelligence to improve the language of the manuscript. After using this tool/service, the author(s) reviewed and edited the content as needed and take(s) full responsibility for the content of the publication.

References

1. P. Dominique, P. Crégo, *Wearables, Smart Textiles & Smart Apparel* (Elsevier, 2018)
2. Growth Market Reports, *Functional Textiles Market Research Report 2033* (2025). https://growthmarketreports.com/report/functional-textiles-market-global-industry-analysis
3. Y. Zhang et al., Functional textiles with smart properties: their fabrications and sustainable applications. Adv. Func. Mater. **33**(33), 2301607 (2023)
4. R. Nayak, Sportswear and activewear: analysis of recent trends from sustainability standpoint, in *Sustainable Sportswear and Activewear*. (Elsevier, 2025), pp.1–22
5. S. Mahmud et al., Comprehensive review on plant fiber-reinforced polymeric biocomposites. J. Mater. Sci. **56**(12), 7231–7264 (2021)
6. T. Shah, A. Rawal, Textiles in filtration, in *Handbook of Technical Textiles*. (Elsevier, 2016), pp.57–110
7. K. Klinkhammer et al., Functionalization of technical textiles with chitosan. Textiles **4**(1), 70–90 (2024)
8. M.K. Singh, *Textiles Functionalization-A Review of Materials, Processes, and Assessment* (IntechOpen, 2021)
9. K.M.F. Hasan et al., Colorful and antibacterial nylon fabric via in-situ biosynthesis of chitosan mediated nanosilver. J. Mater. Res. Technol. **9**(6), 16135–16145 (2020)
10. K.M.F. Hasan et al., Coloration of aramid fabric via in-situ biosynthesis of silver nanoparticles with enhanced antibacterial effect. Inorg. Chem. Commun. **119**, 108115 (2020)
11. H. Jiang et al., Eco-friendly dyeing and finishing of organic cotton fabric using natural dye (gardenia yellow) reduced-stabilized nanosilver: full factorial design. Cellulose **29**(4), 2663–2679 (2022)
12. S.M.R. Billah, Textile coatings, in *Functional Polymers*. (Springer, 2018), pp.1–58
13. W. Fung, *Coated and Laminated Textiles*, vol. 23 (Woodhead Publishing, 2002)
14. M. Joshi, B. Butola, Application technologies for coating, lamination and finishing of technical textiles, in *Advances in the Dyeing and Finishing of Technical Textiles*. (Elsevier, 2013), pp.355–411
15. S. Mahmud et al., In situ synthesis of green AgNPs on ramie fabric with functional and catalytic properties. Emerg. Mater. Res. **8**(4), 623–633 (2019)
16. Y.-P. Huang et al., Effect of PET melt spinning on TiO_2 nanoparticle aggregation and friction behavior of fiber surface. Ind. Eng. Chem. Res. **46**(17), 5548–5554 (2007)
17. K.M.F. Hasan et al., A novel coloration of polyester fabric through green silver nanoparticles (G-AgNPs@PET). Nanomaterials **9** (2019). https://doi.org/10.3390/nano9040569
18. S. Mahmud et al., Multifunctional organic cotton fabric based on silver nanoparticles green synthesized from sodium alginate. Text. Res. J. **90**(11–12), 1224–1236 (2020)
19. K.M.F. Hasan et al., Enhancing mechanical and antibacterial performances of organic cotton materials with greenly synthesized colored silver nanoparticles. Int. J. Cloth. Sci. Technol. **34**(4), 549–565 (2022)

20. S. Mahmud et al., Surface functionalization of "Rajshahi Silk" using green silver nanoparticles. Fibers **5**(3), 35 (2017)
21. K.M.F. Hasan et al., Wool functionalization through AgNPs: coloration, antibacterial and wastewater treatment. Surf. Innov. **9**(1), 25–36 (2020)
22. R. Rehman et al., Effective method development on wool dyeing by using Fl based cotton-reactive dyes. Int. J. Sci. Eng. Res. **6**(6), 505–511 (2015)
23. M.S. Hossain et al., Coloration and functionalization of organic cotton fabric using Chinese fringe leaf extraction: a greener approach. Results Eng. **25**, 104054 (2025)
24. R. Seth, A. Meena, Enzymes-based nanomaterial synthesis: an eco-friendly and green synthesis approach. Clean Technol. Environ. Policy 1–24 (2024)
25. A. Das et al., Optimizing electrical resistivity in flexible conductive films by way of silver nanowire doping. Surf. Innov. **13**(2), 109–116 (2025)
26. J.S. Heo et al., Recent progress of textile-based wearable electronics: a comprehensive review of materials, devices, and applications. Small **14**(3), 1703034 (2018)
27. D. Wang et al., A universal method towards conductive textile for flexible batteries with superior softness. Energy Storage Mater. **36**, 272–278 (2021)
28. S. Riaz et al., Functional finishing and coloration of textiles with nanomaterials. Color. Technol. **134**(5), 327–346 (2018)
29. B. Weise et al., Melt-and wet-spinning of graphene-polymer nano-composite fibres for multifunctional textile applications. Mater. Today Proc. **4**, S135–S145 (2017)
30. S.H. Aboutalebi et al., High-performance multifunctional graphene yarns: toward wearable all-carbon energy storage textiles. ACS Nano **8**(3), 2456–2466 (2014)
31. H. He et al., Mechanically strong globular-protein-based fibers obtained using a microfluidic spinning technique. Angew. Chem. Int. Ed. **59**(11), 4344–4348 (2020)
32. Z. Wang et al., Designer patterned functional fibers via direct imprinting in thermal drawing. Nat. Commun. **11**(1), 3842 (2020)
33. D.H. Ho et al., β-phase-preferential blow-spun fabrics for wearable triboelectric nanogenerators and textile interactive interface. Nano Energy **77**, 105262 (2020)
34. M. Liao et al., Industrial scale production of fibre batteries by a solution-extrusion method. Nat. Nanotechnol. **17**(4), 372–377 (2022)
35. Y. Jing et al., Alginate/gelatin mineralized hydrogel modified by multilayers electrospun membrane of cellulose: Preparation, properties and in-vitro degradation. Polym. Degrad. Stab. **192**, 109685 (2021)
36. Y. Tian et al., Electrospun PVDF-Ag@AgCl porous fiber membrane: stable antifoul and antibacterial surface. Surf. Innov. **9**(2–3), 156–165 (2020)
37. J. Ghosh et al., Grafting modification for textile functionalization: innovations and applications. Discov. Appl. Sci. **7**(1), 49 (2025)
38. D.W. Wei et al., Superhydrophobic modification of cellulose and cotton textiles: methodologies and applications. J. Bioresour. Bioprod. **5**(1), 1–15 (2020)
39. B. Nabil et al., Development of new multifunctional filter based nonwovens for organics pollutants reduction and detoxification: high catalytic and antibacterial activities. Chem. Eng. J. **356**, 702–716 (2019)
40. J. Dave, R. Kumar, H. Srivastava, Studies on modification of polyester fabrics I: alkaline hydrolysis. J. Appl. Polym. Sci. **33**(2), 455–477 (1987)
41. J. Moens, G. Smets, Alkaline and acid hydrolysis of polyvinylamides. J. Polym. Sci. **23**(104), 931–948 (1957)
42. A. Zille, F.R. Oliveira, A.P. Souto, Plasma treatment in textile industry. Plasma Processes Polym. **12**(2), 98–131 (2015)
43. R. Shishoo, *Plasma Technologies for Textiles* (Elsevier, 2007)
44. R.H. Sanatgar, *L'Impression 3D de nanocomposites polymères conducteurs: un nouveau procédé pour la fonctionnalisation de surface de smart textiles. (Doctoral dissertation, Université de Lille, 2029)*
45. M.N. Morshed et al., Surface modification of polyester fabric using plasma-dendrimer for robust immobilization of glucose oxidase enzyme. Sci. Rep. **9**(1), 15730 (2019)

46. M.N. Morshed et al., Stabilization of zero valent iron (Fe^0) on plasma/dendrimer functionalized polyester fabrics for Fenton-like removal of hazardous water pollutants. Chem. Eng. J. **374**, 658–673 (2019)
47. V. Nierstrasz, Enzyme surface modification of textiles, in *Surface Modification of Textiles*. (Elsevier, 2009), pp.139–163
48. S. Khosravi et al., Screen-printed textile-based electrochemical biosensor for noninvasive monitoring of glucose in sweat. Biosensors **13**(7), 684 (2023)
49. J. Qin et al., AI-assisted dyeing optimization of polyimide fibers via backpropagation neural networks. Prog. Org. Coat. **209**, 109611 (2025)
50. Y. Yang et al., Mechanisms and effects of sunlight exposure on disperse-dyed aromatic polysulfonamide fibers. Surf. Innov. **13**(3), 176–186 (2025)
51. Y. Yang et al., Sunlight exposure on vat-dyed polysulfonamide fibers: analysis of mechanisms and effects. ChemistrySelect **10**(37), e01682 (2025)

Chapter 2
Surface Chemistry and Physics in Textile Surface Functionalization

Mohammad Raza Miah, Nahida Akter, Sakil Mahmud, Faisal Ahmed, Abu Saifullah, S. M. Fijul Kabir, Jin Zhu, and Ayub Nabi Khan

Abstract Textile surface functionalization is a dynamic and rapidly advancing discipline that combines the principles of functional materials, surface chemistry, and surface physics to improve fabric performance, durability, and responsiveness. This interdisciplinary approach allows for the creation of sophisticated textiles with customized features such as antimicrobial properties, UV protection, electrical conductivity, water repellency, and responsiveness to external stimuli. Various functional agents, including nanoparticles, conductive polymers, and phase-change materials, are applied to textile surfaces via methods such as coating, grafting, and in situ synthesis. Surface chemistry plays a pivotal role by promoting strong

M. R. Miah (✉) · F. Ahmed · A. N. Khan
Department of Textile Engineering, BGMEA University of Fashion and Technology (BUFT), Dhaka, Bangladesh
e-mail: raza.miah@buft.edu.bd

M. R. Miah · J. Zhu
Key Laboratory of Bio-Based Polymeric Materials Technology and Application of Zhejiang Province, Ningbo Institute of Materials Technology and Engineering, Chinese Academy of Sciences, Ningbo, People's Republic of China

N. Akter
Department of Islamic History and Culture, Government Debendra College Manikganj, National University, Gazipur, Bangladesh

S. Mahmud (✉)
Ivory V. Nelson Center for the Sciences, Department of Chemistry and Physics, Lincoln University, Oxford, PA, USA
e-mail: smahmud@lincoln.edu

Department of Textile Engineering, Faculty of Engineering, Daffodil International University, Dhaka, People's Republic of Bangladesh

A. Saifullah
Portsmouth Centre for Advanced Materials and Manufacturing, Walthamstow, London, UK

S. M. Fijul Kabir
Engineering (Fibers), Procter and Gamble, Iowa City, IA, USA

M. N. Morshed and M. Shahid (eds.), *Advances in Surface Modification and Functionalization in Textiles*, Textile Science and Clothing Technology,
https://doi.org/10.1007/978-3-032-23377-6_2

interfacial bonding and maintaining the durability of functional layers via covalent bonds, electrostatic interactions, and targeted surface functionalization. Concurrently, surface physics influences the textiles' structural, mechanical, and optical properties, affecting aspects such as wettability, surface energy, texture, and adhesion. This chapter presents a general overview of functional materials, surface chemistry, and physics in textile surfaces.

Keywords Textile surface functionalization · Functional materials · Surface chemistry · Surface physics

2.1 Introduction

Textile surface functionalization is crucial for enhancing fabric performance and functionality beyond their natural properties [1]. By integrating functional agents with a comprehensive understanding of surface chemistry and surface physics [2], it is possible to design advanced textile systems that exhibit tailored responsiveness, durability, and multifunctionality for applications spanning protective clothing, biomedical devices, and wearable electronics [3]. In recent years, the textile industry has experienced a significant shift from traditional fabric manufacturing toward the development of multifunctional, high-performance materials [4]. This transformation is driven largely by increasing demand for textiles that provide not only aesthetic and structural value but also added functionalities such as antimicrobial effects, flame retardancy, water resistance, UV protection, and electrical conductivity [5, 6]. At the heart of this progress lies the strategic functionalization of textile surfaces, which involves deliberately altering the outermost fiber layers to introduce desired properties without compromising mechanical strength, flexibility, or breathability [7, 8]. To achieve these enhanced functionalities, a systematic approach to textile surface modification is required, focusing on selecting appropriate functional materials and the mechanisms by which they interact with fiber surfaces. Unlike bulk modification, surface functionalization targets only the outermost layers of textile substrates, allowing the introduction of new properties while preserving intrinsic bulk characteristics such as strength, softness, and comfort. This approach enables precise control over surface-related phenomena and provides a versatile platform for incorporating advanced functionalities through physical, chemical, or physicochemical techniques. Understanding the nature of functional agents and their interaction pathways is therefore essential for the rational design of high-performance textile surfaces. [5, 6].

Functional agents, including nanoparticles, smart polymers, and conductive compounds, are applied to engineer surface features at micro- and nanoscale levels [9–12]. These materials are carefully selected for their ability to respond to external stimuli or to impart specific performance benefits [13, 14]. Consequently, a fundamental understanding of surface chemistry and surface physics is indispensable for predicting performance outcomes and optimizing functionalization processes.

Surface chemistry plays a crucial role in facilitating compatibility between the textile substrate and functional agents [5]. Chemical treatments such as plasma activation, grafting, and coupling agents introduce reactive groups on fiber surfaces, enhancing adhesion, uniformity, and long-term durability of applied coatings [15–17]. Concurrently, principles of surface physics govern the interaction between modified surfaces and their environment, influencing key factors like surface energy, roughness, and wettability [18]. A thorough understanding of the interplay among material characteristics, chemical reactivity, and physical behavior is vital to optimize surface functionalization processes [19–21]. This interdisciplinary strategy not only improves textile performance but also enables the integration of emerging technologies in areas such as wearable electronics, biomedical devices, environmental filtration, and smart apparel systems [22–24]. This chapter provides an overview of functional materials, surface chemistry, and surface physics in textile surface functionalization, emphasizing their collective impact on designing the next generation of functional textiles.

2.2 Understanding Functional Agents in Textile Functionalization

Functional agents are specially engineered substances designed to impart targeted physical or chemical properties to textile surfaces, effectively transforming conventional fabrics into active, responsive systems [5, 25]. In textile surface functionalization, these agents enhance or introduce advanced functionalities such as antimicrobial activity, ultraviolet (UV) protection, electrical conductivity, flame retardancy, and self-cleaning behavior [26]. Functional agents can be broadly classified by chemical nature, origin, mode of interaction, and function as presented in Fig. 2.1. Classification by chemical nature distinguishes agents as inorganic, organic, or biological, reflecting their molecular composition and reactivity. Based on origin, functional agents may be derived from synthetic, natural, or bio-based sources, influencing their sustainability, biocompatibility, and environmental impact. The mode of interaction describes how functional agents attach to textile substrates, which may occur through physical adsorption, chemical bonding, grafting, or encapsulation, directly affecting coating durability and performance stability. Finally, classification by function groups agents according to the specific properties they impart, such as antimicrobial activity, UV protection, flame retardancy, electrical conductivity, water repellency, or stimuli responsiveness. This multidimensional classification enables rational selection and design of functional agents tailored to specific textile applications and performance requirements.

Among the various functional agents from the aforementioned classes, several have found extensive application in textile surface functionalization, including nanoparticles (e.g., silver, titanium dioxide, zinc oxide), conductive polymers (e.g., polyaniline, PEDOT:PSS), carbon-based nanomaterials (e.g., graphene, carbon

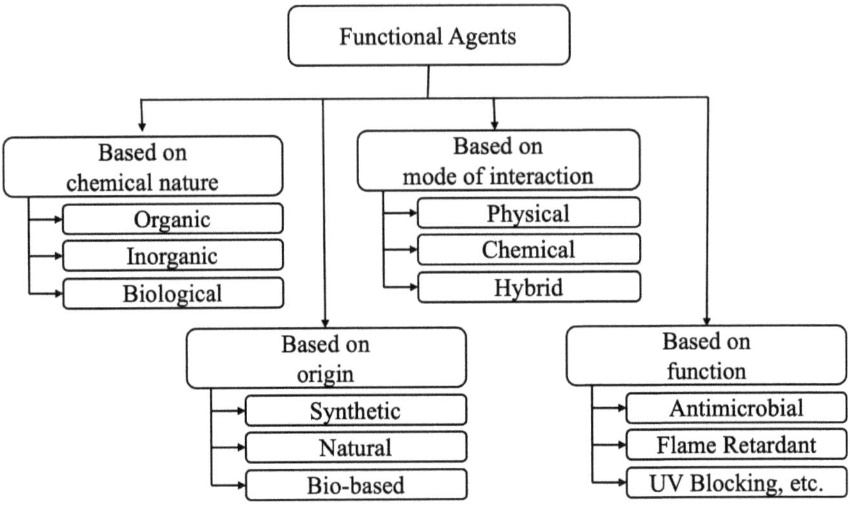

Fig. 2.1 Overview of classification of functional agents

nanotubes), and stimuli-responsive polymers that change their properties in response to environmental triggers such as temperature, pH, or humidity [27, 28] and enzymes [29, 30].

2.2.1 Nanomaterials

Nanoparticles typically range from 1 to 100 nm in size, which is about 1000 times smaller than the width of a human hair. A schematic illustration of 1 nm versus 1 mm is presented in Fig. 2.2. Nanomaterials, particularly metallic and metal oxide nanoparticles such as silver (AgNPs), titanium dioxide (TiO_2), and zinc oxide (ZnO), are extensively employed in textile surface functionalization due to their distinctive physicochemical properties at the nanoscale [31, 32]. Silver nanoparticles are renowned for their potent antimicrobial activity, achieved by disrupting microbial cell membranes and inhibiting DNA replication, making them highly suitable for medical, hygienic, and odor-resistant textiles [33]. Titanium dioxide nanoparticles exhibit strong photocatalytic activity under UV light, enabling self-cleaning and antibacterial functions by decomposing organic pollutants and inactivating pathogens [34]. Zinc oxide nanoparticles offer a dual benefit, combining antimicrobial efficacy with UV-blocking capability, thereby enhancing both hygiene and photoprotection in fabrics [35]. These nanomaterials are commonly applied to textile substrates using methods such as dip-coating, sol–gel processing, or in-situ chemical synthesis. Importantly, their incorporation can achieve long-lasting functionality without compromising the fabric's inherent properties such as breathability, flexibility, and mechanical strength [5].

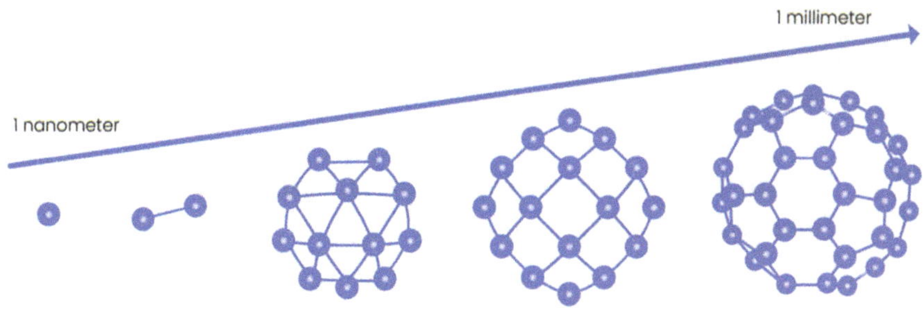

Fig. 2.2 Schematic visualization of 1 nm versus 1 mm

2.2.2 Conductive Polymers and Carbon-Based Materials

Conductive polymers and carbon-based materials play a pivotal role in advancing smart textiles for wearable electronics and sensing technologies [36, 37]. Conductive polymers, particularly poly(3,4-ethylenedioxythiophene): poly(styrenesulfonate) (PEDOT:PSS), are valued for their high electrical conductivity, flexibility, and ease of processing [38]. These properties make them ideal for incorporation into textiles using techniques such as printing, coating, or dipping, enabling real-time monitoring of physiological parameters, temperature, and mechanical strain [39]. Similarly, carbon-based materials like graphene and carbon nanotubes (CNTs) offer exceptional electrical and thermal conductivity, along with superior mechanical strength and chemical stability [40, 41]. Their integration into textile structures results in lightweight, stretchable, and durable systems capable of sensing, energy storage, and wireless data transmission. Together, conductive polymers and carbon nanomaterials form the foundation for next-generation electronic textiles that seamlessly combine comfort, functionality, and advanced performance [42].

2.2.3 Phase-change Materials for Thermoregulation

Phase-change materials (PCMs) are capable of absorbing, storing, and releasing thermal energy during phase transitions, most commonly between solid and liquid states [43]. In textile applications, PCMs such as paraffin waxes, fatty acids, and salt hydrates are integrated into fabrics to deliver thermal regulation by moderating temperature fluctuations [44]. When body or ambient temperatures rise, the PCMs absorb excess heat through melting; as temperatures fall, they solidify and release the stored thermal energy [45]. This reversible process helps maintain a stable microclimate close to the skin, significantly enhancing wearer comfort in changing thermal conditions. PCMs can be incorporated into textiles through methods such as microencapsulation, surface coating, or fiber spinning [46]. They are widely employed in functional clothing, including outdoor apparel, sportswear, military gear, and medical textiles, where thermal adaptability is essential.

2.2.4 *Stimuli-Responsive Polymers for Moisture, Temperature, or pH Sensitivity*

Stimuli-responsive polymers, also referred to as smart or intelligent polymers, are materials that undergo reversible physical or chemical transformations in response to external stimuli such as temperature, moisture, or pH [47, 48]. In textile applications, these polymers allow fabrics to dynamically adjust to changing environmental or physiological conditions [49]. For example, thermo-responsive polymers like poly(N-isopropylacrylamide) can expand or contract with temperature fluctuations, providing adaptive breathability or insulation. Moisture-sensitive polymers enhance wearer comfort by regulating water vapor permeability or triggering moisture-wicking behavior during perspiration [50, 51]. Similarly, pH-responsive polymers alter their charge or solubility in varying pH environments, making them ideal for medical textiles and targeted drug delivery [52]. These functional polymers can be incorporated into textiles through methods such as surface coating, grafting, or electrospinning. Their integration enables the development of advanced, multifunctional fabrics for applications in healthcare, athletic wear, and smart clothing [5].

2.2.5 *Enzymes*

Enzymes represent an important class of biological functional agents in textile surface functionalization due to their high specificity, efficiency, and environmentally benign nature. As biocatalysts, enzymes selectively modify fiber surfaces under mild processing conditions, such as low temperature and neutral pH, thereby minimizing fiber damage and reducing energy and chemical consumption. Commonly used enzymes in textile applications include cellulases, amylases, proteases, lipases, laccases, and catalases, each targeting specific components of textile materials [53, 54]. In textile functionalization, enzymes are primarily employed to tailor surface morphology, enhance wettability, and introduce reactive functional groups that improve subsequent finishing processes. For example, cellulases are widely used in cotton and other cellulosic textiles to hydrolyze surface fibrils, resulting in improved softness, smoothness, and color brightness. Proteases modify protein-based fibers such as wool and silk by partially hydrolyzing surface scales, thereby improving handle, dye uptake, and shrink resistance. Laccases and peroxidases facilitate oxidative reactions that enable surface activation, grafting of functional molecules, and immobilization of bioactive or antimicrobial compounds [54].

Beyond surface modification, enzymes can impart advanced functionalities to textiles, including antimicrobial activity, self-cleaning behavior, and biocompatibility, either directly or through enzyme-assisted immobilization of functional agents. Their biodegradability and low toxicity make enzymes particularly attractive for sustainable textile processing and biomedical applications [55]. However, challenges such as enzyme stability, reusability, and long-term durability on textile surfaces remain areas of ongoing research. Overall, enzyme-based functionalization offers a promising and eco-friendly approach for developing next-generation functional textiles.

2.3 Surface Chemistry and Physics in Textile Functionalization

Surface chemistry and surface physics jointly govern textile functionalization by controlling interactions at the fiber–surface interface, where most functional properties are introduced. Rather than altering the bulk of textile materials, functionalization primarily relies on surface modification, making the chemical composition, physical structure, and reactivity of the outermost fiber layers decisive for the efficiency, durability, and uniformity of functional finishes. A comprehensive understanding of interfacial phenomena enables precise control over adhesion, compatibility, and the long-term stability of functional agents on textile substrates.

Textile fibers possess inherent chemical functionalities—such as hydroxyl groups in cellulosic fibers, amide groups in protein fibers, and ester groups in synthetic polymers—which define their intrinsic surface reactivity and interaction potential with functional agents [56]. However, these native functional groups are often insufficient to form strong or durable bonds with advanced functional systems, necessitating surface activation or modification strategies. At the same time, the physical characteristics of the fiber surface, including roughness, porosity, and morphology, play a critical role in determining wettability, coating uniformity, and adhesion by regulating how liquids, particles, and functional layers interact with the textile surface [57, 58].

Fundamental surface physical principles such as surface energy, interfacial tension, and intermolecular forces, including van der Waals and electrostatic interactions, act in concert with surface chemistry to influence the stability and durability of applied finishes [59]. These interfacial interactions also affect macroscopic fabric properties, including friction, abrasion resistance, tactile comfort, and optical behavior such as reflectivity and color appearance. Consequently, the deliberate engineering of both chemical functionalities and physical surface attributes is essential to achieving textile surfaces that deliver high functional performance while maintaining wearer comfort and long-term durability [60, 61]. The following are basic concepts of surface chemistry and physics that are essential in textile functionalization.

2.3.1 Surface Functional Groups and Surface Energy

The presence and availability of reactive groups on fiber surfaces (e.g., –OH, –COOH, $-NH_2$, –CONH–, –COO– as shown in Fig. 2.3) determine how effectively functional agents can attach to textile substrates. These chemical moieties act as active sites for hydrogen bonding, electrostatic interactions, and covalent coupling with finishes, coatings, enzymes, and nanoparticles. Functionalization strategies, therefore, often aim to increase the density, accessibility, or reactivity of these surface groups through chemical oxidation, plasma treatment, enzymatic activation, or grafting techniques, enabling stronger interfacial bonding and improved durability of the applied functionalities [62, 63].

In parallel, surface energy plays a critical role in governing adhesion, wetting behavior, and coating uniformity. High-energy surfaces promote liquid spreading and intimate contact between the functional agent and the fiber surface, facilitating efficient adsorption and bonding. In contrast, low-surface-energy fibers such as polyester and polypropylene exhibit poor wettability and limited chemical affinity, often resulting in uneven deposition and weak adhesion of functional finishes. Surface activation treatments increase surface energy by introducing polar functional groups and modifying surface chemistry, thereby enhancing wettability and adhesion performance [64, 65]. The combined control of surface functional group chemistry and surface energy is therefore essential for achieving uniform, durable, and high-performance textile functionalization [1].

2.3.2 Wettability, Contact Angle, and Interfacial Bonding Mechanisms

Wettability describes the ability of a liquid to spread and penetrate a textile surface and is a critical parameter in dyeing, printing, and finishing processes. It is governed by both surface chemistry and surface topography, where the presence of polar functional groups and surface roughness together determine liquid–solid interactions. Wettability is commonly quantified through contact angle measurements (Fig. 2.4), with lower contact angles indicating improved wetting and higher affinity between the liquid phase and the fiber surface [66]. Enhanced wettability facilitates uniform chemical uptake, improved coating homogeneity, and increased functional efficiency of applied finishes.

Functional agents interact with textile fiber surfaces through a range of *interfacial bonding mechanisms*. These include physical adsorption driven by van der Waals forces, which provides initial attachment but limited durability; hydrogen bonding between polar functional groups of the fiber and the functional agent; electrostatic interactions arising from surface charge differences, which are particularly relevant in aqueous and colloidal systems; and covalent bonding, which involves the formation of stable chemical linkages at the fiber surface [65]. Among these interactions, covalent

Functional Group	Structure	Properties
Hydroxyl	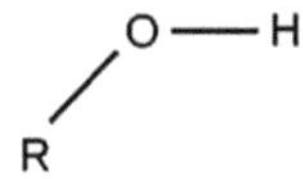	Polar
Methyl		Nonpolar
Carbonyl	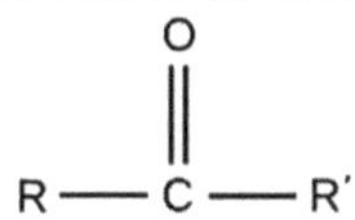	Polar
Carboxyl	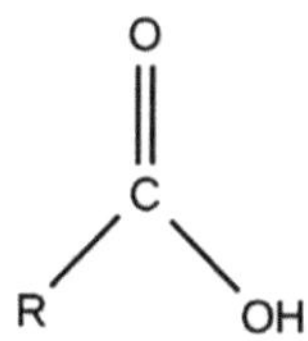	Charged, ionizes to release H^+. Since carboxyl groups can release H^+ ions into solution, they are considered acidic.
Amino	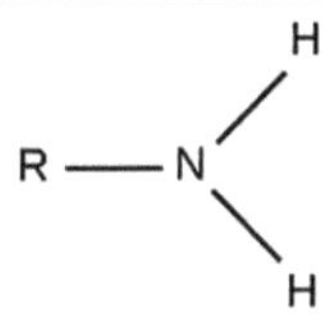	Charged, accepts H^+ to form NH_3^+. Since amino groups can remove H^+ from solution, they are considered basic.
Phosphate	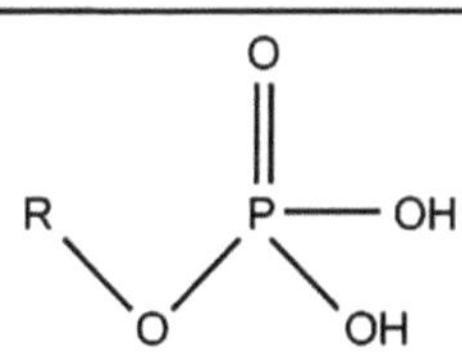	Charged, ionizes to release H^+. Since phosphate groups can release H^+ ions into solution, they are considered acidic.
Sulfhydryl	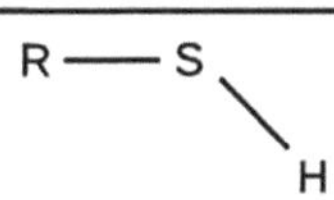	Polar

Fig. 2.3 Examples of functional groups: "R" is the organic molecule

attachment is especially important for achieving wash durability, abrasion resistance, and long-term functional performance, as it minimizes leaching and degradation of the applied functional layer during use and laundering [62].

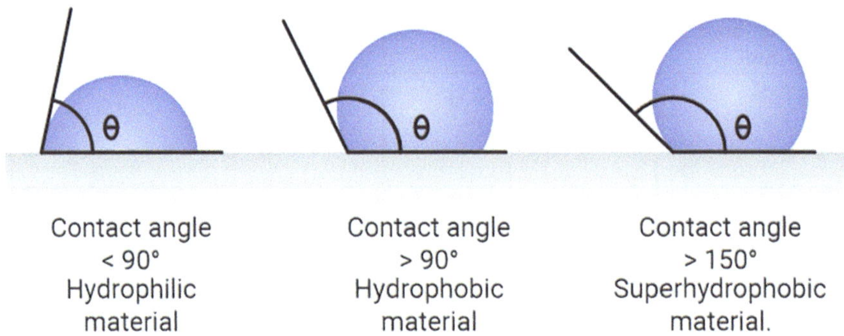

Fig. 2.4 Schematic illustration of wettability and contact angle of liquid over a solid surface

2.3.3 Surface Charge. Zeta Potential and Surface Activation

Surface charge governs interactions between textile fibers and charged species such as dyes, enzymes, nanoparticles, and microorganisms. The sign and magnitude of surface charge, commonly described by *zeta potential*, control electrostatic attraction and repulsion at the fiber–liquid interface and are therefore critical in antimicrobial finishes, enzyme immobilization, and nanoparticle deposition [65, 67]. *Surface activation* is used to modify chemically inert or low-energy fiber surfaces by introducing reactive functional groups and increasing surface energy. Techniques such as plasma treatment, chemical oxidation, enzymatic activation, and UV-induced grafting selectively alter the outer fiber layers, improving wettability, adhesion, and the durability of textile functionalization [68].

2.3.4 Chemical Affinity and Grafting

The chemical compatibility between a textile substrate and a functional agent is crucial for ensuring strong adhesion and long-term stability. When compatibility is poor, functional agents may leach out or distribute unevenly, reducing performance. Grafting offers a solution by covalently attaching polymers or functional molecules directly onto fiber surfaces, creating durable chemical bonds resistant to washing, abrasion, or environmental stress. This method allows textiles to acquire stable and uniform functionalities, such as antimicrobial properties, enhanced hydrophilicity, or stimuli-responsive behavior, making grafting a widely adopted strategy in the development of high-performance functional textiles [69, 70].

2.3.5 *Adsorption Isotherms, Kinetics, and Surface Diffusion and Penetration Depth*

The *adsorption behavior* of functional agents on textile surfaces is commonly described using models such as Langmuir and the Freundlich isotherms, which characterize equilibrium binding and saturation capacity at a given concentration. The Langmuir model assumes monolayer adsorption on homogeneous sites, whereas the Freundlich model describes heterogeneous surfaces and multilayer adsorption, making them useful tools for understanding how molecules distribute between liquid and solid phases during functionalization. *Adsorption kinetics*, often modeled with pseudo-first-order or pseudo-second-order equations, govern how quickly equilibrium is reached and influence processing time and uniformity; for example, rapid kinetics can enhance functional uptake and reduce treatment durations in dyeing systems [71]. During functionalization, transport processes such as *surface diffusion and penetration depth* determine how far agents migrate into fiber surfaces. Functionalization typically occurs in a near-surface region rather than the bulk fiber; controlling penetration depth is crucial to achieve high functional loading without adversely affecting fabric hand or mechanical integrity. Finally, surface chemistry evolves during use due to washing, abrasion, UV exposure, and oxidative aging, altering adsorption sites and affinities; understanding these aging mechanisms is essential for designing finishes with extended durability and sustained performance [72].

2.3.6 *Surface Roughness and Morphology*

Surface roughness and morphology describe the fine-scale texture and shape of textile fiber surfaces, which play a crucial role in determining characteristics such as wettability, friction, and tactile sensation [73]. The Wenzel and Cassie-Baxter models explain how surface roughness influences liquid interaction with fabrics: the Wenzel model applies when liquid completely fills surface grooves, amplifying hydrophilic or hydrophobic effects depending on the fiber's chemistry, whereas the Cassie-Baxter model accounts for air pockets trapped within surface structures, leading to enhanced water repellency and self-cleaning properties [74]. Moreover, surface roughness affects frictional behavior, impacting both durability and the comfort of fabric contact. Through precise control of surface morphology via functionalization methods, textiles can be tailored for superior moisture control, increased longevity, and improved sensory qualities [5].

2.3.7 Interfacial Adhesion Between Coatings and Fibers

Interfacial adhesion, which defines the bonding strength between functional coatings and textile fibers, results from a combination of physical and chemical interactions [75]. Van der Waals forces, weak, short-range molecular attractions, facilitate initial contact and adhesion. Hydrogen bonding provides stronger, more directional connections between polar groups on both the fiber surface and coating molecules, thereby improving bond strength and stability [76]. Electrostatic interactions, generated by opposite charges on the fiber and coating surfaces, further enhance attraction and promote uniform coating coverage. The interplay and balance of these forces are critical in determining the durability, consistency, and overall effectiveness of surface functionalization, ultimately influencing the lifespan and performance of treated textiles [27].

2.3.8 Surface Energy and Charge Distribution

Surface energy and charge distribution are key physical properties that govern the interactions between textiles and their surroundings [77]. Surface energy influences a fabric's affinity for liquids and particles, playing a vital role in processes like dyeing, where higher surface energy facilitates better wetting and stronger dye adhesion to fibers [78]. Meanwhile, the distribution of electrical charges on textile surfaces affects electrostatic interactions, which are essential in applications such as electrostatic filtration, where charged fibers effectively attract and capture airborne particles [79]. By manipulating these properties through surface functionalization, textiles can be engineered to exhibit enhanced features like improved color uptake, superior filtration performance, and minimized static buildup.

2.3.9 Optical Effects Through Nanostructuring

Nanostructuring textile surfaces allows for the manipulation of light at the nanoscale, producing distinctive optical effects such as anti-reflection and iridescence [80]. By engineering precise surface patterns or nanostructures, textiles can reduce light reflection, improve visibility, or minimize glare, beneficial for applications like optical sensors and camouflage [81]. Additionally, these nanostructures can create iridescence, where fabrics exhibit shifting, shimmering colors that vary with the viewing angle, caused by the interference and diffraction of light waves [82]. Beyond aesthetic appeal, these optical properties enable functional applications in smart textiles, signaling, and wearable display technologies.

2.4 Conclusion

Surface chemistry and physics form the critical foundation upon which functional textiles are designed and developed. Unlike bulk material properties, the behavior of fiber surfaces dictates how textiles interact with functional agents, respond to external environments, and retain engineered functionalities over time. Advances in surface science have revealed that fiber surfaces are highly dynamic and heterogeneous, with variations in chemical composition, porosity, and topography that influence adsorption, diffusion, and penetration of molecules. By understanding and controlling these factors, textile scientists can precisely modulate the distribution and activity of functional agents, ensuring both effectiveness and durability. Modern functionalization strategies increasingly move beyond simple surface coating to molecular-level engineering, where fibers are chemically tailored to achieve specific energy landscapes, reactive site distributions, and diffusion pathways. Techniques such as plasma activation, chemical grafting, and polymerization allow the covalent incorporation of functional molecules or polymers onto fiber surfaces, creating long-lasting and stable properties. Such approaches enable textiles to acquire diverse functionalities, ranging from antimicrobial and hydrophilic behavior to stimuli-responsive and smart capabilities, while maintaining essential attributes such as mechanical strength, flexibility, and comfort.

Understanding transport phenomena, including surface diffusion and penetration depth, is critical for controlling how far functional agents migrate into fibers. Near-surface functionalization often provides the desired performance while preserving bulk fiber integrity, but achieving this requires precise knowledge of adsorption kinetics, equilibrium behavior, and diffusion dynamics. Models such as Langmuir and Freundlich isotherms, combined with kinetic analyses, offer predictive insights into how functional agents attach, saturate, and redistribute across fiber surfaces. Integrating these models with experimental methods allows rational design of functionalization processes, minimizing resource use and enhancing reproducibility. Another essential consideration is the dynamic evolution of textile surfaces under real-world conditions. Washing, abrasion, UV exposure, and oxidative environments alter surface chemistry, creating new reactive sites, modifying energy landscapes, and affecting the retention of functional agents. Designing textiles with adaptive durability requires a holistic approach that considers initial surface properties, functionalization strategy, and long-term aging mechanisms, ensuring that performance is maintained throughout the product's service life.

Looking forward, the field of textile functionalization is moving toward multifunctional, sustainable, and high-performance materials. Bio-inspired coatings, stimuli-responsive polymers, and environmentally benign chemistries are increasingly being integrated into textile design, allowing fabrics to respond intelligently to external stimuli while minimizing environmental impact. Computational modeling, machine learning, and advanced characterization techniques offer unprecedented opportunities to predict and control surface behavior at molecular and nanoscale levels. By combining fundamental understanding of surface chemistry and physics with practical engineering solutions, the next generation of functional textiles can achieve robust performance, minimal ecological footprint, and extended operational life, meeting the evolving demands of technical applications, healthcare, wearable electronics, and consumer markets. In conclusion, surface chemistry and physics are not merely supporting considerations in textile functionalization—they are central to designing materials that are durable, effective, and adaptable. A deep understanding of surface interactions, coupled with innovative chemical and physical engineering approaches, enables textiles to transcend conventional limitations, providing tailored functionalities that meet the complex requirements of modern applications. The future of functional textiles will depend on the continued integration of molecular-level control, predictive modeling, and sustainability-driven design, paving the way for textiles that are smarter, longer-lasting, and more environmentally responsible.

Acknowledgements The authors gratefully acknowledge the support provided by the BGMEA University of Fashion and Technology, Dhaka 1230, Bangladesh.

AI-Usage Declaration During the preparation of this chapter, the authors used AI tools to improve the language of the manuscript. After using this tool/service, the authors reviewed and edited the content as needed and take full responsibility for the content of the publication.

References

1. A. Nadi, A. Boukhriss, A. Bentis, E. Jabrane, S. Gmouh, Evolution in the surface modification of textiles: a review. Text. Prog. **50**(2), 67–108 (2018)
2. G. Zhao, X. Li, M. Huang, Z. Zhen, Y. Zhong, Q. Chen, H. Zhu, The physics and chemistry of graphene-on-surfaces. Chem. Soc. Rev. **46**(15), 4417–4449 (2017)
3. M.A. Abtew, D. Atalie, B.K. Dejene, K. McBee-Black, Intelligent and electronic textile materials for adaptive apparel: innovations, functional design, and future directions. J. Ind. Text. **55**, 15280837251346788 (2025)
4. S. Simões, High-performance advanced composites in multifunctional material design: state of the art, challenges, and future directions. Materials **17**(23), 5997 (2024)
5. Y. Zhang, X. Xia, K. Ma, G. Xia, M. Wu, Y.H. Cheung, J.H. Xin, Functional textiles with smart properties: their fabrications and sustainable applications. Adv. Func. Mater. **33**(33), 2301607 (2023)
6. N.R. Dhineshbabu, S. Bose, UV-resistant and fire-retardant properties in fabrics coated with polymer-based nanocomposites derived from sustainable and natural resources for protective clothing application. Compos. B Eng. **172**, 555–563 (2019)
7. K. Althagafy, F. AlhashmiAlamer, Metal-coated textiles in composite materials: advances in manufacturing techniques, properties, and applications. Cellulose 1–25 (2025)
8. M.R. Hossain, M.R. Ahmed, M.S. Alam, Smart textiles. Text. Prog. **55**(2), 47–108 (2023)
9. S. Sharma, P. Sudhakara, A.A.B. Omran, J. Singh, R.A. Ilyas, Recent trends and developments in conducting polymer nanocomposites for multifunctional applications. Polymers **13**(17), 2898 (2021)
10. M.R. Miah, J. Ding, H. Zhao, Q. Chu, H. Wang, J. Wang, J. Zhu, Enhancement of mechanical, thermal, and barrier behavior of sustainable PECF copolyester nanocomposite films using polydopamine-functionalized MXene fillers. Langmuir **41**(15), 9680–9691 (2025)

11. M.R. Miah, S. Mahmud, A.N. Khan, M.A. Jalil, J. Wang, J. Zhu, Recent advances in sustainable thermoplastic polyester elastomers: synthesis, properties and applications. Polymer 128878 (2025)
12. M.R. Miah, J. Wang, J. Zhu, Material advancements in plant/artificial fiber-based woven and non-woven fabrics and their composites, in *Innovations in Woven and Non-woven Fabrics Based Laminated Composites* (Springer Nature Singapore, Singapore, 2025), pp. 17–50
13. J. Ding, H. Wang, H. Zhao, S. Shi, J. Su, Q. Chu, J. Zhu, Large-size ultrathin mica nanosheets: reinforcements of biobased PEF polyester. Giant **18**, 100264 (2024)
14. F. Guo, Z. Guo, Inspired smart materials with external stimuli responsive wettability: a review. RSC Adv. **6**(43), 36623–36641 (2016)
15. M. Varma, S. Chandran, Surface treatment of natural fibers for enhancing interfacial adhesion and mechanical properties in biocomposites-a comprehensive review. Compos. Interfaces 1–37 (2025)
16. M.R. Miah, J. Ding, H. Zhao, H. Wang, Q. Chu, B. Fang, J. Zhu, Enhancing the mechanical and barrier properties of biobased polyester incorporated with carboxylated cellulose nanofibers. Mater. Today Commun. **38**, 108538 (2024)
17. M.R. Miah, J. Ding, H. Zhao, Q. Chu, H. Wang, J. Wang, J. Zhu, Boron nitride-based polyester nanocomposite films with enhanced barrier and mechanical performances. ACS Appl. Polym. Mater. **6**(5), 2913–2923 (2024)
18. C.G. Jothi Prakash, R. Prasanth, Approaches to design a surface with tunable wettability: a review on surface properties. J. Mater. Sci. **56**(1), 108–135 (2021)
19. P. Fabbri, M. Messori, Surface modification of polymers: chemical, physical, and biological routes, in *Modification of Polymer Properties* (William Andrew Publishing, 2017), pp. 109–130
20. H. Wang, J. Ding, H. Zhao, Q. Chu, M.R. Miah, J. Wang, J. Zhu, Preparing strong, tough, and high-barrier biobased polyester composites by regulating interfaces of carbon nanotubes. Mater. Today Nano **25**, 100463 (2024)
21. M.R. Miah, Y. Dong, J. Wang, J. Zhu, Recent progress on sustainable 2, 5-furandicarboxylate-based polyesters: properties and applications. ACS Sustain. Chem. Eng. **12**(8), 2927–2961 (2024)
22. J. Shi, S. Liu, L. Zhang, B. Yang, L. Shu, Y. Yang, X. Tao, Smart textile-integrated microelectronic systems for wearable applications. Adv. Mater. **32**(5), 1901958 (2020)
23. M.A. Taher, X. Wang, K.M. Faridul Hasan, M.R. Miah, J. Zhu, J. Chen, Lignin modification for enhanced performance of polymer composites. ACS Appl. Bio Mater. **6**(12), 5169–5192 (2023)
24. J. Ding, H. Zhao, S. Shi, J. Su, Q. Chu, H. Wang, J. Zhu, High-strength, high-barrier bio-based polyester nanocomposite films by binary multiscale boron nitride nanosheets. Adv. Func. Mater. **34**(1), 2308631 (2024)
25. J. Ghosh, N.S. Rupanty, T. Noor, T.R. Asif, T. Islam, V. Reukov, Functional coatings for textiles: advancements in flame resistance, antimicrobial defense, and self-cleaning performance. RSC Adv. **15**(14), 10984–11022 (2025)
26. L. Jin, C. Ji, S. Chen, Z. Song, J. Zhou, K. Qian, W. Guo, Multifunctional textiles with flame retardant and antibacterial properties: a review. Molecules **28**(18), 6628 (2023)
27. Y. Guo, X. Wei, S. Gao, W. Yue, Y. Li, G. Shen, Recent advances in carbon material-based multifunctional sensors and their applications in electronic skin systems. Adv. Func. Mater. **31**(40), 2104288 (2021)
28. I.A. Rashid, A. Afzal, Smart polymers and composites, in *Advanced Functional Polymers: Synthesis to Applications* (Springer Nature Singapore, Singapore, 2023), pp. 77–99
29. M.N. Morshed, N. Behary, J. Guan, V.A. Nierstrasz, Immobilizing redox enzyme on amino functional group-integrated tailor-made polyester textile: high loading, stability, and application in a bio-fenton system. ACS Sustain. Chem. Eng. **9**(26), 8879–8894 (2021)
30. M. Coradi, M. Zanetti, A. Valério, D. de Oliveira, A. da Silva, S.M.D.A.G. Ulson, A.A.U. de Souza, Production of antimicrobial textiles by cotton fabric functionalization and pectinolytic enzyme immobilization. Mater. Chem. Phys. **208**, 28–34 (2018)

31. M.T. Hossain, S. Reza, M.A. Islam, M. Mohebbullah, T. Islam, Progress and prospects of chemical functionalization of textiles via nanotechnology. ACS Appl. Eng. Mater. **3**(1), 1–20 (2025)
32. S.U. Islam, G. Sun, Biological chemicals as sustainable materials to synthesize metal and metal oxide nanoparticles for textile surface functionalization. ACS Sustain. Chem. Eng. **10**(31), 10084–10104 (2022)
33. D. K. Tripathi, A. Tripathi, Shweta, S. Singh, Y. Singh, K. Vishwakarma, D.K. Chauhan, Uptake, accumulation and toxicity of silver nanoparticle in autotrophic plants, and heterotrophic microbes: a concentric review. Front. Microbiol. **8**, 07 (2017)
34. K. Nagaraj, S. Radha, C.G. Deepa, K. Raja, V. Umapathy, N.P. Badgujar, C. Uthra, Photocatalytic advancements and applications of titanium dioxide (TiO_2): progress in biomedical, environmental, and energy sustainability. Next Res. **2**(1), 100180 (2025)
35. S. Ahmed, A.A.R. Mazumdar, M. Syduzzaman, M.S. Mahmud, UV protection and antimicrobial properties of textiles using nanotechnology, in *Nanotechnology in Textile Finishing: Advancements and Applications* (2024), pp. 191–232
36. J.S. Heo, M.F. Hossain, I. Kim, Challenges in design and fabrication of flexible/stretchable carbon-and textile-based wearable sensors for health monitoring: a critical review. Sensors **20**(14), 3927 (2020)
37. M.R. Miah, M. Yang, M.M. Hossain, S. Khandaker, M.R. Awual, Textile-based flexible and printable sensors for next generation uses and their contemporary challenges: a critical review. Sens. Actuators, A **344**, 113696 (2022)
38. L.V. Kayser, D.J. Lipomi, Stretchable conductive polymers and composites based on PEDOT and PEDOT: PSS. Adv. Mater. **31**(10), 1806133 (2019)
39. M. Azeem, M. Shahid, I. Masin, M. Petru, Design and development of textile-based wearable sensors for real-time biomedical monitoring; a review. J. Text. Inst. **116**(1), 80–95 (2025)
40. M.H. Islam, S. Afroj, M.A. Uddin, D.V. Andreeva, K.S. Novoselov, N. Karim, Graphene and CNT-based smart fiber-reinforced composites: a review. Adv. Func. Mater. **32**(40), 2205723 (2022)
41. J. Ding, H. Wang, H. Zhao, M.R. Miah, J. Wang, J. Zhu, High-compact MXene-based coatings by controllable interfacial structures. Nanoscale **15**(19), 8870–8880 (2023)
42. Y. Tian, R. Ding, S.S. Yoon, S. Zhang, J. Yu, B. Ding, Recent advances in next-generation textiles. Adv. Mater. **37**(8), 2417022 (2025)
43. A. Usman, F. Xiong, W. Aftab, M. Qin, R. Zou, Emerging solid-to-solid phase-change materials for thermal-energy harvesting, storage, and utilization. Adv. Mater. **34**(41), 2202457 (2022)
44. M.T. Hossain, M.A. Shahid, M.Y. Ali, S. Saha, M.S.I. Jamal, A. Habib, Fabrications, classifications, and environmental impact of PCM-incorporated textiles: current state and future outlook. ACS Omega **8**(48), 45164–45176 (2023)
45. M.R. Mohaghegh, Y. Alomair, M. Alomair, S.H. Tasnim, S. Mahmud, H. Abdullah, Melting of PCM inside a novel encapsulation design for thermal energy storage system. Energy Convers. Manag. X **11**, 100098 (2021)
46. K. Yang, M. Venkataraman, X. Zhang, J. Wiener, G. Zhu, J. Yao, J. Militky, Incorporation of organic PCMs into textiles. J. Mater. Sci. **57**(2), 798–847 (2022)
47. A. Balcerak-Woźniak, M. Dzwonkowska-Zarzycka, J. Kabatc-Borcz, A comprehensive review of stimuli-responsive smart polymer materials—recent advances and future perspectives. Materials **17**(17), 4255 (2024)
48. M. Mrinalini, S. Prasanthkumar, Recent advances on stimuli-responsive smart materials and their applications. ChemPlusChem **84**(8), 1103–1121 (2019)
49. D.G. Prajapati, B. Kandasubramanian, A review on polymeric-based phase change material for thermo-regulating fabric application. Polym. Rev. **60**(3), 389–419 (2020)
50. K.H.D. Tang, Advances in thermoregulating textiles: materials, mechanisms, and applications. Textiles **5**(2), 22 (2025)
51. J. Tabor, K. Chatterjee, T.K. Ghosh, Smart textile-based personal thermal comfort systems: current status and potential solutions. Adv. Mater. Technol. **5**(5), 1901155 (2020)

52. J. Singh, P. Nayak, PH-responsive polymers for drug delivery: trends and opportunities. J. Polym. Sci. **61**(22), 2828–2850 (2023)
53. D. Coman, S. Oancea, N. Vrinceanu, Biofunctionalization of textile materials by antimicrobial treatments: a critical overview. Roman. Biotechnol. Lett. **15**(1), 4913–4921 (2010)
54. V.A. Nierstrasz, Enzyme surface modification of textiles, in *Surface Modification of Textiles* (Woodhead Publishing, 2009), pp. 139–163
55. M.N. Morshed, Immobilized enzyme in textile processing, in *Biotechnology Approaches in Textile Technology* (CRC Press, 2025), pp. 156–191
56. R.R. Mather, R.H. Wardman, S. Rana, *The Chemistry of Textile Fibres* (Royal Society of Chemistry, 2023)
57. E. Skrzetuska, A.K. Puszkarz, J. Nosal, Assessment of the impact of the surface modification processes of cotton and polyester fabrics with various techniques on their structural, biophysical, sensory, and mechanical properties. Polymers **14**(4), 796 (2022)
58. B. D'Souza, A.K. Kasar, J. Jones, A. Skeete, L. Rader, P. Kumar, P.L. Menezes, A brief review on factors affecting the tribological interaction between human skin and different textile materials. Materials **15**(6), 2184 (2022)
59. K.J.M. Söderholm, Coatings in dentistry-a review of some basic principles. Coatings **2**(3), 138–159 (2012)
60. M.G. Tadesse, C. Loghin, I. Dulgheriu, E. Loghin, Comfort evaluation of wearable functional textiles. Materials **14**(21), 6466 (2021)
61. S. Motlogelwa, Comfort and durability in high-performance clothing, in High-Performance Apparel (Woodhead Publishing, 2018), pp. 209–219
62. W.D. Schindler, P.J. Hauser, *Chemical Finishing of Textiles* (Elsevier, 2004)
63. M.R. McCoustra, R.R. Mather, Plasma modification of textiles: understanding the mechanisms involved. Text. Prog. **50**(4), 185–229 (2018)
64. K.L. Mittal (Ed.), *Polymer Surface Modification: Relevance to Adhesion*, vol. 2 (CRC Press, 2023)
65. K. Holmberg, *Handbook of Applied Surface and Colloid Chemistry* (2002)
66. A.W. Adamson, A.P. Gast, Physical Chemistry of Surfaces, vol. 150 (Interscience Publishers, New York, 1967), p. 180
67. R.J. Hunter, Zeta Potential in Colloid Science: Principles and Applications, vol. 2 (Academic Press, 2013)
68. H. Höcker, Plasma treatment of textile fibers. Pure Appl. Chem. **74**(3), 423–427 (2002)
69. J. Ghosh, N.S. Rupanty, F. Khan, T. Noor, R. Jahangir, S. Mirmohammadsadeghi, T. Islam, Grafting modification for textile functionalization: innovations and applications. Discov. Appl. Sci. **7**(1), 49 (2025)
70. N. Abidi, Surface grafting of textiles, in *Surface Modification of Textiles* (Woodhead Publishing, 2009), pp. 91–107
71. J. Gao, F. Guan, Y. Tang, X. Li, T. Ma, P. Zheng, Adsorption kinetic and equilibrium studies for maxilon blue GRL dye by cotton fiber in aqueous solutions. Discover Chemistry **2**(1), 224 (2025)
72. M. Amberg, B. Hanselmann, D. Hegemann, Surface functionalization of fibrous material by roll-to-roll low pressure plasma processing. Surf. Coat. Technol. **497**, 131806 (2025)
73. O. Neděla, P. Slepička, V. Švorčík, Surface modification of polymer substrates for biomedical applications. Materials **10**(10), 1115 (2017)
74. A.M. Jonas, R. Cai, R. Vermeyen, B. Nysten, M. Vanneste, D. De Smet, K. Glinel, How roughness controls the water repellency of woven fabrics. Mater. Des. **187**, 108389 (2020)
75. R. Molina, J. Esquena, P. Erra, Interfacial processes in textile materials: relevance to adhesion. J. Adhes. Sci. Technol. **24**(1), 7–33 (2010)
76. T. Zhang, Y. Hu, Y. Dong, S. Jiang, X. Han, The influence of hydrogen bonding in wood and its modification methods: a review. Polymers **17**(15), 2064 (2025)
77. K. Song, J. Lee, S.O. Choi, J. Kim, Interaction of surface energy components between solid and liquid on wettability, and its application to textile anti-wetting finish. Polymers **11**(3), 498 (2019)

78. F. Ferrero, M. Periolatto, Modification of surface energy and wetting of textile fibers. Wetting Wettability 139–168 (2015)
79. B. Le, N. Omran, E. Elnabawy, A.H. Hassanin, K. Mahmoud, N. Shehata, I. Shyha, Exploring advances in nanofiber-based face masks: a comprehensive review of mechanical, electrostatic, and antimicrobial functionality filtration for the removal of airborne particulate matter and pathogens. Emerg. Mater. **7**(3), 765–800 (2024)
80. S. Chattopadhyay, Y.F. Huang, Y.J. Jen, A. Ganguly, K.H. Chen, L.C. Chen, Anti-reflecting and photonic nanostructures. Mater. Sci. Eng. R. Rep. **69**(1–3), 1–35 (2010)
81. L.M. Degenstein, D. Sameoto, J.D. Hogan, A. Asad, P.I. Dolez, Smart textiles for visible and IR camouflage application: state-of-the-art and microfabrication path forward. Micromachines **12**(7), 773 (2021)
82. F. Chen, Y. Huang, R. Li, S. Zhang, B. Wang, W. Zhang, R. Zhang, Bio-inspired structural colors and their applications. Chem. Commun. **57**(99), 13448–13464 (2021)

Chapter 3
Physicochemical Techniques for Textile Surface Modification and Functionalization

Melkie Getnet Tadesse and Jörn Felix Lübben

Abstract Significant progress in surface modification techniques has been promoted by the need for novel fabrics with improved functionality and customized qualities. In order to obtain desired qualities such as enhanced wettability, functional characteristics, antimicrobial activity, self-cleaning qualities, flame retardancy, and dye uptake, surface modification in terms of physicochemical techniques is important. This chapter examines the role of physicochemical approaches in changing and functionalizing textile surfaces. It gives a thorough rundown of methods, stressing their fundamental workings, benefits, and drawbacks. These methods include sol–gel procedures, grafting techniques, plasma technologies, and other surface modification techniques. The chapter explores how each surface modification techniques bring change with respect to physical, mechanical, structural, and chemical properties of textile materials. Advancements and challenges for industrial applications and future research are also briefly discussed.

Keywords Surface modification · Functionalization · Plasma · Grafting

3.1 Introduction

The increasing need for performance [1], sustainability, and multifunctionality in existing applications has driven the perpetual innovation of textile materials. The capacity to modify and functionalize textile fibers' surface characteristics is essential to this advancement since it allows for improvements in wettability, dyeability, adhesion, and added functionalities such as flame retardancy, antimicrobial activity, and other functional qualities without sacrificing the material's bulk properties

M. G. Tadesse (✉) · J. F. Lübben
Sustainable Engineering (STE), Albstadt-Sigmaringen University, Albstadt, Germany
e-mail: tadesse@hs-albsig.de

M. G. Tadesse
Ethiopian Institute of Textile and Fashion Technology, Bahir Dar University, Bahir Dar, Ethiopia

M. N. Morshed and M. Shahid (eds.), *Advances in Surface Modification and Functionalization in Textiles*, Textile Science and Clothing Technology,
https://doi.org/10.1007/978-3-032-23377-6_3

[2]. Surface modification and functionalization techniques can be largely classified into physical and chemical methods. Physical methods include methods such as plasma surface treatment, corona discharge, ozone exposure, and microwave irradiation, which involve modifying the surface's topography or morphology without altering its chemical composition, and the most common mechanisms include etching and blasting. Physical surface modification primarily acts by changing the surface morphology and energy, often increasing roughness and introducing reactive sites that facilitate further functionalization. Therefore, physical methods are preferred when the goal is to measure mechanical properties (e.g., fatigue resistance) or to prepare surfaces for subsequent treatments. Chemical methods such as graft copolymerization, enzymatic treatments, and wet-chemical processes, on the other hand, change the surface's chemistry, for example, by adding extra functional/reactive groups. By doing this, it introduces or alters functional groups on fiber surfaces to impart properties like improved water absorption, antistatic behavior, enhanced chemical affinity, and so on. Hence, they are necessary for applications compelling specific chemical functionalities, such as the biomedical implants (to enhance biocompatibility), electronics (for circuit patterning), or anti-corrosion coatings. The physical and chemical properties of textile materials are transformed by a variety of dynamic processes, including physicochemical approaches for textile surface modification and functionalization. Through the combination of physical and chemical methods, physicochemical methods for textile surface modification provide a practical way to change the surface characteristics of textile fibers while maintaining their inherent bulk qualities, like strength, flexibility, and general structure. These techniques, which include chemical grafting, laser treatment, UV irradiation, and plasma treatment, allow for the precise molecular or nanoscopic modification or functionalization of fiber surfaces. As a result, textiles can be given a variety of new or improved qualities, such as modulated wettability for better moisture management or water repellency, increased surface adhesion to facilitate composite manufacturing, improved electrical conductivity for use in smart textiles and wearable electronics, and improved flame retardancy for improved fire safety. A particular physicochemical approach is chosen based on the fiber type, the planned end-use, and the desired surface properties. Because natural fibers have a lot of reactive groups, they might need different approaches than synthetic polymers, which frequently have no surface activity and have problems with wettability and adhesion. High-performance composites, biomedical textiles, and smart textiles have all benefited from the integration of various surface modification techniques. This chapter offers a thorough summary of the most recent physicochemical methods used for functionalization and surface modification of textiles, such as ozone and plasma treatment. It examines the fundamental ideas, real-world applications, and the improvements in textile qualities that emerge, laying the groundwork for creative uses and further advancements in the area.

3.2 Physicochemical Techniques for Textile Surface Modification and Functionalization

Physicochemical techniques for textile surface modification fundamentally alter the textile surface so that its original properties, such as hydrophilicity, corrosion resistance, biocompatibility, or adhesion, can be altered without affecting the bulk properties. This alteration could be practical by various mechanisms, such as:

- *Etching*: Selectively removes material/waste from the surface through chemical reactions, enabling the creation of intricate patterns or designs. Etching is commonly used in electronics (e.g., PCB fabrication), decorative arts, and microfabrication, offering high precision and customization.
- *Surface Functionalization*: Involves introducing specific chemical groups onto the surface to impart new properties, such as hydrophilicity, biocompatibility, or anti-fouling behavior. This can be achieved via plasma treatments, wet-chemical processes, or grafting techniques.
- *Coating and Film Deposition*: Applies thin layers of materials (such as polymers, metals, or ceramics) onto the substrate to enhance surface properties like corrosion resistance, wear resistance, or biological compatibility.
- *Chemical Grafting*: Covalently attaches molecules or polymers to the surface, providing tailored functionalities for specific applications, such as medical implants or sensors.
- *Thermochemical Treatments*: Techniques like carburizing, nitriding, and boriding diffuse elements into the surface, altering its chemical composition and improving properties such as hardness and corrosion resistance.

The schematic view of physiochemical treatment methods is shown in Fig. 3.1.

3.2.1 *Plasma Treatment*

Plasma, often termed "the fourth state of matter," was invented by Sir William Crookes in 1879. Plasma is characterized as a visible glow discharge with colors ranging from blue, white, to dark purple depending on the type of gas (Fig. 3.2).

Plasma treatment of textile materials is considered just a pretreatment process that improves the wettability of fibers, yarns, or fabrics, which further allows chemicals/dyes to bond well and also enhances durability using two techniques.

(1) *Etching* is a surface modification process that involves the controlled removal of a thin layer from the textile surface [3]. This technique alters the surface morphology and roughness, typically by exposing the material to reactive plasma species or chemical agents. In the context of plasma treatment, etching occurs through physical or chemical interactions between energetic ions or radicals and the textile substrate. This results in micro- or nanoscale texturing of the fibers, which can improve properties such as surface area, adhesion, dye

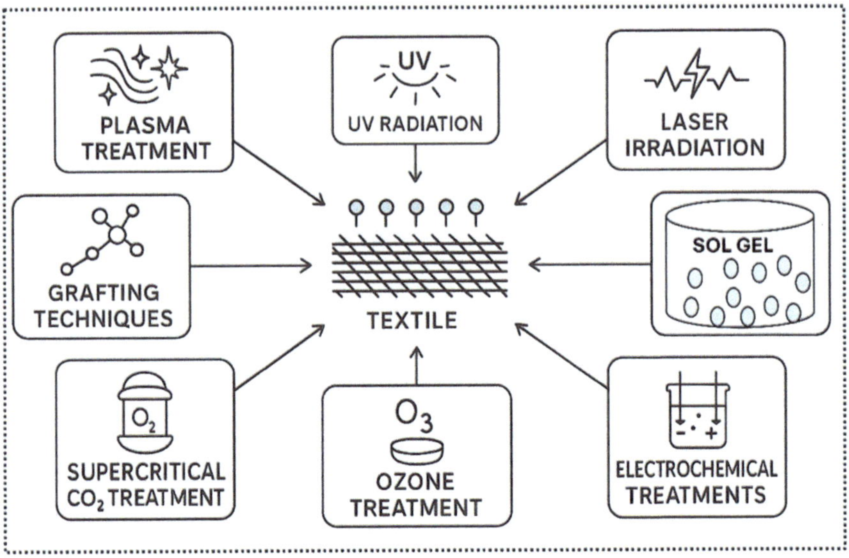

Fig. 3.1 Illustration of surface modification and functionalization methods of textile materials

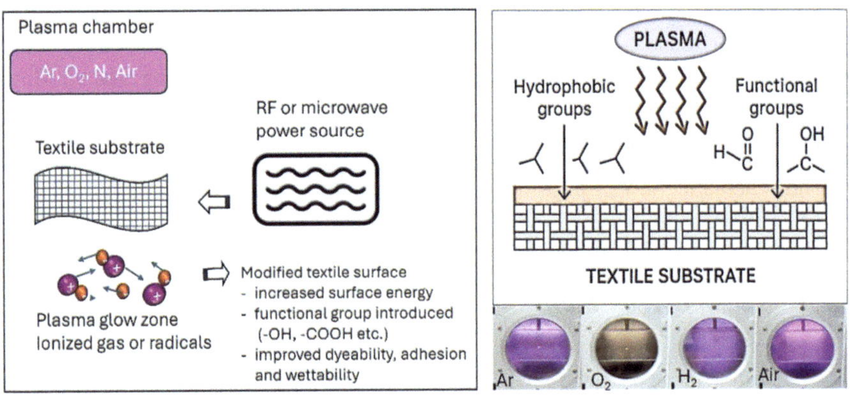

Fig. 3.2 Plasma surface modification and its effect on textile material

uptake, and hydrophilicity. Etching is particularly valuable for preparing textile surfaces for subsequent functional coatings or treatments without compromising the overall integrity or flexibility of the fabric. In this case, removing the surface matter, including fiber finishes, size, and other contaminants, might be performed simultaneously, which could be taken as a pretreatment for the next process. For instance, it could improve wettability and can clean the textile material [4].

(2) *Surface Functionalization*: Fiber surface rendering with recrystallization can occur when reactive gases, such as nitrogen, are employed during surface modification processes [3].

During plasma treatment, the use of reactive gases like nitrogen can initiate complex surface interactions, leading not only to chemical functionalization but also to physical changes at the microscopic level. One such phenomenon is surface recrystallization, where the molecular structure of the fiber surface is reorganized into more ordered crystalline regions. This can enhance the mechanical strength, thermal stability, and chemical resistance of the textile fibers. Additionally, nitrogen-containing plasma treatments can introduce amine or other nitrogen-based functional groups onto the fiber surface, further improving compatibility with coatings, dyes, or biomolecules. The combined effects of recrystallization and chemical modification contribute to the creation of high-performance, functional textiles tailored for advanced applications.

Plasma is an environmentally friendly process [3]. It can be used to enhance dyeability, adhesion, and wettability for both natural and synthetic fibers. Plasma process is controlled by gas pressure (Ar, O_2, Air, etc.) separately using a combination of gases. The gas pressure, type of gas, time of exposure, and treatment type (etching or functionalization) can be optimized to get optimum results. Plasma treatment not only modifies textiles but also related materials such as PET sheets for adhesion improvement [5]. Particularly when it comes to textiles, plasma techniques have shown themselves to be extremely adaptable and mild ways of surface alteration [6]. Without changing the material's bulk properties, these methods enable exact customization of surface attributes like wettability, adhesion, and chemical activity. For applications ranging from functional coatings to smart textile integration, their non-invasive nature makes them especially appropriate for delicate textile fibers, guaranteeing the preservation of structural integrity while improving performance qualities.

Plasma treatment might bring a different property based on the types of textiles. For instance, it improves consistency, paint strength, coatability, and portability for silk fiber [7]. Table 3.1 compares the use of plasma for different textile materials.

Although the classification of plasma technology can differ based on factors such as its intended application, the types of gases utilized, and various technical criteria, for the purposes of this chapter, we have chosen to organize plasma technology into two primary categories: low-pressure plasma and atmospheric plasma. This approach provides a clear and practical framework for discussion, allowing us to address the fundamental differences, operational principles, and unique advantages associated with each category [13]. Atmospheric plasma is further divided into: corona discharge (CD), dielectric barrier discharge (DBD), atmospheric-pressure glow discharge (APGD), and atmospheric-pressure plasma jet (APPJ). In the following section, each category is discussed.

Table 3.1 Comparison of the effects and characteristics of plasma surface modification

Textile	Plasma-induced surface changes	Key property changes	References
Silk	Minimal topographic change; increased N and O content	Improved cell adhesion, enhanced biocompatibility, and portability	[7, 8]
Wool	Enhanced surface roughness; removal of fatty layer, oxidation	Improved dyeability at lower temperatures; increased hydrophilicity; higher breaking strength; better wrinkle recovery	[4, 9, 10]
Cotton	Enhanced hydrophilicity; cracks/grooves; increased surface energy	Improved wickability; dye uptake, and water vapor permeability; reduced water vapor resistance	[10, 11]
Polyamide	Improved wettability; stable surface modification; new functional groups	Enhanced dyeability, wettability, and comfort properties	[10, 11]
Polyester	Introduction of polar groups; increased roughness and surface area	Enhanced dyeing performance, wickability, and comfort	[11, 12]
Viscose	Removal of hydroxyl groups; increased hydrophobicity (with N_2)	Modified comfort properties; can become more hydrophobic depends on plasma gases	[8, 11]
Acrylic	Removal of unwanted surface matter, new amide, and carboxyl groups	Improved dyeing and shrink resistance; tailored surface chemistry	[9]

3.2.1.1 Low-Pressure Plasma

Low-pressure plasma treatment in textiles modifies fabric properties such as hydrophobicity, hydrophilicity, and dirt repellency [14]. For low-pressure plasma treatment of textiles, the pressure limit is generally in the range of 1–100 Pa and usually 70 Pa. By improving these properties, cleaning and activation, adhesion improved, dyeability and wettability improved, and hair removal. Low-pressure plasma is not only used to modify the surface but also to coat a thin polymer on the textile surface [15]. This is the polymerization of an organic gas/monomer gas or vapor to deposit on the surface of the textile material. Low-pressure plasma uses argon, oxygen, air, or nitrogen gas for cleaning, activation, or crosslinking functions. Two gases can be combined in the ratio to optimize the process. Low-pressure plasma is also used to increase the adhesion properties of polyester films [5]. Low-pressure plasma surface modification has determining parameters such as gas pressure, time of exposure, and pre-cleaning activities. After etching the surface with plasma, coating of the substrate will be easy as the adhesion of the substrate is improved. Post-treatment using oxygen plasma can also be used to enhance the hydrophobic nature of textile materials [16]. In addition to superhydrophobic properties, antimicrobial properties can also be added. Various functionalities have been explored through atmospheric plasma treatment, including enhanced hydrophilicity and hydrophobicity, improved adhesion, surface cleaning, better dyeing and printing performance, sterilization and antimicrobial effects, anti-felting and shrink resistance in wool fabrics, wrinkle resistance, increased flame retardancy, ultraviolet (UV) protection, self-cleaning capabilities, and asymmetric wettability. However, slight changes in the physical and mechanical properties of the textiles may occur as a result of the treatment [14]. In addition, low-pressure plasma treatment on textiles has several advantages in terms of sustainability, such as reducing the use of water and chemicals, shortening the processing time when compared to conventional surface modification techniques, and energy savings.

When compared to other plasma-based methods, low-pressure plasma treatment generally stands out as an effective and sustainable way to functionalize textile materials. In line with ecologically conscious processing objectives, it permits accurate surface modification by adding functional groups or changing surface shape without the use of water or dangerous chemicals. Furthermore, low-pressure plasma maintains the material's original mechanical strength, flexibility, and structural integrity because of its surface-selective nature, which has no effect on the bulk characteristics of the textile substrate. This makes it especially appropriate for delicate and high-performance textiles where preserving the fundamental properties of the material is crucial.

3.2.1.2 Atmospheric Pressure Plasma

Surface functionalization is a crucial procedure in the textile industry for improving material performance and transferring particular qualities, including adhesion, hydrophilicity, hydrophobicity, antibacterial activity, and dyeability. A versatile, environmentally friendly, and successful technique for surface functionalization among the many modification methods is atmospheric plasma treatment. Atmospheric plasma treatment functions at ambient pressure settings and provides a dry, solvent-free option with less of an impact on the environment than conventional wet chemical processes, which can call for significant amounts of water, energy, and dangerous chemicals. The process of creating atmospheric plasma entails creating a partially ionized gas made up of ions, electrons, neutral species, and reactive radicals that can interact molecularly with the textile surface. The high-energy plasma causes physical and chemical alterations in textiles, such as etching, crosslinking, and the addition of new functional groups, when they are subjected to this reactive environment. Because these surface changes take place without substantially changing the material's bulk characteristics, atmospheric plasma treatment is particularly well-suited for delicate or high-performance textiles. In the following section, various atmospheric plasma treatment techniques are discussed.

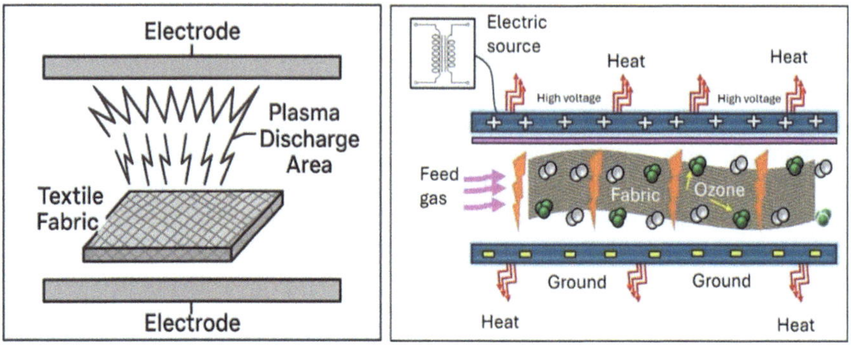

Fig. 3.3 Textile surface functionalization using the corona discharge principle. Ozone is formed via electric discharge and diffused over the area using a dielectric to create corona discharge. For example, oxygen (feed gas) passed through this corona is converted to ozone

(i) Corona Discharge

Corona discharge treatment is a surface modification technique that uses high-voltage electrical discharges to alter the properties of textiles, enhancing their suitability for industrial and commercial applications [17]. This process generates ozone through ionization of ambient air, which interacts with the textile surface to improve adhesion, wettability, and functionality (Fig. 3.3).

Corona discharge improves not only wettability but also modifies bonding, interfacial adhesion, and morphology [18]. Below is a detailed breakdown of the mechanism, effects, and applications:

Mechanism of corona discharge treatment:

1. *Electrical Discharge and Ozone Generation*

   A high-voltage electrode creates an electrical discharge (corona) across the textile surface, ionizing surrounding air molecules. This splits oxygen (O_2) into reactive oxygen atoms (O), which combine with O_2 to form ozone (O_3)[19]. This ozone helps to functionalize the textile surface, and hence it changes its physical properties.
2. *Surface Activation*

   The corona discharge introduces polar functional groups (e.g., –OH, –COOH) onto the textile surface by oxidizing polymer chains or natural fibers [20]. This increases surface energy, making materials more receptive to coatings, adhesives, or dyes.
3. *Physical and Chemical Changes*

   - *Chemical Modification*: Oxygen incorporation alters surface chemistry, improving hydrophilicity (e.g., in polypropylene or wool) [21].
   - *Topographical Changes*: Mild etching occurs on fiber surfaces, such as smoothing wool scales or increasing roughness in synthetic polymers like low-density polyethylene LDPE [22].

Effects on Textiles: Corona discharge is commonly used to improve textile wettability, adhesion toward chemicals/dyeing, and even biocompatibility of the textile surface without affecting its bulk properties. In general, the following basic properties can be modified using corona discharge [17].

- *Enhanced Adhesion*: Low-surface-energy materials like fluoropolymers (e.g., PTFE) or polyolefins (e.g., PP) become printable and paintable after treatment.
- *Improved Wettability*: Natural fibers like wool show reduced hydrophobicity, aiding dye absorption and reducing shrinkage [23].
- *Durability Considerations*: Benefits may diminish with washing or prolonged storage, necessitating post-treatment stabilization (e.g., hydrogen peroxide).

Corona discharge can also be used to coat nanoparticles on the cotton fabric [24]. In this work, ZnO and CU nanoparticles were coated separately before being treated with corona discharge, and hence, the treated cotton fabric might have different functionalities based on the properties of each nanoparticle. Corona discharge brings different changes for different fibers (Fig. 3.4).

Corona discharge has been shown to be a scalable, environmentally benign, and successful method for altering textile surfaces. It improves important qualities, including wettability, dyeability, adhesion, and printability, without sacrificing the material's bulk properties by adding polar functional groups and changing surface energy. It is especially appealing for environmentally friendly textile production due to its compatibility with continuous processing and lack of hazardous chemical residues. However, more study and development are still needed to determine the long-term stability of the adjustments and optimization for various fiber types.

(ii) Dielectric Barrier Discharge

Dielectric barrier discharge (DBD) is a type of non-thermal, non-local thermodynamic equilibrium (non-LTE) plasma that operates at atmospheric pressure [25]. In this technique, a moderate to high voltage is applied across two electrodes, at

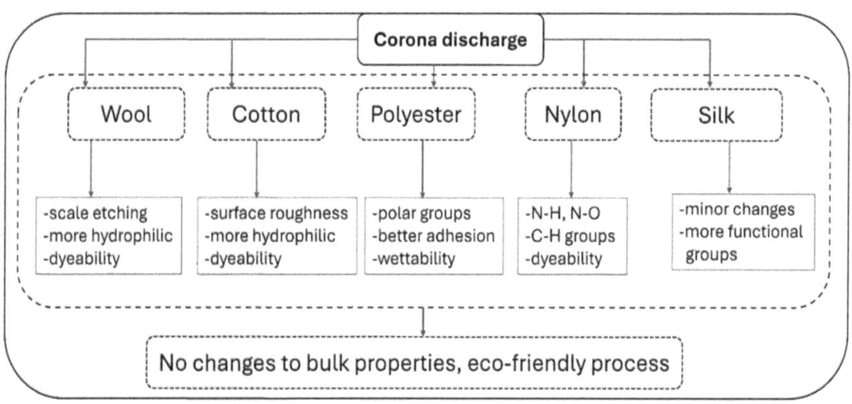

Fig. 3.4 Effect of corona discharge on different textile fabrics

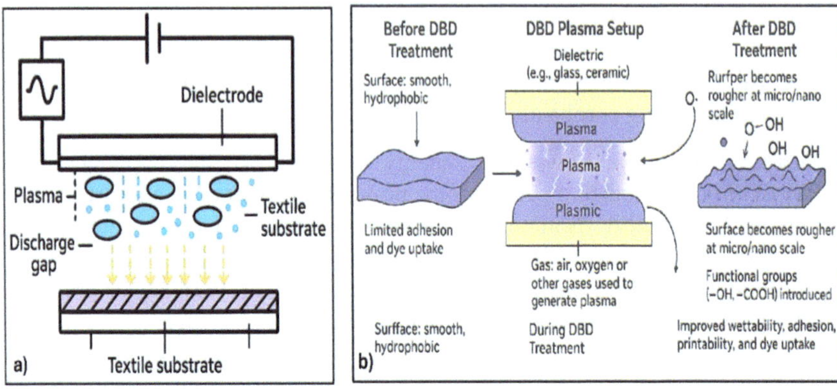

Fig. 3.5 Mechanism (**a**) and effect (**b**) of dielectric barrier discharge surface modification on textile materials

least one of which is covered by a dielectric (insulating) material. The presence of the dielectric barrier prevents the formation of a continuous arc, ensuring that the plasma remains diffuse and non-thermal. This allows for effective surface treatment of materials, such as textiles, without significant heating or damage to the substrate. Dielectric barrier discharge brought the following changes to the textiles surface [26] (Fig. 3.5):

- Enhances wettability, allowing dyes to be absorbed more effectively, resulting in an increased K/S value.
- Leads to a slight improvement in tensile strength.
- Improves the washing fastness of the dyed garment.
- Alters the surface topography of the textile material, increasing its roughness.

Plasma technology, particularly dielectric barrier discharge (DBD), holds significant potential for a wide range of applications, especially in surface modification and functionalization of materials. However, several critical challenges must be overcome to enable its successful integration into industrial processes. One of the primary limitations is the lack of commercially available, scalable machinery suited for large-scale industrial applications. This gap highlights a broader issue: the disconnect between applied research and industrial implementation. Although promising results have been demonstrated at the laboratory scale, the translation of these findings into practical, commercially viable systems remains slow and underdeveloped.

Moreover, specific technical limitations hinder widespread adoption. These include the shallow penetration depth of plasma treatment, which restricts its effectiveness to surface-level modifications, and the potential for non-uniform treatment, especially over large or complex surfaces. This inconsistency can compromise the reproducibility and quality of treated materials. Another concern is the aging effect, where the benefits of plasma treatment diminish over time, reducing the long-term stability of the modified surfaces. In addition, plasma processes can be highly material-sensitive; delicate substrates such as silk or certain synthetic fabrics may suffer from degradation due to exposure to high-energy discharges.

From an economic and operational standpoint, the complexity and prohibitive cost of plasma equipment present further barriers to adoption. These systems often require precise control and maintenance, which can be resource intensive. Furthermore, many plasma treatment setups offer low throughput, particularly in applications involving continuous or high-speed processing of large textile rolls, making them unsuitable for high-volume manufacturing without significant system redesign. Addressing these challenges—through the development of robust industrial-scale plasma systems, enhanced process control for uniformity, deeper and more durable treatment effects, and cost-effective equipment—will be essential for realizing the full potential of dielectric barrier discharge and similar plasma technologies in commercial settings.

(iii) Atmospheric-Pressure Glow Discharge

Atmospheric-pressure glow discharge (APGD) plasma treatment is an advanced surface modification technique widely employed to enhance the functional properties of textile materials [27]. Operating at atmospheric pressure, this method eliminates the need for expensive vacuum systems, making it more practical and cost-effective for industrial use. Through the generation of a stable, low-temperature plasma, APGD treatment enables the activation and functionalization of textile surfaces without significantly affecting the bulk properties of the material. The reactive species produced during the discharge—such as ions, electrons, and radicals—interact with the textile surface to introduce polar functional groups (e.g., hydroxyl, carboxyl, and amino groups). These chemical changes lead to several beneficial outcomes, including improved surface wettability, which enhances the fabric's ability to absorb water and other liquids.

Additionally, APGD treatment significantly increases dye uptake by promoting stronger interactions between dye molecules and the modified surface, resulting in more vibrant and uniform coloration. The treatment also enhances adhesion properties, making it particularly valuable for subsequent finishing processes, such as coating, laminating, or printing, where strong bonding between layers is critical. Due to its non-thermal nature and ability to provide uniform treatment across a range of materials, APGD plasma is suitable for a variety of textile types, including natural fibers, synthetics, and blends. Its versatility, combined with its environmental friendliness (no water or chemical waste), positions APGD as a promising technology for sustainable textile processing.

(iv) Atmospheric-Pressure Plasma Jet

An atmospheric-pressure plasma jet (APPJ) is an advanced device used to generate a plasma plume under standard atmospheric conditions, eliminating the need for vacuum systems and making it suitable for a wide range of practical applications. The basic structure of an APPJ typically includes a narrow nozzle or tubular chamber through which a working gas—such as argon, helium, or air—is continuously flowed. When a high-voltage electric field is applied across the electrodes surrounding this

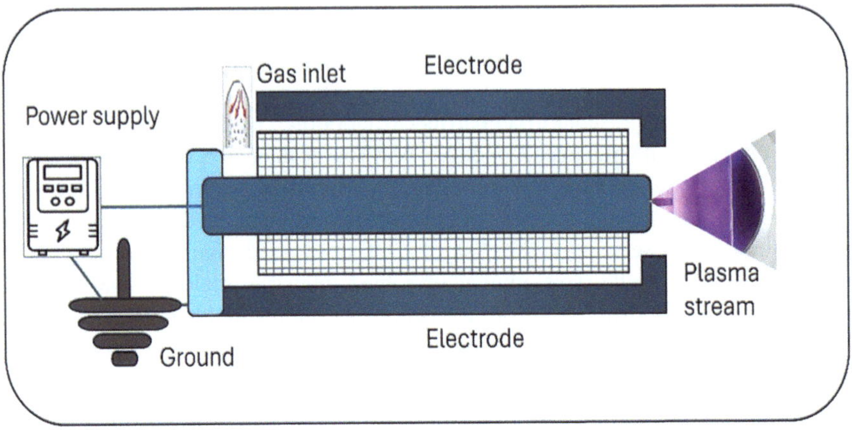

Fig. 3.6 Mechanisms of textile functionalization using atmospheric-pressure plasma jet. Adapted from [13] under CC

gas flow, the gas becomes ionized, producing a visible and highly energetic plasma plume that extends beyond the nozzle (Fig. 3.6).

The plasma produced by APPJs is characterized not only by elevated temperatures but also by a rich mixture of reactive species, such as ions, electrons, radicals, and excited molecules. These reactive components are capable of initiating chemical reactions at the surface of materials without significantly affecting their bulk properties. This makes APPJs particularly advantageous in the treatment and functionalization of the textile materials, where surface modifications—such as enhancing wettability, adhesion, dye uptake, or imparting antimicrobial or flame-retardant properties—are desired. Furthermore, because APPJ treatments are contactless, localized, and can be applied at atmospheric pressure and low temperatures, they are well-suited for use on heat-sensitive and flexible substrates like textiles. As a result, APPJ technology has emerged as a versatile and environmentally friendly alternative to traditional wet chemical processing in the textile industry.

Plasma treatment is a favorable method for textile surface modification and functionalization. There are numerous ways to create plasma in a vacuum or atmosphere. The type of plasma and the conditions of the process have a significant impact on how the plasma affects the substrate. Cleaning, activation, etching, and polymerization (grafting) are among the numerous effects that can be achieved by carefully choosing the apparatus, process, gas, and parameters. In summary, atmospheric plasma treatment significantly influences the surface characteristics of textile materials, offering a versatile and efficient approach to surface modification without affecting the bulk properties. The effectiveness of this treatment depends on several factors, including the type of plasma source used, the uniformity of plasma exposure across the material surface, and the intended final application. Each plasma system presents its own set of advantages and limitations—for instance, in terms of energy efficiency, treatment speed, scalability, or compatibility with specific textile substrates. Table 3.2

Table 3.2 Comparison of atmospheric plasma treatment methods

Technology	Plasma source	Plasma uniformity	Applications	Advantages	Limitations	References
CD	• High voltage • Generates plasma at sharp points	• Nonuniform • Localized	• Surface activation • Ozone generation • Textile modification	• Simple setup • Low inception voltage	• Limited to surface • Low density	[28, 29]
DBD	• High voltage AC across electrode • Micro discharges	Diffusion can be made more uniform	• Surface modification • Ozone generation • Textile finishing	• Large area treatment • Scalable	• Requires dielectric • Higher voltage • Filamentary	[30]
APGD	Sustained glow discharge b/n electrode	• Uniform • Diffuse	Surface modification	• Homogenous plasma • Uniform treatment	Challenge to scale at large	[31]
APPJ	Plasma generated inside a tube and ejected jet into open air	• Localized • Focused jet	Localized surface treatment	• Treats complex shapes • Operates in open air	• Small treatment area • Requires gas flow	[32]

provides a comparative overview of various atmospheric plasma surface modification techniques, highlighting their distinct features, operational parameters, and practical implications for textile processing.

3.2.2 *UV (Ultraviolet) Radiation (Irradiation)*

UV (ultraviolet) radiation is one of the physicochemical methods used to modify or functionalize textile to change the inherent or added nature of textiles to improve the performance of the next process. UV radiations can be used in pre-or post-treatments for the surface modification of textile [33]. UV radiation can help to improve the dyeability, improve pilling resistance, and water swelling properties of cotton fibers. Table 3.3 shows the properties that could be affected due to UV radiation for different fibers.

In the textile material, UV radiation treatment should not only achieve the desired functional effects but also meet several key criteria: (i) ease of application without

Table 3.3 Effect of UV radiation treatment on different fibers

Fiber type	Effect of UV radiation	Key changes	References
Cotton	Enhanced reactivity, dyeability, water swelling, pilling resistance	Impurity removal, fault correction, increased surface reactivity	[33]
Wool	Decreased strength, increased crystallinity, improved dyeability	Structural degradation, dye affinity enhancement	[33, 34]
Silk	Reduced strength, crosslink breakage	Mechanical degradation, structural changes	[35]
Nylon	Strength loss, molecular weight reduction	Weakening, viscosity decrease	[34]
Acrylic	High UV stability	Minimal degradation, retained mechanical properties	[34]
Polyester	Superior resistance to UV	Loss of tensile strength, decolorization, reduce gloss in coatings on intense UV	[36]

the need for costly or complex tools [37]; (ii) durability and fastness of the treatment to ensure long-term performance [38]; (iii) cost-effectiveness, allowing for broad market accessibility; (iv) compliance with environmentally friendly practices to support sustainable textile [39]. Numerous studies have explored the effects of UV radiation on both natural and synthetic fibers [33, 40, 41], with a significant focus on surface modification of cotton, wool, polyester, and nylon has been investigated. Li et al. [42] reported the effect of UV radiation on polyester fabric. The surface modified polyester fabric using UV radiation exhibited a water contact angle reduced to 0°, with water absorption increasing from 85 to 104%. Capillary rise height significantly improved from 0.2 to 6.2 cm, while the spray rating declined from 4 to 1, indicating enhanced wettability. Crease recovery angles showed only a slight reduction of 1.2–8.8%, and the mechanical and physical properties of the fabric remained well-preserved.

UV light has been shown to be a successful, environmentally responsible, and adaptable method for altering textile surfaces. It makes it possible to precisely modify surface qualities including wettability, adhesion, and functionalization without sacrificing the material's bulk properties. UV treatment can improve dyeability, hydrophilicity, antibacterial activity, and compatibility with functional coatings by stimulating photochemical processes or surface grafting. It is a promising technique for developing high-performance and sustainable textile applications because of its low energy consumption, solvent-free nature, and scalability.

3.2.3 Laser Irradiation

Laser irradiation is the process of exposing on a material for instance on a textile substrate to a laser light [43]. For instance, exposing laser irradiation on a textile surface can cause adhesion and wetting properties [44] (Fig. 3.7).

Laser irradiation is regarded as an effective and versatile technique for modifying the surface of textile materials, enabling precise control over surface properties such as roughness, wettability, chemical functionality, and microstructure without significantly affecting the bulk characteristics of the fabric. Bahners [45], The authors conducted a comprehensive review on the application of laser irradiation as a surface modification technique for the functionalization of synthetic fibers. They highlighted that laser treatment enables precise alteration of the fiber surface morphology and topography without compromising the bulk properties of the material. This modification enhances the interaction between the fiber surface and external agents, resulting in significantly improved dye uptake, better fixation, and increased absorption of various chemical substances. The review also emphasized the potential of laser irradiation to tailor surface functionalities for advanced textile applications, including improved coloration, wettability, and chemical reactivity.

Due to laser irradiation treatment (surface modification), the typically smooth surface of synthetic fibers can be transformed into a regular, roll-like structure using this non-contact technique, significantly influencing the fiber's overall properties—such as particle and coating adhesion, wetting behavior, and optical appearance [46]. Morgan et al. [47] treated wool fabric using CO_2 laser irradiation and claimed that the dyeability and color difference value. CO_2 laser irradiation treatment for cotton brought a profoundly effect on the cotton fabric [48]. However, the change on the cotton fabric is not always positive. For instance, Fabric weight and strength decreased with increase of resolution and pixel time and yellowing has been occurred. laser irradiation is a strong and adaptable technique for surface modification and textile functionalization. Laser-based methods provide notable benefits in terms of

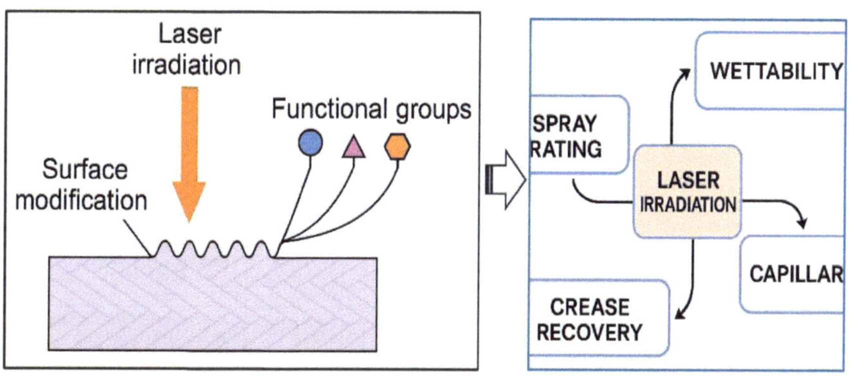

Fig. 3.7 Mechanisms of surface modification of textile using laser irradiation and its effect on the properties of the material

process control, scalability, and environmental sustainability by allowing precise, localized changes without the need for chemical treatments or physical contact. To increase textile performance or add new capabilities like hydrophobicity, antimicrobial behavior, or improved dyeability, surface attributes including roughness, wettability, adhesion, and optical features can be precisely adjusted. The creation of high-performance, multipurpose textiles that meet the needs of contemporary applications is anticipated as laser technologies continue to progress and are incorporated into textile production.

3.2.4 Grafting Techniques

Surface grafting is another way of functionalization of textile materials which import new functionalities or improve the existed functional properties [49]. It improves hydrophilicity, enhances dyeability, adds new functions such as antimicrobial/antifoaming, improve adhesion and increase chemical resistance of textile materials [50] (Fig. 3.8).

Grafting techniques are further divided into chemical grafting polymerization (CGP), radiation-induced grafting (RIG), plasma-induced grafting (PIG), and light-induced grafting (LIG) [50]. Table 3.4 summarizes the properties imparted and modified due to these grafting techniques.

Chemical grafting is a versatile and widely employed technique used for imparting various functionalities and has a minimal impact on air permeability when using very small molecules. While radiation-induced grafting is suitable for inert or synthetic fibers. Furthermore, plasma-induced grafting is a dry, eco-friendly process that add some functionality on the textile surface without affecting the bulk properties. Green processing recently got a strong focus [57]. Finally, light-induced grafting allows for mild introduction of functionalities and hence suitable for advanced and sensitive

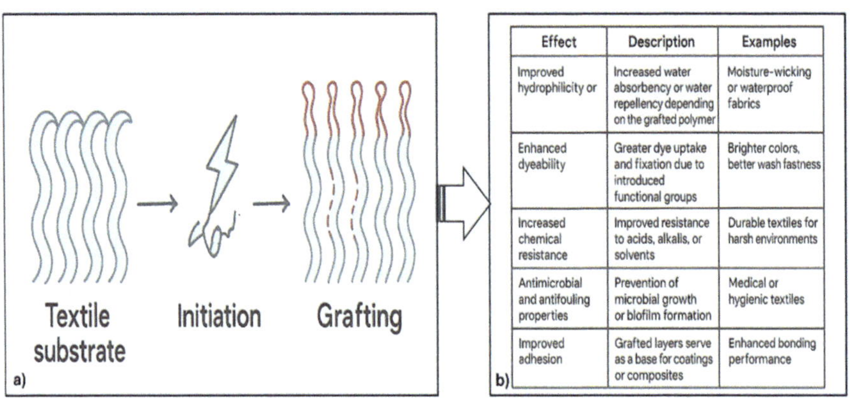

Effect	Description	Examples
Improved hydrophilicity or	Increased water absorbency or water repellency depending on the grafted polymer	Moisture-wicking or waterproof fabrics
Enhanced dyeability	Greater dye uptake and fixation due to introduced functional groups	Brighter colors, better wash fastness
Increased chemical resistance	Improved resistance to acids, alkalis, or solvents	Durable textiles for harsh environments
Antimicrobial and antifouling properties	Prevention of microbial growth or biofilm formation	Medical or hygienic textiles
Improved adhesion	Grafted layers serve as a base for coatings or composites	Enhanced bonding performance

Fig. 3.8 Mechanism (**a**) and (**b**) effect of grafting techniques on the textile material

Table 3.4 Properties imparted, enhanced and applications areas of textile materials using different grafting techniques

Grafting method	Properties imparted	Application areas	Limitations	References
CGP	Hydrophilicity, water wicking, antimicrobial, flame retardant, dyeability, adhesion	Medical textiles, moisture management, fire protection, dyeing and printing textiles	Harsh chemicals, environmental impact	[50, 51]
RIG	Hydrophobic, oleophilic, flame retardant, UV-resistance, antimicrobial	Air/water filtration, medical textile, protective clothing	Requires radiation infrastructure	[52, 53]
PIG	Surface energy increased, hydrophilicity, wettability, dyeability, adhesion, antimicrobial, antistatic, shrink resistance, anti-felting	Technical textiles, UV-protection, fire protection, medical and hygienic, smart textile (sensor)	Limited penetration depth	[54, 55]
LIG	Photo-responsive, pH-responsive, antimicrobial, covalent bonding	Color changing, pH measurement, biosensors, self-cleaning and medical	Dependent on photoactive initiators	[56]

textile applications. In summary, grafting techniques are newly introduced methods not only for modifying the surface of textiles but also for imparting new functionalities—such as antimicrobial, flame-retardant, oil-repellent, and water-repellent properties—enabling the treated textiles to be used in various technical applications.

3.2.5 *Sol–Gel Coating*

Sol–gel coating is a process that changes a liquid material the so called sol into a solid gel structure, which can then be used as a coating on textile substrates [58]. The sol–gel process starting from preparing precursor solutions (e.g., titanium tetraisopropoxide ($Ti(OiPr)_4$, zinc nitrate) and then sol materials are prepared from these precursor solutions to obtain gel-like materials; dip-coating, spin coating, and electrodeposition are followed for initiation and finally the deposited materials are cured on the surface of the textile material (Fig. 3.9).

The sol–gel process is straightforward to carry out, requiring neither extreme conditions nor high temperatures [60]. It involves a sequence of basic hydrolysis and condensation reactions. The sol–gel technique presents extensive opportunities for surface modification across a wide range of materials, including but not limited to textiles [61]. This versatile approach enables the functionalization of textiles using

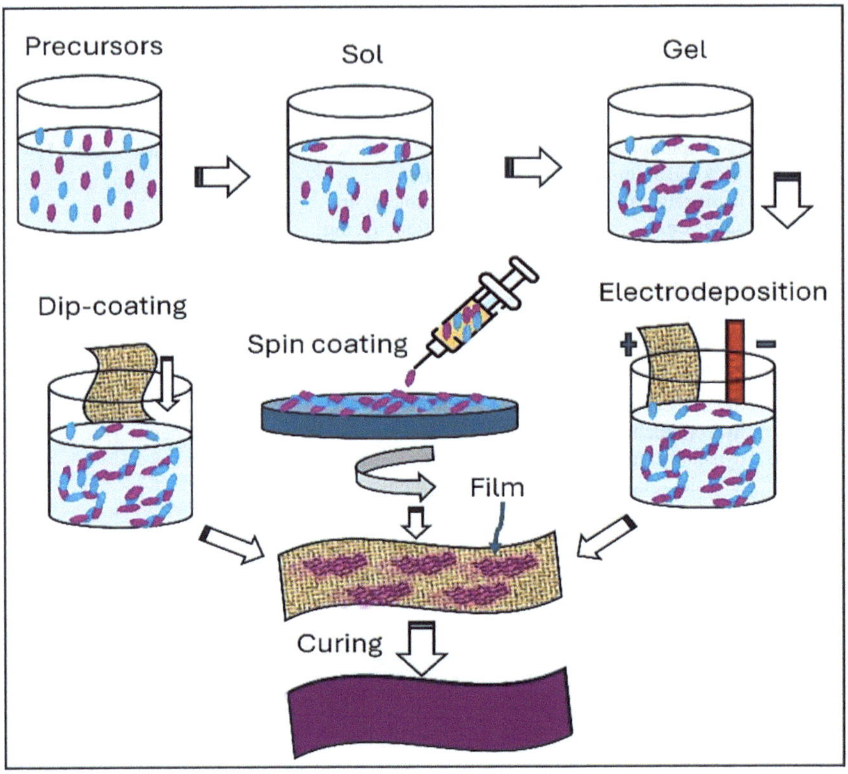

Fig. 3.9 Processing route to obtain sol–gel coatings. Adapted from [59] under CC

conventional application methods under mild conditions, while also allowing for the integration of multiple functionalities into a single material. Yang et al. [62] uses sol–gel technique to impart hydrophobic nature to cotton textile. Sol–gel technique impart various functionalities to textile materials such as phase change materials [63], photo-induced [64], flame retardancy [65], water repellency [66], and multifunctional durable properties [67] to textile materials. Flame retardancy treatment has got recent focuses for textile treatment [68].

A flexible and effective technique for surface modification and functionalization of textile materials is the sol–gel coating process. Under mild circumstances, metal alkoxides or organosilanes undergo hydrolysis and condensation processes to form inorganic or hybrid organic–inorganic networks. By using this method, functional coatings can be evenly applied to textile surfaces without substantially changing their bulk characteristics. Numerous cutting-edge properties, including hydrophobicity, flame retardancy, antibacterial activity, UV protection, and electrical conductivity, can be added to textiles by sol–gel processing. Furthermore, the technique makes it possible to combine several functional agents into a single coating, which makes it easier to create multipurpose textiles for use in protective gear, smart clothes, medical textiles, and technical textiles. The sol–gel technique is a promising tool in the development of high-performance and sustainable textile products because of its versatility, scalability, and compatibility with traditional textile finishing procedures.

3.2.6 *Ozone Treatment*

Ozone is a very strong oxidant which can be produced artificially, as well as naturally available in the stratosphere between 15 and 30 km above the Earth's surface [69]. Artificially ozone can be generated using photochemically, electrolytic, radiochemical, and corona discharge methods [70]. The generated ozone has two modes of application that is in gas phase where porous subtracted absorbed this phase and aqueous phase where it can be solubilized in water and hence can be applied for textile substrate. Ozone treatment of textile materials has been used to modify its dyeability [71]. In addition to improving the dyeability of textile materials, ozone treatment is dry process and hence it is environmentally friendly process [72] as ozone treatment conducted at room temperature [73]. Furthermore, ozone treatment does not affect the bulk properties of textile material beyond the enhancement of wettability and dyeability [74]. In general, textile properties such as surface wettability (adhesion and hydrophobicity hence dyeability) [75], whiteness (bleaching) [76], surface roughness [77], antimicrobial properties [78], color fading (jeans fabric) [73], chemical functionalization [79], shrinkage resistance [80], electrostatic behavior [81], and other properties has been modified using ozone treatment.

Ozone treatment has emerged as a promising and eco-friendly method for the surface modification of textile materials. This technique supports sustainable textile processing by offering several environmental advantages, including minimal use of harmful chemicals, reduced water consumption, and significantly lower energy requirements compared to conventional treatments. As a powerful oxidizing agent, ozone facilitates efficient surface activation without leaving behind toxic residues, thereby promoting cleaner production practices. In recent years, ozone treatment has gained recognition as a cutting-edge advancement in textile engineering. It enables the enhancement of key textile properties such as wettability, dyeability, printability, and adhesion, while also improving the overall functionality of fibers. Its versatility makes it suitable for a wide range of fiber types, including natural, synthetic, and blended materials. Given its environmental benefits and performance-enhancing capabilities, ozone treatment is increasingly being adopted as a green alternative in modern textile processing.

3.2.7 *Supercritical CO_2 Treatment*

Most textile treatments are water-intensive and therefore pose numerous environmental pollution problems. For this reason, most researchers are now giving attention

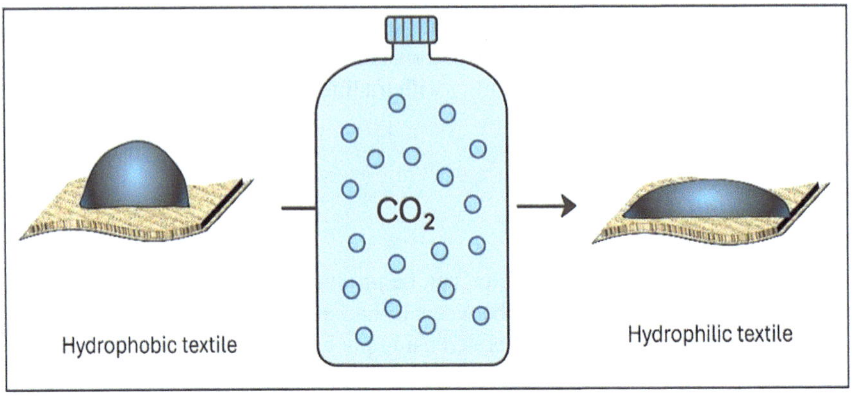

Fig. 3.10 Supercritical carbon dioxide treatment of textiles

to processes that requires less water. Supercritical CO_2 ($scCO_2$) treatment is one of the waterless surface-modification methods in textile pretreatment. More specifically, supercritical CO_2 dyeing is getting more attention in recent days. Supercritical CO_2 treatment offers environmentally friendly treatments which are more efficient when compared to traditional treatment methods. Supercritical CO_2 treatment utilizes CO_2 in a supercritical state (above 31.1 °C and 73 atmosphere) and act as a solvent which can dissolve dyes and also helps to modify contaminants without the need of water and hence no drying needed [82]. Molla et al. [83] have used supercritical CO_2 activation of polyester fabric for easy dyeability. No need of mordant and any pretreatment for natural dyeing of polyester. The adhesion, surface function, and wettability issues has been resolved using $scCO_2$. Simple representation of $scCO_2$ treatment on textile is shown in Fig. 3.10.

Supercritical carbon dioxide treatment of textile brought various changes to the substrate material. For example, it helps to incorporate various active principles into a polymeric matrix, such as medicines, functional finishing agents, dyes, and other agents [84]. Supercritical carbon dioxide treatment therefore acts as not only physical modification but also add functions into textiles such as antimicrobial, water and oil resistance, UV protection and flame-resistant fabrics [85]. The interfacial adhesion of textile substrate has been improved using $scCO_2$ treatment [86]. In addition, $scCO_2$ treatment on textile enhances surface wettability and moisturization efficiency [87]. In summary, supercritical carbon dioxide treatment not only brings surface modification to textile surface but also brings functionalization of the substrate. In addition, $scCO_2$ is a water-less process and hence encourages sustainability. Finally, $scCO_2$ is more efficient in which diffusivity, low surface tension, and simplicity of solvent removal at the end of the process is possible by the process [84].

3.3 Future Perspectives and Conclusion

Surface modification of textile materials is essential for two main reasons: the inherent characteristics of textiles—such as the hydrophobicity of synthetic fibers—and the presence of impurities in natural fibers, both of which can hinder the penetration of finishing chemicals like dyes and printing pastes. Therefore, surface modification plays a crucial role in enhancing these properties. Additionally, it enables the functionalization of textiles, allowing the introduction of new surface functionalities. The evolution of surface modification methods is poised to address growing demands of sustainable production, multifunctionality, and scalability. Furthermore, these developments solve the problems occurred during traditional surface modification techniques. Supercritical CO_2 treatment, while traditionally used for dyeing, has the potential to be expanded for a variety of advanced textile applications, including imparting antimicrobial properties, water repellency, and even the deposition of nanomaterial-based finishes. Ozone treatment offers a precise method for surface activation and can be effectively utilized to enhance specific functionalities without significantly altering the bulk properties of the fabric. Grafting techniques, particularly when combined with high-energy methods such as electron beam irradiation, must be scaled up to meet industrial production demands.

Furthermore, sol–gel techniques present a promising route for incorporating nanocomposites into textiles, enabling improved durability, UV resistance, and thermal stability. Plasma-based methods, both for surface modification and for plasma-assisted deposition of biodegradable nanolayers, are crucial for achieving scalable and environmentally friendly textile finishing—especially within the context of a circular economy.

Looking ahead, hybrid approaches that combine multiple surface modification techniques could offer synergistic effects, optimizing textile performance across various functional requirements. Importantly, all of these technologies must be developed and implemented with strong consideration for sustainability, energy efficiency, and alignment with circular economy principles to ensure environmentally responsible innovation in textile manufacturing.

Surface modification techniques have the ability to transform conventional textiles into advanced functional materials by altering or enhancing their surface properties without compromising their inherent bulk characteristics. Through the application of physical, chemical, or biological treatments, these techniques can impart a wide range of functionalities—such as antibacterial activity, UV protection, water and oil repellency, flame retardancy, and electrical conductivity. As a result, surface-engineered textiles are increasingly being used in high-performance sectors including healthcare, sportswear, military, wearable electronics, and smart textiles. This transformation not only adds value to textile products but also supports the development of sustainable and multifunctional materials for next-generation applications.

Acknowledgements This work was assisted financially by the Alexander Von Humboldt Foundation at Albstadt Sigmaringen University via the Georg Forster scheme (return fellowship from AvH)

Conflicts of Interest The authors declare no conflict of interest.

AI-Usage Declaration During the preparation of this chapter, the authors used AI tools to improve the language of the manuscript. After using this tool/service, the authors reviewed and edited the content as needed and take full responsibility for the content of the publication.

References

1. M.G. Tadesse, J.F. Lübben, Heat-and energy-harvesting technology for smart textiles and applications, in *Smart Textiles from Natural Resources*. (Elsevier, 2024), pp.741–767
2. M.J. John, R. Anandjiwala, Surface modification and preparation techniques for textile materials, in *Surface Modification of Textiles*. (Elsevier, 2009), pp.1–25
3. L. Ammayappan, L. Nayak, D. Ray, G. Basu, Plasma treatment on textiles. Asian Dyer **8**(6), 34–37 (2012)
4. H. Barani, A. Haji, Comprehensive plasma-enhanced wool advancements: cleaning, dyeing, and finishing, in *Advances in Plasma Treatment of Textile Surfaces*. (Elsevier, 2024), pp.13–36
5. M.G. Tadesse, N. Simon, J. Felix Lübben, Highly conductive films through PEDOT-PSS ink formulation via doping using spontaneous wicking of liquids for supercapacitor applications. Adv. Energy Sustain. Res. **5**(7), 2400006 (2024)
6. S. Periyasamy, D. Gupta, M. Parvathi, S.B. Chaudhari, Surface modification techniques in textiles: a review, in *International Conference on Functional Textiles & Clothing*. (Springer, 2023)
7. S. Abrar, S. Kiran, S. Iqbal, B. Munir, A. Rasool, Recent advances in plasma modification of silk, in *Advances in Plasma Treatment of Textile Surfaces* (2024), pp. 37–56
8. M.R. McCoustra, R.R. Mather, Plasma modification of textiles: understanding the mechanisms involved. Text. Prog. **50**(4), 185–229 (2018)
9. M. Naveed, Plasma treatment as green technology for dyeing of textile fabrics. Trends Text. Eng. Fashion Technol **2**, 168–173 (2018)
10. A. Zille, F.R. Oliveira, A.P. Souto, Plasma treatment in textile industry. Plasma Processes Polym. **12**(2), 98–131 (2015)
11. B.B. Yilma, J.F. Luebben, M.G. Tadesse, Effect of plasma surface modification on comfort properties of polyester/cotton blend fabric. Mater. Res. **24**, e20210021 (2021)
12. A. Zille, Plasma technology in fashion and textiles, in *Sustainable Technologies for Fashion and Textiles*. (Elsevier, 2020), pp.117–142
13. A. Haji, H. Barani, Introduction of plasma technology, in *Advances in Plasma Treatment of Textile Surfaces*. (Elsevier, 2024), pp.1–12
14. R.A. Jelil, A review of low-temperature plasma treatment of textile materials. J. Mater. Sci. **50**(18), 5913–5943 (2015)
15. M. Hossain, G. Grabher, Applications of low-pressure plasma in textile finishing. Melliand Int. **17**, 149–151 (2011)
16. G. Shanmugavelayutham, T. Anupriyanka, P. Bhagyashree, P. Premasudha, Plasma surface modification of cotton fabric by using low pressure plasma. IEEE Trans. Plasma Sci. **49**(2), 497–501 (2020)
17. S. Brzezinski, S. Polowinski, D. Kowalczyk, G. Malinowska, Effect of corona discharge treatment on the surface strength and performance properties of synthetic fibre textiles. Fibres Text. East. Eur. **17**(5), 76 (2009)
18. G. Wypych, *PVC Degradation and Stabilization* (Elsevier, 2025)
19. Y. Zhu, C. Chen, J. Shi, W. Shangguan, A novel simulation method for predicting ozone generation in corona discharge region. Chem. Eng. Sci. **227**, 115910 (2020)
20. W. Xu, X. Liu, Surface modification of polyester fabric by corona discharge irradiation. Eur. Polym. J. **39**(1), 199–202 (2003)

21. X. Wang, Y. Zhao, W. Li, H. Wang, Effect of surface modifications on the thermal and moisture behavior of wool fabric. Appl. Surf. Sci. **342**, 101–105 (2015)
22. N. Le Moigne, B. Otazaghine, S. Corn, H. Angellier-Coussy, A. Bergeret, N. Le Moigne, B. Otazaghine, S. Corn, H. Angellier-Coussy, A. Bergeret, Modification of the interface/interphase in natural fibre reinforced composites: treatments and processes, in *Surfaces and Interfaces in Natural Fibre Reinforced Composites: Fundamentals, Modifications and Characterization* (2018), pp. 35–70
23. G. Ke, W. Yu, W. Xu, W. Cui, X. Shen, Effects of corona discharge treatment on the surface properties of wool fabrics. J. Mater. Process. Technol. **207**(1–3), 125–129 (2008)
24. S. Nourbakhsh, H. Sepehrnia, E. Akbari, Novel corona discharge treatment of cotton fabric with Cu and ZnO nanoparticles. J. Text. Inst. **111**(9), 1269–1276 (2020)
25. A. Valinattajomran, *Characterization and the Study of the Behavior of Transporting Cold Plasmas in Dielectric Capillary Tubes and Their Applications* (Université Pierre et Marie Curie (Paris 6); Université de Mazandaran (Iran), 2016)
26. H. Ahmed, K. Ahmed, H. Mashaly, A. El-Halwagy, Treatment of cotton fabric with dielectric barrier discharge (DBD) plasma and printing with cochineal natural dye. Indian J. Sci. Technol. **10**(10), 1–10 (2017)
27. F. Cheng, H. Yibo, J. Chenggang, W. Xuemei, W. Wenli, The effect of atmospheric pressure glow discharge plasma treatment on the dyeing properties of silk fabric. Plasma Sci. Technol. **22**(1), 015503 (2019)
28. Z. Rahman, H. Rahman, A. Rahman, Classification and generation of atmospheric pressure plasma and its principle applications. Int. J. Math. Phys. Sci. Res. **2**, 127–146 (2014)
29. A. Yehia, Characteristics of the dielectric barrier corona discharges. AIP Adv. **9**(4) (2019)
30. F. Gasi, G. Petraconi, E. Bittencourt, S.R. Lourenço, A.H.R. Castro, F.d.S. Miranda, A.M. Essiptchouk, L. Nascimento, A. Petraconi, M.A. Fraga, Plasma treatment of polyamide fabric surface by hybrid corona-dielectric barrier discharge: material characterization and dyeing/washing processes. Mater. Res. **23**, e20190255 (2019)
31. C. Tendero, C. Tixier, P. Tristant, J. Desmaison, P. Leprince, Atmospheric pressure plasmas: a review. Spectrochim. Acta, Part B **61**(1), 2–30 (2006)
32. A. Schutze, J.Y. Jeong, S.E. Babayan, J. Park, G.S. Selwyn, R.F. Hicks, The atmospheric-pressure plasma jet: a review and comparison to other plasma sources. IEEE Trans. Plasma Sci. **26**(6), 1685–1694 (1998)
33. F. Ferrero, G. Migliavacca, M. Periolatto, UV treatments on cotton fibers. Cotton Res. 235–256 (2016)
34. R.L. Shishoo, *Degradation by Ultraviolet Light of Wool and Other Textile Fibres* (University of Glasgow, United Kingdom, 1963)
35. H. Liu, S. Zhao, Q. Zhang, T. Yeerken, W. Yu, Secondary structure transformation and mechanical properties of silk fibers by ultraviolet irradiation and water. Text. Res. J. **89**(14), 2802–2812 (2019)
36. H. Gu, Degradation of glass fibre/polyester composites after ultraviolet radiation. Mater. Des. **29**(7), 1476–1479 (2008)
37. R. Schwalm, *UV Coatings: Basics, Recent Developments and New Applications* (Emerald Publishing, 2006)
38. V. Rubeziene, S. Varnaite, J. Baltusnikaite, I. Padleckiene, Effects of light exposure on textile durability, in *Understanding and Improving the Durability of Textiles* (Woodhead Publishing, Oxford, 2012), pp. 104–125
39. S.F. Kabir, S. Chakraborty, S.A. Hoque, K. Mathur, Sustainability assessment of cotton-based textile wet processing. Clean Technol. **1**(1), 232–246 (2019)
40. M. Moezzi, M. Ghane, The effect of UV degradation on toughness of nylon 66/polyester woven fabrics. J. Text. Inst. **104**(12), 1277–1283 (2013)
41. M. Sadeghi-Kiakhani, S. Safapour, F. Sabzi, A.R. Tehrani-Bagha, Effect of ultra violet (UV) irradiation as an environmentally friendly pre-treatment on dyeing characteristic and colorimetric analysis of wool. Fibers Polym. **21**, 179–187 (2020)

42. M. Li, T. Deng, S. Liu, F. Zhang, G. Zhang, Superhydrophilic surface modification of fabric via coating with nano-TiO_2 by UV and alkaline treatment. Appl. Surf. Sci. **297**, 147–152 (2014)
43. K.W. Kolasinski, M.C. Gupta, L.V. Zhigilei, Plume and Nanoparticle Formation During Laser Ablation in *Encyclopedia of Interfacial Chemistry* (Elsevier, 2018)
44. T. Bahners, W. Kesting, E. Schollmeyer, Designing surface properties of textile fibers by UV-laser irradiation. Appl. Surf. Sci. **69**(1–4), 12–15 (1993)
45. T. Bahners, Excimer laser irradiation of synthetic fibres as a new process for the surface modification of textiles—a review. Opt. Quant. Electron. **27**, 1337–1348 (1995)
46. D. Knittel, E. Schollmeyer, Surface structuring of synthetic polymers by UV-laser irradiation. Part IV. Applications of excimer laser induced surface modification of textile materials. Polym. Int. **45**(1), 110–117 (1998)
47. L. Morgan, J. Tyrer, F. Kane, The effect of CO_2 laser irradiation on surface and dyeing properties of wool for textile design, in *ICALEO 2014: 33nd International Congress on Laser Materials Processing, Laser Microprocessing and Nanomanufacturing* (AIP Publishing, 2014)
48. Y.F. Chow, A. Chan, C.-W. Kan, Effect of CO_2 laser irradiation on the properties of cotton fabric. Text. Res. J. **82**(12), 1220–1234 (2012)
49. J. Ghosh, N.S. Rupanty, F. Khan, T. Noor, R. Jahangir, S. Mirmohammadsadeghi, T. Islam, Grafting modification for textile functionalization: innovations and applications. Discov. Appl. Sci. **7**(1), 1–29 (2025)
50. N. Abidi, Surface grafting of textiles, in *Surface Modification of Textiles*. (Elsevier, 2009), pp.91–107
51. W. Yang, B. Rånby, Radical living graft polymerization on the surface of polymeric materials. Macromolecules **29**(9), 3308–3310 (1996)
52. C. Brendlé, R. El Hage, J.-L. Clément, S. Rouif, R. Sonnier, B. Otazaghine, Flame retardancy of short flax fibers modified by radiation-induced grafting of phosphonated monomers: comparison between pre-and simultaneous irradiation grafting. Molecules **29**(5), 1176 (2024)
53. J. Taibi, S. Rouif, B. Ameduri, R. Sonnier, B. Otazaghine, Radiation induced graft polymerization of fluorinated monomers onto flax fabrics for the control of hydrophobic and oleophobic properties. Polymer **281**, 126132 (2023)
54. E. El-Sayed, A.G. Hassabo, Recent advances in the application of plasma in textile finishing (A review). J. Text. Color. Polym. Sci. **18**(1), 33–43 (2021)
55. S. Yoshida, K. Hagiwara, T. Hasebe, A. Hotta, Surface modification of polymers by plasma treatments for the enhancement of biocompatibility and controlled drug release. Surf. Coat. Technol. **233**, 99–107 (2013)
56. J. Zhuo, G. Sun, Light-induced surface graft polymerizations initiated by an anthraquinone dye on cotton fibers. Carbohydr. Polym. **112**, 158–164 (2014)
57. M.G. Tadesse, J.F. Lübben, Green nanomaterials: processing, characterization and applications, in *Mechanics of Nanomaterials and Polymer Nanocomposites*. (Springer, 2023), pp.87–104
58. A.P. Periyasamy, M. Venkataraman, D. Kremenakova, J. Militky, Y. Zhou, Progress in sol–gel technology for the coatings of fabrics. Materials **13**(8), 1838 (2020)
59. Y.A. Attia, *Sol–Gel Processing and Applications* (Springer Science & Business Media, 2012)
60. W.N.W. Ismail, Sol–gel technology for innovative fabric finishing—a review. J. Sol–Gel Sci. Technol. **78**, 698–707 (2016)
61. T. Textor, Modification of textile surfaces using the sol–gel technique, in *Surface Modification of Textiles*. (Elsevier, 2009), pp.185–213
62. M. Yang, W. Liu, C. Jiang, S. He, Y. Xie, Z. Wang, Fabrication of superhydrophobic cotton fabric with fluorinated TiO_2 sol by a green and one-step sol–gel process. Carbohydr. Polym. **197**, 75–82 (2018)
63. X. Liu, Y. Lou, Preparation of microencapsulated phase change materials by the sol–gel process and its application on textiles. Fibres Text. East. Eur. **23**(2(110)) (2015)
64. E. Pakdel, W.A. Daoud, X. Wang, Assimilating the photo-induced functions of TiO_2-based compounds in textiles: emphasis on the sol–gel process. Text. Res. J. **85**(13), 1404–1428 (2015)

65. A. Bentis, A. Boukhriss, A.M. Grancaric, M. El Bouchti, M. El Achaby, S. Gmouh, Flammability and combustion behavior of cotton fabrics treated by the sol gel method using ionic liquids combined with different anions. Cellulose **26**, 2139–2153 (2019)
66. A. Bentis, A. Boukhriss, S. Gmouh, Flame-retardant and water-repellent coating on cotton fabric by titania–boron sol–gel method. J. Sol–Gel Sci. Technol. **94**, 719–730 (2020)
67. J. Foksowicz-Flaczyk, J. Walentowska, M. Przybylak, H. Maciejewski, Multifunctional durable properties of textile materials modified by biocidal agents in the sol–gel process. Surf. Coat. Technol. **304**, 160–166 (2016)
68. M.G. Tadesse, E. Kasaw, B.F. Adamu, Thermal properties of wool: thermal degradation studies and fire-retardant properties, in *The Wool Handbook*. (Elsevier, 2024), pp.327–340
69. J.C. Van der Leun, The ozone layer. Photodermatol. Photoimmunol. Photomed. **20**(4), 159–162 (2004)
70. A. Körlü, Use of ozone in the textile industry, in *Textile Industry and Environment* (IntechOpen, 2018)
71. R.S. Gabardo, D.S. de Carvalho Cotre, M.J. Lis Arias, M.P. Moisés, B.T. Martins Ferreira, R.B. Samulewski, J.P. Hinestroza, F.M. Bezerra, Surface modification of polyester fabrics by ozone and its effect on coloration using disperse dyes. Materials **14**(13), 3492 (2021)
72. S. Ibrahim, N. El-Zaher, M. Micheal, Characterization and evaluation of physico-chemical properties of polymeric fabrics treated with UV/ozone. Res. J. Text. Appar. **14**(2), 59–71 (2010)
73. H.A. Eren, İ. Yiğit, S. Eren, O. Avinc, Ozone: an alternative oxidant for textile applications, in *Sustainability in the Textile and Apparel Industries: Production Process Sustainability* (2020), pp. 81–98
74. F. Fattahi, H. Izadan, A. Khodami, Investigation into the effect of UV/Ozone Irradiation on the dyeing behaviour of poly(lactic acid) and poly(ethylene terephthalate) substrates. Prog. Color Colorants Coatings **5**(1), 15–22 (2012)
75. E.M. Osman, M.N. Michael, H. Gohar, The effect of both UV\Ozone and chitosan on natural fabrics. Int. J. Chem. **2**(2), 28 (2010)
76. H.A. Eren, O. Avinc, B. Erişmiş, S. Eren, Ultrasound-assisted ozone bleaching of cotton. Cellulose **21**, 4643–4658 (2014)
77. H.S. Maqsood, U. Bashir, J. Wiener, M. Puchalski, S. Sztajnowski, J. Militky, Ozone treatment of jute fibers. Cellulose **24**, 1543–1553 (2017)
78. F. Özyildiz, S. Karagönlü, G. Basal, A. Uzel, O. Bayraktar, Micro-encapsulation of ozonated red pepper seed oil with antimicrobial activity and application to nonwoven fabric. Lett. Appl. Microbiol. **56**(3), 168–179 (2013)
79. J. Rahmatinejad, A. Khoddami, Z. Mazrouei-Sebdani, O. Avinc, Polyester hydrophobicity enhancement via UV-Ozone irradiation, chemical pre-treatment and fluorocarbon finishing combination. Prog. Org. Coat. **101**, 51–58 (2016)
80. W. Thorsen, W. Ward, M. Millard, Wool shrinkage control and surface modification by ozone. J. Appl. Polym. Sci. **24**(2), 523–546 (1979)
81. Y. Dong, J. Jang, The enhanced cationic dyeability of ultraviolet/ozone-treated meta-aramid fabrics. Color. Technol. **127**(3), 173–178 (2011)
82. M.T. Abate, M.G. Tadesse, Airflow, foam, and supercritical carbon dioxide dyeing technologies, in *Innovative and Emerging Technologies for Textile Dyeing and Finishing* (2021), pp. 137–164
83. M.T. Abate, A. Ferri, J. Guan, G. Chen, V. Nierstrasz, Colouration and bio-activation of polyester fabric with curcumin in supercritical CO_2: Part I-Investigating colouration properties. J. Supercrit. Fluids **152**, 104548 (2019)
84. A.G. Hassabo, M. Zayed, M. Bakr, E. Abd El-Aziz, H. Othman, Applications of supercritical carbon dioxide in textile finishing: a review. J. Text. Color. Polym. Sci. **19**(2), 179–187 (2022)
85. L.S. Abd-Elaal, A.N. El-Wekil, A.G. Marey, L.N. Allam, A.G. Hassabo, Supercritical carbon dioxide as an impregnation medium for producing functional materials in textile finishing. J. Text. Color. Polym. Sci. **22**(1), 187–195 (2025)
86. M.C. Seghini, F. Touchard, L. Chocinski-Arnault, V. Placet, C. François, L. Plasseraud, M.P. Bracciale, J. Tirillò, F. Sarasini, Environmentally friendly surface modification treatment of flax fibers by supercritical carbon dioxide. Molecules **25**(3), 438 (2020)

87. W.X. Ma, C. Zhao, S. Okubayashi, I. Tabata, K. Hisada, T. Hori, A novel method of modifying poly(ethylene terephthalate) fabric using supercritical carbon dioxide. J. Appl. Polym. Sci. **117**(4), 1897–1907 (2010)

Chapter 4
Oxidative Surface Modification of Textiles

Tariful Islam, Saiful Islam, Badhon C. Mazumder, Md. Himel Mahmud, Sakil Mahmud, and Md. Reazuddin Repon

Abstract This chapter describes how oxidative surface modification plays an essential role in advanced textile processing, along with the sustainable, efficient, and performance-enhancing approaches across diverse fiber types. Chemical oxidation, ozone treatment, plasma activation, photocatalytic processes, and electrochemical and enzymatic-assisted oxidation are different methods for textile surface oxidation. These techniques introduce oxygenated functional groups and micro- to nanoscale roughness onto the textile surface. These approaches improve the wettability, dyeability, adhesion, and chemical reactivity of surfaces. Unlike conventional oxidative routes, which tend to be heavy on chemicals, high in temperature, and require large volumes of water, advanced oxidation techniques offer cleaner, less expensive alternatives. Studies were conducted using cellulose, polyester, wool, and technical fibers to improve dye uptake, fiber-matrix interfacial adhesion, and fastness properties, thereby maintaining or enhancing overall fabric quality. Ozone–ultrasound

T. Islam
Department of Textile Engineering, University of Scholars, Dhaka, Bangladesh

S. Islam
Department of Textile Engineering, Atish Dipankar University of Science and Technology, Dhaka, Bangladesh

B. C. Mazumder · Md. Himel Mahmud
Department of Textile Engineering, Northern University Bangladesh, Ashkona, Dhaka, Dakshinkhan, Bangladesh

S. Mahmud
Ivory V. Nelson Center for the Sciences, Department of Chemistry and Physics, Lincoln University, Oxford, PA, USA

Department of Textile Engineering, Faculty of Engineering, Daffodil International University, Dhaka, Bangladesh

Md. Reazuddin Repon (✉)
Department of Bioproducts and Biosystems, School of Chemical Engineering, Aalto University, Espoo, Finland
e-mail: reazmbstu.te@gmail.com

M. N. Morshed and M. Shahid (eds.), *Advances in Surface Modification and Functionalization in Textiles*, Textile Science and Clothing Technology,
https://doi.org/10.1007/978-3-032-23377-6_4

treatment and control plasma modification are also employed to enhance surface activation without compromising fiber integrity. This chapter identifies oxidative modification as a promising and sustainable route for textile pretreatment and functional finishing, offering significant advantages in efficiency, environmental compatibility, and durability.

Keywords Surface modification · Textile functionalization · Sustainable textile processing · Plasma treatment · Eco-friendly finishing

4.1 Introduction

Textile finishing methods, which often use hazardous chemicals and energy-intensive processes, have been practiced for a long time and have several limitations, which is why sustainability has been a significant concern for the textile industry. Both environmental and human health are affected by these practices in the textile industry. For example, textile production houses use large volumes of water, and the effluent often contains a high level of contaminants, which affect the environment tremendously [1]. This situation is worsened by the use of synthetic dyes, which can disrupt aquatic ecosystems, decrease oxygen levels, and produce potentially carcinogenic byproducts [2]. Apart from this, the carbon footprint of the overall process is also high [3]. Altogether, the evidence clearly shows that conventional textile finishing practices are unsustainable, highlighting the need for cleaner and more eco-friendly alternatives [4].

Nowadays, surface modification has become one of the critical approaches that is used to enhance not only the performance but also the functionality of textiles without changing the basic structure or bulk properties [5]. Widely used methods, i.e., polymerization, nanotechnology, plasma treatment, and grafting, provide textiles with some advanced features, e.g., water repellency, flame resistance, antimicrobial activity, and UV protection [6]. It has been found that surface modification can significantly improve the quality and performance of textile materials. For example, surface layers modification, adhesion, wettability, and chemical reactivity can be significantly improved by plasma-based treatments [7]. Similarly, the dyeability, color fastness, and overall fabric quality can be modified using surface modifications [8]. Overall, surface modification has become an effective and sustainable way to improve the performance and quality of textile materials [5].

Oxidation, by introducing chemical and structural modifications, has become a significant surface modification tool in enhancing fabric functionality. Even new functional groups, e.g., carboxyl and hydroxyl, can be introduced on the fiber through a controlled oxidation process, enabling dyes and finishing chemicals to form stronger bonds with fibers that have gained improved reactivity through oxidative modification. The surface charge is also altered, enabling fibers to interact with various chemicals and achieve the intended functionality. Furthermore, this treatment roughens the surface and increases moisture sorption, thereby improving wettability and, consequently, comfort in textiles. This technique also transforms the fiber surface by activating, cleaning, and etching. This transformation later on helps in subsequent coating, printing, or other processes. For this purpose, techniques such as chemical oxidants, plasma treatment, or nanoparticle-assisted reactions are widely used. Although oxidation can cause a minor drop in fiber strength, it is compensated for by greatly enhanced surface activity [9, 10].

In this chapter, the use of oxidants to modify and improve textile surfaces has been discussed, along with various oxidation methods, namely chemical, plasma, and enzymatic approaches. This chapter also covers how oxidative surface modification influences the surface structure, chemistry, and overall performance of fabrics, e.g., dyeability, wettability, adhesion, and antimicrobial activity, while supporting sustainable, eco-friendly textile processing.

4.2 Fundamentals of Oxidative Surface Modification

4.2.1 Chemical Reactions Involved

Generally, chain scission in polymers occurs when the main polymer is broken down into smaller molecular fragments, primarily through oxidative processes used in oxidative surface modification. The reactive radicals, which are produced from oxygen, attack the polymer chains, ultimately breaking them down into smaller fragments [11]. This process reduces the material's molecular weight and alters its properties. Environmental factors, i.e., light intensity and oxygen concentration, further accelerate this process. Light exposure, for instance, can rapidly decrease the molecular weight of polymers [12]. Although oxidative chain scission commonly results in material degradation, which includes embrittlement and loss of mechanical strength [13, 14]. It can also be beneficial if used in controlled treatments, e.g., plasma exposure, which introduces hydrophilic functional groups on the polymer surface and improves surface reactivity [14]. In general, C–C bonds are among the easiest to cleave by oxidation, as they are abundant in complex organic structures such as lignin. These bonds are easily degraded when exposed to oxidative conditions [15, 16]. Also, when nearby C–H bonds become activated, C–C bonds are even more likely to break, ultimately accelerating the reaction. However, C–O bonds are more stable and need explicit conditions to be cleaved, while C–H bonds are usually the toughest, mainly helping nearby C–C bonds break more easily [15, 17].

Radical-chain autoxidation produces hydroxyl, carbonyl, carboxyl, and peroxide groups. The C–H bonds in polymers are initially targeted by hydroxyl radicals, which are frequently produced when H_2O_2 degrades. Alkyl radicals are produced as a result. These alkyl radicals then combine with oxygen to form hydroperoxides and alkyl peroxides [18]. Alkyl peroxides and hydroperoxides are produced when these alkyl radicals react with oxygen. Oxidation faults are produced when these unstable intermediates break down to produce carbonyl and carboxylic acid groups [19]. Similarly, reactive oxygen species can attack the molecular backbone to form hydroxyl and epoxide groups [20]. pH of the medium is another important factor that basically controls the role of peroxide as a nucleophile or an electrophile during the process [21]. The surface etching in this type of modification is a combination of chemical and physical impact. Radicals, such as hydroxyl (•OH) and oxygen (O•), typically remove H-atoms and break a carbon–carbon (C–C) bond. This allows oxygen to be inserted and initiate the chain scission process, ultimately producing lower molecular weight oxidized compounds and progressively eroded surface [22, 23]. Physically, the neutral species and the energetic ions collide with the polymer, which supplements oxidation and helps to remove degraded fragments [24, 25].

The effective interfacial area created by surface etching also influences mechanical interlocking and adhesive strength, thereby promoting strong anchoring between the adhering layer and the polymers. As a result of this interaction, oxidative modification produces microscopic holes, as well as surface roughness, which helps in strengthening the bonds [26]. As the modification progresses, surface etching slows because the surface is reorganized during oxidative modification and becomes more stable. A typical surface etching process comprises the following [25, 27]:

- An initial surface oxidation stage;
- The penetration into the bulk;
- The removal of materials.

4.2.2 *Factors Influencing Modification Efficiency*

The following factors influence chain scissoring [28, 29]:

- Temperature: Higher temperature induces thermal degradation, hence reducing molecular weight.
- Oxidant concentration: With high oxidant levels, especially •OH radicals, accelerate scission in some polymers like PEG, whereas lower levels may start crosslinking.
- Exposure time: If the exposure time is high, it increases the amount of damage done by oxidative modification, which in turn results in more fragmented and enhanced biodegradable polymers.

Usually, chain scission is accompanied by disruption of molecular alignment and reduced molecular weight, resulting in reduced strength, elasticity, and strain-induced crystallization. The mechanical integrity also gets weakened due to the modification of surface morphology, which causes changes in microstructural uniformity and crystallinity [30–32]. Hydroxyl (•OH) and ozone (O_3) are the primary oxygen species that are responsible for generating oxygenated functional groups. Among them, •OH can initiate strong oxidation, which may produce functional groups such as C–OH, C=O, C–O–O, and O–C=O [22, 33]. On the other hand, O_3 produces carbonyl (C=O) and hydroxyl groups (OH). On the contrary, $O_2{\bullet}^-$ and H_2O_2 play a role as precursors and hence as a secondary source to produce hydroxyl radicals in surface modification [34]. Often, the formation of the oxygenated group and the severity depend on the type of oxidizing agent. For example:

- Hydroxyl radical (•OH) induces a sequential oxidation on polymers like polystyrene [33]. This process generates carbonyl (C=O) and carboxyl (–COOH) groups.
- Ozone (O_3) induces more extensive oxidation, generating significant amounts of carbonyl (C=O) and hydroxyl groups (OH) [34].
- Chlorine dioxide radicals (ClO_2•) oxidize specific polymer sites, such as polypropylene methyl groups, into carboxylic acid and hydroxyl groups [35].
- Oxygen plasma rapidly activates surfaces, contributing to C–OH formation [36], while UV/oxygen systems create peroxides and carbonyls through radical chain reactions [37].

4.3 Methods of Oxidative Surface Modification

4.3.1 Chemical Treatments (Wet Processes)

4.3.1.1 Alkaline Hydrolysis

Alkaline hydrolysis modifies polyester fibers primarily via surface-selective cleavage of ester bonds. It is a regulated erosion process that creates hydrophilic terminal groups by hydroxide-ion attack on ester linkages, leading to endwise peeling of the PET surface and removing the molecular overlayer while the bulk remains intact [38]. As evidenced by SEM analysis, this reaction generates uneven nanoroughness [39] and reduces fiber diameter [40]. FTIR spectra and contact angle measurements generally show higher amounts of –COOH and –OH groups [41], indicative of increased dye uptake and improved wettability. Therefore, with the correct parameters, alkaline hydrolysis can simultaneously alter the surface morphology and chemistry of polyesters, thereby improving interfacial performance while retaining bulk integrity.

Polyester fibers treated by sodium hydroxide undergo a mainly surface alkaline hydrolysis (Fig. 4.1). Hydroxide ions play the role of nucleophiles in this reaction, and the hydroxyl groups attack the ester linkages along the PET backbone, in turn, breaking the –COO– to make carboxylate ($–COO^-Na^+$) and hydroxyl (–OH) end groups. The reaction proceeds via an endwise peeling pathway [38], which refers to the fact that most of the reaction only occurs at the molecular end with little diffusion into the fiber core. As hydrolysis continues and hydrophobic ester segments get replaced by more hydrophilic terminals, the surface becomes more polar, and wettability is thus enhanced [42]. The process remains surface-controlled mainly at approximately 80 °C, allowing for a controlled modification of the surface without

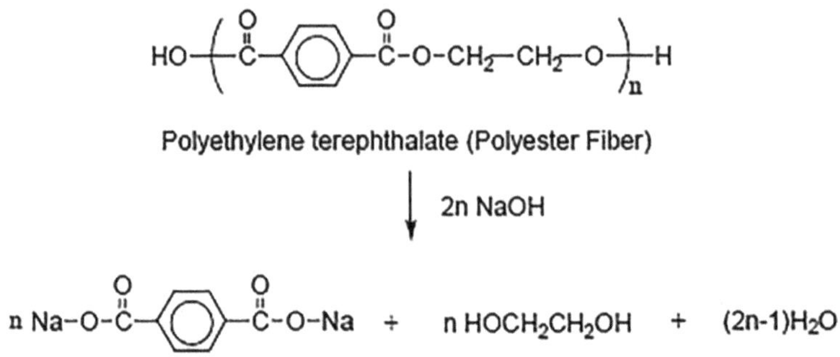

Fig. 4.1 The alkaline hydrolysis of polyester fiber (reproduced under the terms of CC BY [46])

affecting the integrity of the bulk material [41]. Thus, NaOH tailored the chemistry of polyester by cleavage of ester bonds, resulting in –COO$^-$ and –OH entities to regulate adhesion, dyeing, and subsequent chemical modifications. Alkaline hydrolysis of polyester fibers is typically evaluated using SEM, FTIR, and tensile testing to assess changes in both surface morphology and structural integrity. FTIR identifies the formation of carboxylate and hydroxyl groups and tensile testing to assess mechanical effects, with up to ~ 20% strength reduction reflecting the delicate balance between effective surface activation and the preservation of fiber integrity [43]. Controlled alkaline hydrolysis simultaneously modifies the surface chemistry and morphology of polyester fabrics, enhancing their dyeability and wettability. The treatment with NaOH lyses esters partially on the surface of the fiber, leading to the generation of hydroxyl and carboxyl functional groups on the fiber and therefore increasing its surface polarity and hydrophilicity [44]. Simultaneously, the reaction causes nanoscale etching, resulting in micro- and nanoroughness that increases surface area and disrupts crystalline regions. This leads to improved dye uptake, lower contact angles, and faster water absorption because the newly created polar sites strengthen hydrogen bonds with dyes and water molecules [45]. The result is a technique that updates polyester materials without compromising bulk integrity and is both eco-friendly and repeatable.

4.3.1.2 Acidic Oxidation

Nitric acid oxidation modifies the fiber surface by introducing nitro, carbonyl, and carboxyl groups, resulting from C–H and C=C bond cleavage and subsequent transformation into higher-oxidized species. Chromic acid is another potent oxidizing agent that enhances surface oxygen level due to the oxidation of hydroxyl and alkyl sites, and also exists in the forms of carbonyl, carboxylic acid, ester, and hydroperoxide groups [47]. Regarding peroxydisulfate, the sulfates ($SO_4{\bullet}-$) attack the polymer chains, causing partial degradation of the surface and the introduction of sulfonic or oxygenated groups [48]. In general, these oxidants convert fiber surfaces via bond cleavage and the addition of oxygen or sulfur-based moieties, but the stepwise mechanisms differed depending on fiber chemistry.

Oxidants change the surface properties of fibers, introducing polar functional groups to the fiber surface through the cleavage of former chemical bonds such as C–H, C–C, or glycosidic linkages, and replacing them with oxygen-rich functional groups. In this regard, TEMPO (2,2,6,6-Tetramethylpiperidine-1-oxyl) preferentially oxidizes the primary alcoholic groups of cellulose to form carboxylate ($-COO^-$) groups that increase the surface charge and hydrophilicity [49]. Likewise, ozone treatment generates reactive oxygen species that may cleave double bonds or hydroxyl sites, resulting in carbonyl groups on cellulose-based fibers [50]. More potent oxidants, like potassium permanganate or hydrogen peroxide, break intermolecular forces and partially degrade non-cellulosic fractions, allowing for additional carbonyl or hydroxyl groups to be added [50]. Collectively, these methodical oxidative procedures render the fiber face more polar and, consequently, more chemically reactive, thereby aiding subsequent finishing or modification processes.

Oxidant concentration, treatment temperature, and reaction time are critical factors controlling fiber surface oxidation. These contribute to the fiber's structural and functional aspects. Higher oxidant concentrations and higher temperatures lead to the faster clearance of the non-cellulosic phase, thus leading to more reactive cellulose and an increase in crystallinity as observed in the case of the jute fibers subjected to 2M peracetic acid at 50–60 °C [51]. In contrast, because excessive oxidation (caused by prolonged reactions) results in chain scission, which in turn decreases fiber dimensions and consequently, in some cases, compromises mechanical strength, despite increased surface charge and thermal stability [52]. Optimal oxidation and dyeable conditions strike a compromise between reactivity and structural integrity, with enhanced polarity that facilitates dyeability and prevents over-degradation. On the other hand, overly aggressive treatments compromise the tensile properties of composites; therefore, all reaction parameters must be well-regulated.

4.3.1.3 Bleaching of Natural Fibers

The bleaching of cotton with hydrogen peroxide primarily proceeds via radical-coupled oxidation. Hydrogen peroxide decomposes to hydroxyl radicals (HO•) and superoxide radicals (O_2•–) under alkaline conditions and in the presence of catalysts or activators. Hydroxyl radicals are involved in reducing color over a broader range [53]. They target the chromophoric groups present in natural pigments of the cotton fiber. Consequently, the whiteness index is higher [54]. Perhydrolysis reactions yield peracetic acid, a precursor that promotes the oxidative degradation of the pigments [54]. Although the bleaching process is highly effective for cotton, polyester components in cotton–polyester blends are less reactive due to their hydrophobic and crystalline nature. As a result, the bleaching efficiency of cotton–polyester blends depends on fiber accessibility. Cotton undergoes radical-driven oxidation, while polyester remains largely unaffected unless aided by carriers or higher temperatures.

The decomposition of peroxide into an active radical concentration is strongly dependent on pH, time, and temperature, indicating that these factors heavily influence hydrogen peroxide bleaching. Although peroxide itself is stable and releases radicals slowly at pH ~ 7.5–8 (slightly alkaline), a decrease in pH causes rapid degradation and a loss of bleaching ability; this is evident in the literature [55]. Temperature accelerates radical production; it is also harmful to fibers, but the whiteness value of activator-assisted systems is high at a low temperature (80 °C) [56]. Both temperature and treatment time need to be optimized, as long treatment times provide marginal additional benefit while aggravating degradation. These parameters need to be optimized for efficient bleaching and for maintaining fabric strength, both in cotton and in cotton–polyester blends.

4.3.2 Advanced Oxidative Processes

4.3.2.1 Ozone Treatment

Surface modification with ozone is based on radical-driven oxidation. Ozone decomposes into hydroxyl radicals in aqueous systems, which attack the most reactive sites on the fiber surface, causing molecular chains to be cleaved and the introduction of various oxidized groups, including carbonyl, carboxylic acid, ester, and hydroperoxide [57]. This conversion is confirmed by an increase in surface oxygen content seen from XPS analysis [47]. Fibers with more lignin or hemicellulose are less likely to be cleaved, with each molecule undergoing ~ 0.2–5.3 chain scission events, to show substrate chemistry underwrites oxidation efficiency [50]. Overall, ozone treatment changes surface chemistry by substituting weaker ionic bonds with polar oxygen-functional groups, thereby balancing reactivity and wettability, and, most importantly, the fiber core remains intact when the parameters are controlled. The ozone treatment results are sensitive to many parameters, including concentration, exposure time, ambient humidity, and the balance between surface activation and structural stability. More specifically, higher concentrations of ozone or longer reaction time can improve the oxidation efficiency, resulting in more carboxyl and hydroxyl groups, such as carbon fibers with carboxyl groups increased from 1.27 to 8.12% after 60 min in 5 L/min of ozone [58]. But overexposure can reduce mechanical strength, particularly in natural fibers. The catalytic action of humidity arises because it allows ozone to decompose into hydroxyl radicals, which promote oxidation at aqueous or high-moisture sites. Controlled exposure of flax fibers to ozone under humid conditions decreased lignin content and degree of polymerization, and increased the crystallinity and roughness of flax fibers [59]. On the other hand, too-dry conditions limit radical formation and inhibit functionalization. Ozone treatment, when applied for 30–60 min, can also enhance fiber–matrix bonding in synthetic fibers like polyethylene and hence increase the tensile strain (1.3–2.5 times) in the treated fibers without damage to their structure [60]. The oxidation efficiency is higher with higher ozone flux, longer exposure times, and sufficient humidity; however, due to the rapid nature of oxidation, overdosing must be avoided, requiring a tailored dose for each fiber type.

4.3.2.2 Plasma Treatment (Low-Pressure and Atmospheric Pressure)

Plasma discharge produces a spectrum of highly reactive oxygen species that oxidize the fiber surface. In the dielectric barrier discharge, the energetic collisions between the gas molecules generate ozone (O_3), atomic oxygen (O), superoxide anion (O_2^-), hydrogen peroxide (H_2O_2), and hydroxyl radicals (•OH), which affect the surface activation [61]. This species cleaves surface bonds and absorbs oxygen to create carbonyl and hydroxyl groups, which causes micro-cracks and surface roughness [62]. The plasma species and conditions determine the extent and intensity of plasma oxidation. In low-pressure plasma, due to negligible diffusion of gas species to the solid, reactive species are only present on the surface, whereas in atmospheric-pressure plasma, reactive species can penetrate the inner porosity of the structure [63]. The number of radicals and their mobility depend on the discharge power, the gas composition, and the treatment distance. This prevents the formation of surface radicals, which limits the extent of surface modification [7]. The quantity of fiber surface oxides has a direct correlation with the underlying physical and chemical mechanisms that generate reactive species, which are clearly controlled by the plasma process parameters. When the discharge power is higher, the radical concentration increases and the oxygen absorption strengthens, creating features such as micro-cracks and grooves [62]. Chamber pressure regulates reaction depth. Low pressures are conducive to bulk-scale functionalization, whereas high pressures result in surface-constrained modification. Since gas composition identifies the type of reaction pathway, for instance, oxygen plasmas enhance the generation of carboxyl and carbonyl groups, while inert gases like argon provide only physical etching [64]. Excessive exposure can lead to material loss. Notably, alterations remain confined to the outer nanolayer, preserving bulk integrity while enabling tunable surface activation [64, 65].

4.3.2.3 Electrochemical Oxidation

Electrochemical oxidation at the fiber–electrode interface involves sequential redox reactions that generate reactive oxygen species, thereby modifying the fiber surface. The functional groups, including hydroxyl (•OH), carboxyl (–COOH), and carbonyl (C=O), can be controlled and are introduced by anodic oxidation, and simultaneous oxygen evolution occurs at the electrode surface [66]. Electrode material, electrolyte composition, and applied potential all strongly influence oxidation efficiency and the type of functional groups formed. Water oxidation generates surface-bound hydroxyl radicals at moderate potentials, resulting in mild functionalization. In other words, at high potentials, carbon sites are directly oxidized to generate carboxyl carbonyl functionalities. It proceeds with a low and then a high intensity phase, wherein the former seems to be specific to defect regions or grain boundaries, while the latter ruptures the aromatic carbon rings and enhances the surface roughness [67]. These electrochemical reactions are known to amplify fiber wettability, surface energy, and interfacial bonding efficiency by introducing oxygen-containing functional groups across the oxidized surface.

4.3.2.4 Fenton's Reagent and Advanced Oxidation Technologies

Fenton's reagent produces hydroxyl radicals that unequivocally oxidize the fiber surface for pervasive non-selective oxidation. This generates free radicals that break the chemical bonds in the fiber's surface layers. In this process, functionalities containing oxygen, such as hydroxyl, carbonyl, carboxyl, and peroxy groups, are introduced [22]. Oxidation usually proceeds from the outside to the inner part of the fiber. This results in a continuous gradient of chemical changes that enhance surface polarity and wettability. Cellulosic fibers are surface-treated or coated to enhance their ability to attract and retain water. It additionally induces fiber morphological alteration, such as an increase in surface roughness and microcrack formation [68]. These changes may enhance dyeability and fiber attachment. For a longer treatment, however, fiber strength and thermal stability decrease.

4.3.2.5 Photocatalytic Oxidation

Titanium dioxide (TiO_2) has been widely studied as a photocatalyst for light-driven surface oxidation of textile fibers via photoexcited charge carriers and redox reactions. TiO_2 absorbs photons with energy equal to or larger than its bandgap and forms electron–hole pairs, which diffuse towards the surface of the material. In this process, the photoexcited electrons react with atmospheric or dissolved O_2 molecules, producing superoxide radicals ($O_2{\bullet}-$), while producing photogenerated holes, oxidizing surface hydroxyl groups or water molecules, producing hydroxyl radicals (•OH) [69, 70]. These reactive oxygen species act in concert to drive the oxidative degradation of organic impurities, pigment residues, and surface contaminants, while also introducing oxygen-containing functional groups that improve fiber hydrophilicity and chemical reactivity. Exciting advances in this field, such as metal-ion doping, heterojunction formation, and TiO_2 nanostructuring, have significantly enhanced the photocatalytic performance of materials. This makes it easier for them to collect more light and separate charges more effectively. This leads to the formation of greater amounts of reactive species and prolonged photocatalytic activity under visible light [71]. Based on the above overview, this technique has served as a guideline for developing modern photocatalytic surface treatment for textiles, resulting in cleaner, low-energy-consumption technology.

4.3.3 *Emerging Green Oxidation Techniques*

4.3.3.1 Enzyme-Assisted Oxidation

There are several oxidative enzymes, e.g., laccase and peroxidase, which transfer electrons, thereby oxidizing textile fibers and producing reactive radicals. One-electron oxidation of phenolic and non-phenolic substrates is enabled by multicopper oxidases, such as laccase, which convert molecular oxygen to water. At the fiber surface, this reaction produces phenoxy radicals that can participate in coupling, grafting, or crosslinking reactions. The fiber gains additional oxygen-based groups like hydroxyl, carbonyl, and carboxyl as a result of these processes, which covalently bind functional molecules [72, 73]. To initiate electron transfer reactions, peroxidase enzymes use hydrogen peroxide as an oxidizing agent. These reactions generate radicals that can modify or polymerize molecules on the fiber surface. These reactions take place under gentle, water-based conditions. The modification of fiber improves properties such as wettability, dyeability, UV resistance, and antioxidant activity, as demonstrated in the grafting of phenolic compounds onto cotton using laccase [74]. These enzymatic oxidation routes collectively provide an environmentally friendly, regulated method for surface functionalization, consistent with sustainable textile finishing practices.

4.3.3.2 Photocatalytic and Sonochemical Oxidation

Reactive oxygen species (ROS) are generated in combination with light and ultrasonic energy. This can be used to modify textile fibers. Cavitation phenomena generate a localized zone of high temperature and pressure in an ultrasonic treatment. It cleaves water molecules and forms hydroxyl radicals and hydrogen atoms [75]. Catalysts like Zn(II) and Pd(II) porphyrins can be used to enhance the ROS yield by singlet oxygen generation, which improves the oxidation efficiency [76]. Sonophotocatalysis is the process where sonication is combined with light irradiation. In this method, not only photon-induced but also cavitation-induced processes function simultaneously, whereas photonic excitation generates electron–hole pairs to initiate redox reactions, and on the other side, ultrasonic agitation promotes dispersion and inhibits charge recombination [77]. This results in hydroxyl, perhydroxyl, and superoxide radicals, which basically play a significant role in oxidizing impurities and/or introducing polar functional groups, ultimately enhancing surface wettability, dyeability, interfacial adhesion, etc. This proves the effectiveness of sonophotocatalysis in advanced fiber surface oxidation [78].

The process parameters, i.e., treatment duration and catalyst concentration, are vital in achieving effective results in photocatalytic and sonochemical oxidation. This is because-

- The extent of oxidation or the level of effect is achieved by the duration of treatment, as well as the intensity of energy. Generally, intense exposure may increase the amount of radicals generated, at the expense of over-oxidation and fiber damage. Typically, the process is optimized by balancing these, with the temperature maintained at a moderate level (20–40 °C) and the ultrasonic frequency kept low (20–200 kHz). This condition maintains the optimum cavity formation as well as reactive species formation without damaging the fibers [79].

- Catalyst concentration should be maintained at a lower level, i.e., 0.1–100 μM. This prevents fiber damage while maximizing radical yield and surface activity.
- On the other hand, the reaction medium, especially the pH, is another factor to consider. A mildly acidic condition helps the process by maintaining other desirable outputs.

Process kinetics mainly depend on substrate concentration and temperature, affecting the rate of radical diffusion and surface interaction [80]. Hence, well-optimized systems can balance oxidation efficiency with minimal fiber degradation. A substantial functionalization is possible by optimizing the process while consuming 80% lower energy [80].

4.3.3.3 Bio-Based Oxidizing Agents

Textile surface treatments are being updated to include more sustainable alternatives derived from naturally occurring compounds, such as plant extracts, organic acids, and biogenic peroxides. Biodegradable materials, such as organic acids (e.g., citric acid, tartaric acid, and malic acid), provide a gentle oxidation process that modifies fiber surfaces without damaging their structure [4]. Phenolic and flavonoid compounds, found in plant extracts and essential oils, can also serve as mild oxidants and as eco-friendly sources of colorants and fiber-finishing agents [81]. On the contrary, traditional oxidants like hydrogen peroxide (H_2O_2) or chromic acid offer superior environmental compatibility and safety [82].

4.4 Effects of Oxidative Modification on Textile Properties

4.4.1 Changes in Surface Morphology (Roughness, Porosity, Microstructure)

Oxidation can yield surface changes, such as roughness and porosity at the micro- or nanoscale. For instance, Fang et al. discovered the etching effect of air plasma on the surface of polyester fabric. As a result, the surface resembled rough and pit-like features at a micro level due to oxygen species, whereas the untreated polyester surface was smooth [83]. In another study, Xiao et al. noted the limitations of air plasma for etching the carbon fiber surface. Oxygen-containing functional groups were attached to the exposed carbon sites on the graphite plane, along with being etched [84]. In wool, Thakker et al. found that scales were unfastened and stripped after air plasma treatment. In other words, the surface became more scaly and grooved [85].

4.4.2 Alteration of Surface Chemistry (Functional Groups, Wettability, Charge)

Oxidative modification mainly changes two things, i.e., polarity and surface energy. In a study, the polyester fabric surface showed the existence of oxygenated species, i.e., hydroxyl and carboxyl groups, following plasma treatment. Therefore, polyester exhibited radial increment in terms of wettability [86]. In another study, these groups were identified by chemical analysis in the case of ozone oxidized sample, through the FTIR spectroscopy, where –C=C– bonds belong to the band at 1640 cm^{-1} and 1537 cm^{-1} [87]. Similarly, the increase in C–O, C=O, and COOH peaks indicates a polar surface that is chemically active. As a result, air plasma-treated polyester fabric absorbs water in 503 ms compared to 90 s for untreated polyester fabric, proving its significant improvement in wettability [83]. The oxygen-to-carbon (O/C) ratio can indicate the level of oxidation of a surface. For example, Fang et al. reported the increment of this ratio in air plasma-treated and air/argon-treated (50%), respectively, in a polyester fabric [83].

4.4.3 Impact on Dyeability and Printability

Oxidative modification yields better dyeability and printability by generating more accessible chemical sites. For instance, Lou et al. used oxidized raffinose, which improved dyeability and anti-wrinkle properties. Usually, both properties don't complement each other. However, targeted chemical reduction via oxidized raffinose overcame the issue [88]. Moreover, oxidative agents like ozone tend to show affinity for chromophores [89]. In another study, Tertyshnaya et al. reported color change in nonwoven polylactide-natural rubber fibers due to ozone oxidation [87]. It is observed that printability increases with oxidative modifications. By employing air plasma, Fang et al. got darker and more vivid colors due to more roughening and higher color strength (K/S) while inkjet printing on polyester fabric [83].

4.4.4 Improvement in Adhesion (Coatings, Composites, Laminates)

Oxidative pretreatment with ozone or oxygen plasma increases adhesion between the fiber surface and applied coatings, such as chitosan. This is an outcome of enhanced surface energy and chemical compatibility. In a study, physical anchoring of fibers was observed when chitosan was applied to a polyester fabric [90].

4.4.5 Influence on Mechanical Properties (Strength, Flexibility)

Chain scission after advanced oxidative modification is a risk for the mechanical performance of the fibers. In that case, protective coatings can overcome the matter. For example, Bušac et al. recorded ozone- and plasma-treated polyester fabric to have an optimum protective layer due to the following chitosan treatment [90]. In a study, electrospun PHB fabrics underwent ozone treatment, and tensile strength increased with exposure time up to 7 min. This happened due to the enhancement in the degree of crystallinity [91]. In another study, silk underwent photooxidation, resulting in a decrease in mechanical strength due to peptide bond breakage. Along with this, aromatic compounds get excited and oxidized under UV irradiation, causing yellowing [92]. Moreover, Xiao reported the increase in the interlaminar shear strength (ILSS) from 64.9 to 80.0 MPa owing to 60 s of atmospheric plasma treatment. This effect was gradually growing and varied with the time of treatment [84].

4.4.6 Durability and Fastness Behavior

To sustain the oxidative modification, robust chemical anchoring and crosslinking can precede oxidation. In a study, Thakker et al. reported that air-plasma-treated wool and cotton achieved excellent wash, staining, and rubbing fastness. In terms of color change, wash fastness properties for wool and cotton were labeled as very good and moderate, respectively [85].

4.5 Applications of Oxidative Surface Modification

4.5.1 Pre-treatment for Dyeing and Printing

Oxidative modifications can facilitate pretreatment and subsequent dyeing and printing. For instance, Perincek et al. discovered that a combination of ozone and ultrasound treatment can provide a significant level of whiteness in woven cotton fabric. That means the bleaching role was unique, along with a better hydrophilic nature [93]. In another study, Gashti et al. pretreated cellulose fibers with ozone gas to perform grafting of fluoromonomer. That caused surface oxidation, thereby improving reactivity. As a result, the fluoromonomer got readily anchored to the surface [94]. Surprisingly, Erin et al. reported a significant effect of the ultrasound method combined with ozone treatment in the homogenizer bath. While achieving a better bleaching effect, conventional properties were reported to remain unharmed. This method has the full potential to replace the conventional bleaching with hydrogen peroxide [95]. Benli and Bahtiyari reported similar results from their experiment for both the ozone-ultrasound method and the conventional methods, such as desizing, scouring, and bleaching. After that, they observed a similar dyeing quality in both cases, with only minor adjustments to the greige fabrics [95].

4.5.2 Enhancing Adhesion in Technical Textiles and Composites

Some surface needs to be modified to enhance adhesion. For example, Netravali and Bahners reported UV treatment of polyester in atmospheric oxygen, yielding a polar surface due to the introduction of polar groups. That's why binder joining became easier subsequently. Particularly, this feature can help make composites [96].

4.5.3 Improvement of Hydrophilicity and Wettability

Wettability and hydrophilicity usually depend on water contact angle and surface energy. In a study, UV-treated polyester within air demonstrated better wetting property due to the decrease in water contact angle [96]. In another study, Thakker et al. reported an increase in the hydrophilicity of air-plasma-treated wool and cotton. Contact angle was reduced from 145.84° to 41.35° for wool, whereas cotton's contact angle came lower from 146.35° to 80.43° [85].

4.5.4 Antimicrobial and Functional Finishing

Textiles can undergo pretreatment, such as chemical oxidation, to better accommodate subsequent finishes. In other words, oxidative modifications can indirectly impart functionality to textiles, which is then enhanced by proper finishing. In a study, Kováčik et al. reported the chemical grafting of acrylic acid onto a polypropylene nonwoven fabric via plasma polymerization. There, the fabric got more wetting and adhesive property owing to more carboxyl groups [97].

4.5.5 Applications in Filtration, Medical Textiles, and Smart Textiles

Alexeeva reported the use of electrospinning to impart antimicrobial properties to nonwoven fabric from PLA fibers and ozonized oil. Hence, these nonwovens show potential for biomedical textiles, wound dressings, and packaging materials. Here, 3% ozonized oil also serves other functions, such as plasticization, improved fiber uniformity, higher porosity, higher water absorption, better mechanical strength and flexibility, etc. [91].

4.5.6 Enhancing Coating and Lamination Adhesion

Oxidative surface modification can improve adhesion, resulting in a better coating. For instance, Gao et al. reported the fabrication of carbon fiber (CF)-reinforced polyaryletherketone (PAEK) composite. There, adhesion improved owing to electrochemical oxidation. They found that textiles include pretreatment and dyeing at an applied current of 200 A. They also pointed out that an excess carboxyl group leads to less adhesion, as it hampers the structure [98].

4.5.7 Facilitating Composites Manufacturing

Pre-oxidation of the fiber surface plays a vital role in the manufacturing of composites. For example, Li et al. reported the electrochemical oxidation of recycled carbon fibers, resulting in improved fiber-matrix bonding. The final composition included anodized fiber and alkali-activated matrix, which were chemically bonded due to prior modifications [99].

4.6 Environmental and Sustainability Aspects

Traditional surface modification techniques in textiles include pretreatment, dyeing, and finishing processes. For example, H_2O_2 bleaching is prevalent for whitening textile materials, with significant environmental impacts, including high chemical and energy use, toxic effluent, etc. Conventional oxidative modifications require high temperature treatment, like 98 °C for desirable H_2O_2 bleaching in a large volume water bath, for which the pressure on electricity and water remains consistently higher [100]. In other contexts, higher water use yields greater wastewater volumes, which require prior treatment. In this case, higher energy consumption is found while further treatment of effluents [101].

4.6.1 Effluent Issues: High COD, BOD, Salt Content, and Toxic By-Products (E.G., from Chromic Acid)

Traditional oxidation systems use a significant amount of types of chemicals and water. Hence, those create more effluents, including higher BOD, COD, and toxic by-products. Often, effluent from textile industries has color, along with dyes and chemicals that are poorly degradable, resulting in higher COD and BOD values. In a study, Afanga et al. reported a 325 mg/L COD value and 35 mg/L in the case of untreated textile wastewater [102]. According to Tanveer et al., traditional technologies that are not sustainable carry 100 times more water footprint [103]. Moreover, toxic by-products, such as intermediates, can be more harmful than the original compound. This happens due to the complex oxidation chemistry, where pollutants are transformed into more toxic compounds [104] (Fig. 4.2).

4.6.2 Advantages of Dry Processes

Dry processes like plasma are recognized as environmentally friendly as they consume less electricity, water, and chemical auxiliaries. In the process of cold plasma, there is only the utilization of electricity and simple gases, i.e., air, oxygen, argon, causing far less chemical and liquid waste [106]. In a study, Bušac et al. mentioned microplastic release can be mitigated due to the physical anchoring of fibers in chitosan-coated polyester fabric, which is a significant byproduct here, causing less pollution [90]. Dry processes, such as plasma, can help achieve sustainable development goals. Because it becomes easier to avoid carbon emissions and toxic effluents, causing responsible production and consumption, along with benefiting both nature and human beings [85].

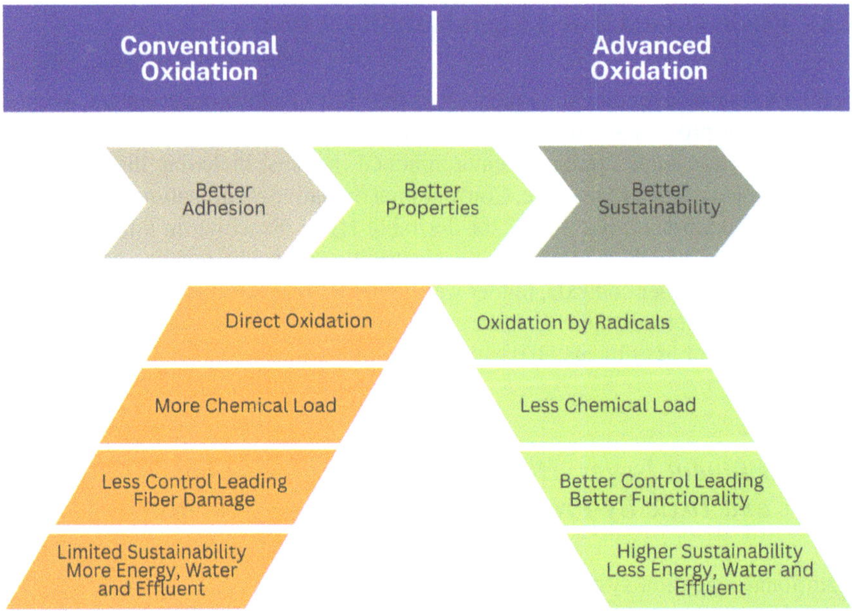

Fig. 4.2 Comparison between conventional and advanced oxidation for surface modifications [105]

4.6.3 *The Challenge of Treatment Longevity*

The phenomenon of "hydrophobic recovery" in plasma-treated polymers and strategies to mitigate it. Despite achieving hydrophilicity, the textile surface tends to revert to its initial hydrophobic state over time. As a result, this functionality doesn't last long, yielding hydrophobic recovery [107]. In other words, Watson et al. reported this thing as the result of the ageing of the finishing [108]. There are several techniques to mitigate this phenomenon, such as combined treatment, a two-step oxygen-diffusion process, and surface crosslinking. In a study, Primc and Mozetič reported the successive plasma treatment with O_2 and CF_4 to sustain the hydrophilicity of PET fabric [109]. Similarly, Behnisch reported the retardation of hydrophobic recovery through a hydrogen plasma treatment of polyethylene [110]. In addition, Mozetič reported a two-step approach: subsurface enrichment with O_2 and cooling, followed by oxygen plasma. Due to prior subsurface, hydrophobic recovery gets suppressed [111].

4.6.4 *Cost–Benefit Analysis*

Among oxidative modifications, ozone treatment requires less time, water, and chemicals than conventional methods. Simultaneously, this can offer shared applications with UV, plasma, and ultrasound [112]. In a study, the total costs for a Jet, Pad-steam, and DBD were calculated. There, DBD was reported at 0.013 euros per kg, which is lower than both conventional systems. As a result, this outcome proved plasma's economic advantage while pretreating cotton [113].

Oxidation modification includes not only toxic chemicals but also more toxic by-products, volatile organic compounds, and harmful emissions such as nitrogen dioxide and chlorine dioxide. As a result, compulsory initiatives are taken to control emissions and purify effluents [114]. Li et al. reported lower energy consumption in their one-step scouring/bleaching approach. They found that 82.3% of energy consumption was lower than that of a conventional bath. In terms of liquor ratio, the one-step approach needs a half M:L ratio compared to the conventional bath [100].

4.7 Challenges and Future Perspectives

Madhushree et al. reported that controlling uniformity and the degree of oxidation is possible when reagents and catalysts overcome diffusion limitations, thereby avoiding either under- or over-oxidation. Again, advanced catalysts are required to ensure the precise location of newly generated functional groups. Otherwise, oxidation will be uneven, where reaction conditions also influence to a great extent [114]. Li et al. reported a one-step process for scouring/bleaching cotton/spandex fabric. They faced a decrease in tensile strength of less than 10% in the UV/H_2O_2 approach [100]. In a study, Madhushree et al. reported cellulose degradation due to high levels of oxidation. This makes cellulose weak and brittle [114]. Lab experiments and industrial applications vary depending on several criteria, such as the uniformity of methods, which yields consistency in properties across textile batches. In industry, standardization is essential for achieving several properties. Consequently, this will be implemented in industries for bulk production [115]. Nanoparticles can be incorporated into textiles to impart multifunctional properties, such as antibacterial and self-cleaning properties. For this, plasma modification is crucial, as it etches the fabric surface, enabling nanoparticles to adhere more easily. Again, plasma parameters play a vital role in improving adhesion. However, agglomeration of nanoparticles and uneven nano-coating can make the entire work unsuccessful [116]. In the future, the focus will be on sustainable oxidative modifications, such as plasma. To ensure that, LCA and improved functionality are the main areas to work on using air plasma [85]. Also, oxidative modification can act as an oxidative clearing agent in place of reduction clearing in PC-blended textiles. In the study, the effect of disperse dyeing has been explored before and after. A better result was found after reactive dyeing, as reported in the study [117]. Moreover, Madhushree et al. concluded that novel, environmentally friendly oxidants are poised to be the next game-changer, offering a smaller environmental footprint and better energy and water efficiency [114].

4.8 Conclusion

Oxidation has been a prevalent surface modification technique in textiles to enhance functionality. Basically, this involves fiber chemistry and its physicochemical behavior, which varies from fiber to fiber. In general, alterations in fiber reactivity through chemical and physical change impart better functionality to the fibers. In particular, chemical modifications include generating new reactive functional groups, such as hydroxyl and carboxyl groups, which influence surface energy. Similarly, physical changes, such as surface roughness and microstructural alterations, affect fiber reactivity. In both cases, the fiber's reactivity improves significantly, facilitating better adhesion, dye uptake, wettability, etc. Adaptability and sustainability make a considerable difference in the adoption of conventional and advanced techniques for oxidative modifications. Traditional oxidation methods, such as bleaching, acidic oxidation, and alkaline hydrolysis, lack sustainability in terms of energy and water consumption, effluent issues (COD, BOD, toxicity), etc. In contrast, plasma, ozone, Fenton's reagent, electrochemical, and photocatalytic techniques are advanced oxidative techniques. Ways to overcome the shortcomings of conventional approaches include significantly reducing energy, water, and effluent use. Surprisingly, low-pressure air plasma has been the only dry process in this matter. As a result, it outperforms in terms of chemical and water issues prominently. In essence, the purpose of modifying textile surfaces is to improve fiber reactivity in subsequent processes to achieve durable functionality, such as hydrophilicity, hydrophobicity, flame retardancy, etc. In other words, oxidative modifications serve as a pretreatment, improving the surface's interaction with subsequent chemicals. As a result, the fiber-matrix bond is stronger and more adhesive, yielding durable finishes. While doing that, it is essential not to compromise fiber's inherent properties and sustainability. For this, it is expected to promote green chemistry and the circular economy in line with the Sustainable Development Goals (SDGs). With continual development in this regard, new developments may include choosing natural oxidants, achieving economic balance in plasma, improving system control, and improving other processes to reduce the environmental footprint. Finally, LCA plays the most crucial role, on which sustainability in any process can enjoy gradual improvements if followed.

AI-Usage Declaration During the preparation of this chapter, the authors used AI tools to improve the language of the manuscript. After using this tool/service, the authors reviewed and edited the content as needed and take full responsibility for the content of the publication.

References

1. G.A. Kallawar, B.A. Bhanvase, A review on existing and emerging approaches for textile wastewater treatments: challenges and future perspectives. Environ. Sci. Pollut. Res. **31**(2), 1748–1789 (2024). https://doi.org/10.1007/s11356-023-31175-3

2. M. Berradi et al., Textile finishing dyes and their impact on aquatic environs. Heliyon **5**(11) (2019). https://doi.org/10.1016/j.heliyon.2019.e02711
3. A. Azanaw, B. Birlie, B. Teshome, M. Jemberie, Textile effluent treatment methods and eco-friendly resolution of textile wastewater. Case Stud. Chem. Environ. Eng. **6**, 100230 (2022). https://doi.org/10.1016/j.cscee.2022.100230
4. B.M. Eid, N.A. Ibrahim, Recent developments in sustainable finishing of cellulosic textiles employing biotechnology. J. Cleaner Prod. **284**, 124701 (2021). https://doi.org/10.1016/j.jclepro.2020.124701
5. A. Nadi, A. Boukhriss, A. Bentis, E. Jabrane, S. Gmouh, Evolution in the surface modification of textiles: a review. Text. Prog. **50**(2), 67–108 (2018). https://doi.org/10.1080/00405167.2018.1533659
6. J. Ghosh et al., Grafting modification for textile functionalization: innovations and applications. Discov. Appl. Sci. **7**(1), 49 (2025). https://doi.org/10.1007/s42452-024-06364-5
7. J. Peran, S. Ercegović Ražić, Application of atmospheric pressure plasma technology for textile surface modification. Text. Res. J. **90**(9–10), 1174–1197 (2019). https://doi.org/10.1177/0040517519883954
8. Y. Zhang, I. ul Shahid, L.J. Rather, Q. Li, Recent advances in the surface modification strategies to improve functional finishing of cotton with natural colourants—a review. J. Cleaner Prod. **335**, 130313 (2022). https://doi.org/10.1016/j.jclepro.2021.130313
9. M. Knežević et al., Influence of potassium permanganate oxidation on structure and properties of cotton. J. Nat. Fibers **19**(2), 403–415 (2022). https://doi.org/10.1080/15440478.2020.1745120
10. C. Gleissner, J. Landsiedel, T. Bechtold, T. Pham, Surface activation of high performance polymer fibers: a review. Polym. Rev. **62**(4), 757–788 (2022). https://doi.org/10.1080/15583724.2022.2025601
11. T. Ishida, Y. Doi, T. Uneyama, Y. Masubuchi, Modeling for heterogeneous oxidative aging of polymers using coarse-grained molecular dynamics. Macromolecules **56**(21), 8474–8483 (2023). https://doi.org/10.1021/acs.macromol.3c01708
12. B. Louis, S. Caubergh, P.-O. Larsson, Y. Tian, I.G. Scheblykin, Light and oxygen induce chain scission of conjugated polymers in solution. Phys. Chem. Chem. Phys. **20**(3), 1829–1837 (2018). https://doi.org/10.1039/C7CP07347J
13. S. Konica, T. Sain, A reaction-driven evolving network theory coupled with phase-field fracture to model polymer oxidative aging. J. Mech. Phys. Solids **150**, 104347 (2021). https://doi.org/10.1016/j.jmps.2021.104347
14. R.D. Boyd, A.M. Kenwright, J.P.S. Badyal, D. Briggs, Atmospheric nonequilibrium plasma treatment of biaxially oriented polypropylene. Macromolecules **30**(18), 5429–5436 (1997). https://doi.org/10.1021/ma960940x
15. E. Xu, F. Xie, T. Liu, J. He, Y. Zhang, Photocatalytic, oxidative cleavage of C−C bond in lignin models and native lignin. Chem. Eur. J. **30**(24), e202304209 (2024). https://doi.org/10.1002/chem.202304209
16. S.-H. Shi, Y. Liang, N. Jiao, Electrochemical oxidation induced selective C–C bond cleavage. Chem. Rev. **121**(1), 485–505 (2021). https://doi.org/10.1021/acs.chemrev.0c00335
17. W.M. Zhang, Y. Li, Efficient water reforming of biomass to h2 via well-organized redox-neutral cleavage of C−C, O−H and C−H bonds. Angew. Chem. **137**(4), e202416867 (2025)
18. J. Helberg, D.A. Pratt, Autoxidation vs. antioxidants—the fight for forever. Chem. Soc. Rev. **50**(13), 7343–7358 (2021). https://doi.org/10.1039/D1CS00265A
19. Y. Ahn, X. Colin, G. Roma, Atomic scale mechanisms controlling the oxidation of polyethylene: a first principles study. Polymers **13**(13). https://doi.org/10.3390/polym13132143
20. C.M. Spickett, Formation of oxidatively modified lipids as the basis for a cellular epilipidome. Front. Endocrinol. Mini Rev. **11** (2020). (Online). Available: https://doi.org/10.3389/fendo.2020.602771
21. A. More, T. Elder, Z. Jiang, A review of lignin hydrogen peroxide oxidation chemistry with emphasis on aromatic aldehydes and acids. Holzforschung **75**(9), 806–823 (2021). https://doi.org/10.1515/hf-2020-0165

22. H. Du, A. Komuro, Y. Seki, M. Kobayashi, R. Ono, Mechanism behind polypropylene surface modifications by OH radicals: an experimental study. Appl. Surf. Sci. **648**, 159086 (2024). https://doi.org/10.1016/j.apsusc.2023.159086
23. H. Guo et al., The quantum chemistry research on the polyethylene surface modification of He atmospheric pressure plasma jet. J. Phys. D Appl. Phys. **58**(28), 285206 (2025). https://doi.org/10.1088/1361-6463/adea83
24. G. Memos, E. Lidorikis, E. Gogolides, G. Kokkoris, A hybrid modeling framework for the investigation of surface roughening of polymers during oxygen plasma etching. J. Phys. D Appl. Phys. **54**(17), 175205 (2021). https://doi.org/10.1088/1361-6463/abdb0b
25. A. Hassan, S. A. Abd El Aal, M.M. Shehata, A.A. El-Saftawy, Plasma-etching and modification of polyethylene for improved surface structure, wettability and optical behavior. Surf. Rev. Lett. **26**(07), 1850220 (2018). https://doi.org/10.1142/S0218625X18502207
26. S. Siau, A. Vervaet, S. Nalines, E. Schacht, A. Van Calster, Kinetic study of wet chemical treatments on the surface roughness of epoxy polymer layers for buildup layers: I. sweller influence. J. Electrochem. Soc. **15**(12), C816 (2004). https://doi.org/10.1149/1.1813991
27. P. Luan, G.S. Oehrlein, Stages of polymer transformation during remote plasma oxidation (RPO) at atmospheric pressure. J. Phys. D Appl. Phys. **51**(13), 135201 (2018). https://doi.org/10.1088/1361-6463/aaaf60
28. S. Ray, R.P. Cooney, Chapter 9—thermal degradation of polymer and polymer composites, in *Handbook of Environmental Degradation of Materials*, 3rd edn, ed. by M. Kutz (William Andrew Publishing, 2018), pp. 185–206
29. K. Kleemann et al., Photochemical chain scissions enhance polyethylene glycol biodegradability: from probabilistic modeling to experimental demonstration. Environ. Sci. Technol. **59**(33), 17773–17784 (2025). https://doi.org/10.1021/acs.est.5c03567
30. S. Taourit, P.-Y. Le Gac, B. Fayolle, Relationship between network structure and ultimate properties in polyurethane during a chain scission process. Polym. Degrad. Stab. **201**, 109971 (2022). https://doi.org/10.1016/j.polymdegradstab.2022.109971
31. T.A. Will, Y. Lu, K. Wakabayashi, Effects of polymer properties on solid-state shear pulverization: thermoplastic processability and nanofiller dispersibility. ACS Appl. Polym. Mater. **5**(3), 1848–1858 (2023). https://doi.org/10.1021/acsapm.2c01932
32. N. Chongvimansin, T.C. O'Connor, The work of mechanical degradation in elongating polymer melts. Macromolecules 58(4), 1787–1794 (2025). https://doi.org/10.1021/acs.macromol.4c02063
33. A. Vesel, R. Zaplotnik, G. Primc, M. Mozetič, Evolution of surface functional groups and aromatic ring degradation upon treatment of polystyrene with hydroxyl radicals. Polym. Degrad. Stab. **218**, 110582 (2023). https://doi.org/10.1016/j.polymdegradstab.2023.110582
34. R. Zafar, S.Y. Park, C.G. Kim, Surface modification of polyethylene microplastic particles during the aqueous-phase ozonation process. Environ. Eng. Res. **26**(5) (2021)
35. K. Ohkubo, H. Asahara, T. Inoue, Photochemical C–H oxygenation of side-chain methyl groups in polypropylene with chlorine dioxide. Chem. Commun. **55**(32), 4723–4726 (2019). https://doi.org/10.1039/C9CC01037H
36. Y. Fukunaga et al., Interaction of oxygen with polystyrene and polyethylene polymer films: a mechanistic study. J. Appl. Phys. **127**(2), 023303 (2020). https://doi.org/10.1063/1.5127863
37. G. Grause, M.-F. Chien, C. Inoue, Changes during the weathering of polyolefins. Polym. Degrad. Stab. **181**, 109364 (2020). https://doi.org/10.1016/j.polymdegradstab.2020.109364
38. J. Bengtsson, A. Peterson, A. Idström, H. de la Motte, K. Jedvert, Chemical recycling of a textile blend from polyester and viscose, part II: mechanism and reactivity during alkaline hydrolysis of textile polyester. Sustainability 14(11). https://doi.org/10.3390/su14116911
39. S. Lee, Superhydrophobicity and conductivity of polyester-conductive fabrics using alkaline hydrolysis. RSC Adv. **12**(35), 22911–22921 (2022)
40. A. Karimi, H. Izadan, A. Khoddami, S.A. Hosseini, Modifying the surface of poly(ethylene terephthalate) nanofibrous materials by alkaline treatment and TiO_2 nanoparticles. J. Ind. Text. **47**(8), 1944–1958 (2017). https://doi.org/10.1177/1528083717716164

41. I. Čorak, A. Tarbuk, D. Đorđević, K. Višić, L. Botteri, Sustainable alkaline hydrolysis of polyester fabric at low temperature. Materials **15**(4). https://doi.org/10.3390/ma15041530
42. R. Ghamarpoor, M. Jamshidi, M. Sayyadian, M. Razavizadeh, Chemical/photochemical functionalization of polyethylene terephthalate fabric: effects on mechanical properties and bonding to nitrile rubber. Sci. Rep. **13**(1), 14533 (2023). https://doi.org/10.1038/s41598-023-41432-7
43. S. Nourbakhsh, M. Montazer, Z. khandaghabadi, Zinc oxide nano particles coating on polyester fabric functionalized through alkali treatment. J. Ind. Text. **47**(6), 1006–1023 (2016). https://doi.org/10.1177/1528083716657819
44. J.A. Kumar, M. Senthil Kumar, A study on improving dyeability of polyester fabric using lipase enzyme. Autex Res. J. **20**(3), 243–249 (2020)
45. S. Youn, C. Hee Park, Development of breathable Janus superhydrophobic polyester fabrics using alkaline hydrolysis and blade coating. Text. Res. J. **89**(6), 959–974 (2018). https://doi.org/10.1177/0040517518760750
46. Z. Ömeroğulları Başyiğit, Effects of chemical and surface modification on mechanical and chemical properties of polyester fabrics. Duzce Univ. J. Sci. Technol. **6**(4), 1344–1353 (2018) (in en). https://doi.org/10.29130/dubited.423516
47. O.C. Grimm, E.A. Welsh, R. Pang, J. E. Whitten, Comparison of oxidative treatments of polypropylene films and fibers to promote zinc oxide nanoparticle adhesion. J. Adhes. Sci. Technol. **36**(3), 302–316 (2022). https://doi.org/10.1080/01694243.2021.1918917
48. P.V. Nidheesh, G. Divyapriya, F. Ezzahra Titchou, M. Hamdani, Treatment of textile wastewater by sulfate radical based advanced oxidation processes. Sep. Purif. Technol. **293**, 121115 (2022). https://doi.org/10.1016/j.seppur.2022.121115
49. M. Asgarpour Khansary, P. Pouresmaeel-Selakjani, M.A. Aroon, A. Hallajisani, J. Cookman, S. Shirazian, A molecular scale analysis of TEMPO-oxidation of native cellulose molecules. Heliyon **6**(12) (2020). https://doi.org/10.1016/j.heliyon.2020.e05776
50. C. Valls, O. Cusola, M.B. Roncero, Evaluating the potential of ozone in creating functional groups on cellulose. Cellulose **29**(12), 6595–6610 (2022). https://doi.org/10.1007/s10570-022-04694-4
51. M.M. Islam et al., Effect of time, temperature and chemical concentration of peracetic acid treatment on enhancement of crystallinity of jute fibers. J. Ind. Text. **55**, 15280837251321136 (2025). https://doi.org/10.1177/15280837251321136
52. M. Marwanto et al., Effect of oxidation time on the properties of cellulose nanocrystals prepared from balsa and kapok fibers using ammonium persulfate. Polymers **13**(11). https://doi.org/10.3390/polym13111894
53. Q. Li, R. Lu, H. Jiang, Y. Zhou, Efficient and sustainable bleaching of cotton/spandex integrating UVA radiation and peroxide activator: a study focus on the degradation mechanism of natural pigments in fibres. Arab. J. Chem. **16**(12), 105348 (2023). https://doi.org/10.1016/j.arabjc.2023.105348
54. Y. Zhou et al., An eco-friendly approach to low-temperature and near-neutral bleaching of cotton knitted fabrics using glycerol triacetate as an activator. Cellulose 28(12), 8129–8138 (2021). https://doi.org/10.1007/s10570-021-04030-2
55. F.N. Gumy et al., The decomposition rate and bleaching efficacy of in-office bleaching gels with different pHs: a randomized controlled trial. Clin. Oral Invest. **28**(8), 440 (2024). https://doi.org/10.1007/s00784-024-05821-0
56. L. Küster, B. Porto, C.R. de Aguiar, M.A. Granato, Performance of low-temperature bleaching techniques for cotton fabrics using hydrogen peroxide activators. ACS Omega **10**(26), 28167–28177 (2025). https://doi.org/10.1021/acsomega.5c02605
57. L. Dabasinskaite et al., The effect of ozone treatment on the physicochemical properties and biocompatibility of electrospun poly(ε)caprolactone scaffolds. Pharmaceutics **13**(8). https://doi.org/10.3390/pharmaceutics13081288
58. Z. Saka Dinç, S. Eren, Y. Öz, H.A. Eren, Influence of surface modifications on interfacial and mechanical properties in fibre-reinforced composites. J. Reinf. Plast. Compos., 07316844241259870 (2024). https://doi.org/10.1177/07316844241259870

59. L. Zhang et al., Effect of ozone treatment on the chemical and mechanical properties of flax fibers. Ind. Crops Prod. **189**, 115694 (2022). https://doi.org/10.1016/j.indcrop.2022.115694
60. Z. Lu, R. Yin, J. Yao, C.K.Y. Leung, Surface modification of polyethylene fiber by ozonation and its influence on the mechanical properties of strain-hardening cementitious composites. Compos. Part B Eng. **177**, 107446 (2019). https://doi.org/10.1016/j.compositesb.2019.107446
61. E.S. Massima Mouele, O.O. Fatoba, O. Babajide, K.O. Badmus, L.F. Petrik, Review of the methods for determination of reactive oxygen species and suggestion for their application in advanced oxidation induced by dielectric barrier discharges. Environ. Sci. Pollut. Res. **25**(10), 9265–9282 (2018). https://doi.org/10.1007/s11356-018-1392-9
62. C.-W. Kan, W.-S. Man, Surface characterisation of atmospheric pressure plasma treated cotton fabric—effect of operation parameters. Polymers **10**(3). https://doi.org/10.3390/polym1003 0250
63. G. Primc, R. Zaplotnik, A. Vesel, M. Mozetič, Mechanisms involved in the modification of textiles by non-equilibrium plasma treatment. Molecules **27**(24). https://doi.org/10.3390/mol ecules27249064
64. M. Saleem et al., Plasma-fabric interaction for surface activation and functionalization: a review. IOP Conf. Ser. Mater. Sci. Eng. **863**(1), 012036 (2020). https://doi.org/10.1088/1757-899X/863/1/012036
65. A. Zille, 6—Plasma technology in fashion and textiles, in *Sustainable Technologies for Fashion and Textiles*, ed. by R. Nayak (Woodhead Publishing, 2020), pp. 117–142
66. M. Bauer, S. Beratz, K. Ruhland, S. Horn, J. Moosburger-Will, Anodic oxidation of carbon fibers in alkaline and acidic electrolyte: quantification of surface functional groups by gas-phase derivatization. Appl. Surf. Sci. **506**, 144947 (2020). https://doi.org/10.1016/j.apsusc.2019.144947
67. C. Huang et al., Effect of electrochemical anodic oxidation modification on the interfacial properties of carbon fiber reinforced polyimide composites. Polym. Compos. **46**(3), 2390–2403 (2025). https://doi.org/10.1002/pc.29112
68. E. Meurs, M.N. Morshed, M. Kahoush, N. Kadi, Study on Fenton-based discoloration of reactive-dyed waste cotton prior to textile recycling. Sci. Rep. **14**(1), 24536 (2024). https://doi.org/10.1038/s41598-024-75450-w
69. C. Liao, Y. Li, S.C. Tjong, Visible-light active titanium dioxide nanomaterials with bactericidal properties. Nanomaterials **10**(1). https://doi.org/10.3390/nano10010124
70. C. Tao, Q. Jia, B. Han, Z. Ma, Tunable selectivity of radical generation over TiO_2 for photocatalysis. Chem. Eng. Sci. **214**,115438 (2020). https://doi.org/10.1016/j.ces.2019.115438
71. M.M. Rashid, B. Simončič, B. Tomšič, Recent advances in TiO_2-functionalized textile surfaces. Surf. Interfaces **22**, 100890 2021. https://doi.org/10.1016/j.surfin.2020.100890
72. X. Li, S. Li, X. Liang, D.J. McClements, X. Liu, F. Liu, Applications of oxidases in modification of food molecules and colloidal systems: laccase, peroxidase and tyrosinase. Trends Food Sci. Technol. **103**, 78–93 (2020). https://doi.org/10.1016/j.tifs.2020.06.014
73. X. Wang, B. Yao, X. Su, Linking enzymatic oxidative degradation of lignin to organics detoxification. Int. J. Mol. Sci. **19**(11). https://doi.org/10.3390/ijms19113373
74. S. Kim et al., Laccase-mediated grafting of polyphenols onto cationized cotton fibers to impart UV protection and antioxidant activities. J. Appl. Polym. Sci. **135**(6), 45801 (2018). https://doi.org/10.1002/app.45801
75. Y. Wang et al., Ultrasonic activation of inert poly(tetrafluoroethylene) enables piezocatalytic generation of reactive oxygen species. Nat. Commun. **12**(1), 3508 (2021). https://doi.org/10.1038/s41467-021-23921-3
76. F. Giuntini et al., Insight into ultrasound-mediated reactive oxygen species generation by various metal-porphyrin complexes. Free Radical Biol. Med. **121**, 190–201 (2018). https://doi.org/10.1016/j.freeradbiomed.2018.05.002
77. M. Moradi, A. Elahinia, Y. Vasseghian, E.-N. Dragoi, F. Omidi, A. Mousavi Khaneghah, A review on pollutants removal by Sono-photo -Fenton processes. J. Environ. Chem. Eng. **8**(5), 104330 (2020). https://doi.org/10.1016/j.jece.2020.104330

78. G. Wang, H. Cheng, Application of photocatalysis and sonocatalysis for treatment of organic dye wastewater and the synergistic effect of ultrasound and light. Molecules **28**(9). https://doi.org/10.3390/molecules28093706
79. M. Pirsaheb, N. Moradi, A systematic review of the sonophotocatalytic process for the decolorization of dyes in aqueous solution: synergistic mechanisms, degradation pathways, and process optimization. J. Water Process Eng. **44**, 102314 (2021). https://doi.org/10.1016/j.jwpe.2021.102314
80. J.C. García-Prieto, L.A. González-Burciaga, J.B. Proal-Nájera, M. García-Roig, Study of influence factors in the evaluation of the performance of a photocatalytic fibre reactor (TiO_2/SiO_2) for the removal of organic pollutants from water. Catalysts **12**(2). https://doi.org/10.3390/catal12020122
81. T.L. Nguyen et al., Innovative extraction technologies of bioactive compounds from plant by-products for textile colorants and antimicrobial agents. Biomass Convers. Biorefinery **14**(20), 24973–25002 (2024). https://doi.org/10.1007/s13399-023-04726-4
82. G. Natarajan, T.P. Rajan, S. Das, Application of sustainable textile finishing using natural biomolecules. J. Nat. Fibers **19**(11), 4350–4367 (2022). https://doi.org/10.1080/15440478.2020.1857895
83. K. Fang, C. Zhang, Surface physical–morphological and chemical changes leading to performance enhancement of atmospheric pressure plasma treated polyester fabrics for inkjet printing. Appl. Surf. Sci. **255**(17),7561–7567 (2009). https://doi.org/10.1016/j.apsusc.2009.04.028
84. J. Xiao, X. Zhang, Z. Zhao, J. Liu, Q. Chen, X. Wang, Rapid and continuous atmospheric plasma surface modification of PAN-based carbon fibers. ACS Omega **7**(13), 10963–10969 (2022). https://doi.org/10.1021/acsomega.1c06818
85. A.M. Thakker, D. Sun, D. Bucknall, Inkjet printing of plasma surface–modified wool and cotton fabrics with plant-based inks. Environ. Sci. Pollut. Res. **29**(45), 68357–68375 (2022). https://doi.org/10.1007/s11356-022-20659-3
86. M. Ayesh, A.R. Horrocks, B.K. Kandola, The impact of atmospheric plasma/UV laser treatment on the chemical and physical properties of cotton and polyester fabrics. Fibers **10**(8). https://doi.org/10.3390/fib10080066
87. Y.V. Tertyshnaya, S.G. Karpova, M.V. Podzorova, Effect of ozone on nonwoven polylactide/natural rubber fibers. Polymers **17**(15). https://doi.org/10.3390/polym17152102
88. J. Lou, J. Zhang, D. Wang, X. Fan, Improving the dyeability and anti-wrinkle properties of cotton fabric via oxidized raffinose. Appl. Sci. **11**(10). https://doi.org/10.3390/app11104641
89. S. Eren, İ. Yiğit, K. Kutlay, Z. Kaya, C. Basrık, H.A. Eren, A new sight of ozone usage in textile: improving flame retardant properties. Polymers **16**(6). https://doi.org/10.3390/polym16060735
90. T. Bušac, M. Čurlin, T. Pušić, and S. Ercegović Ražić, The influence of low-pressure plasma and ozone pretreatment on the stability of polyester/chitosan structure in the washing process—part 1. Coatings **15**(9). https://doi.org/10.3390/coatings15091030
91. A.A. Olkhov, V. Siracusa, A.L. Iordanskii, O. Alexeeva, M.L. Konstantinova, A.A. Berlin, Ozonation of non-woven ultrathin fibrous biomaterials for medical and packaging implementations, in *Ozonation—New Aspects*, ed. by M. Eyvaz, A. Albahnasawi, E. Gürbulak, E. Yüksel (IntechOpen, London, 2022)
92. C. Solazzo, J.M. Dyer, S. Deb-Choudhury, S. Clerens, P. Wyeth, Proteomic profiling of the photo-oxidation of silk fibroin: implications for historic tin-weighted silk. Photochem. Photobiol. **88**(5), 1217–1226 (2012). https://doi.org/10.1111/j.1751-1097.2012.01167.x
93. S. Perincek, I. Bahtiyari, A. Korlu, K. Duran, New techniques in cotton finishing. Text. Res. J. **79**(2), 121–128 (2009). https://doi.org/10.1177/0040517508092022
94. M.P. Gashti, A. Pournaserani, H. Ehsani, M.P. Gashti, Surface oxidation of cellulose by ozone-gas in a vacuum cylinder to improve the functionality of fluoromonomer. Vacuum **91**, 7–13 (2013). https://doi.org/10.1016/j.vacuum.2012.10.015
95. H.A. Eren, O. Avinc, B. Erişmiş, S. Eren, Ultrasound-assisted ozone bleaching of cotton. Cellulose **21**(6), 4643–4658 (2014). https://doi.org/10.1007/s10570-014-0420-2

96. A.N. Netravali, T. Bahners, Adhesion promotion in fibers and textiles using photonic surface modifications. J. Adhes. Sci. Technol. **24**(1), 45–75 (2010). https://doi.org/10.1163/016942409X12538865055999
97. D. Kováčik et al., Plasma-induced polymerization and grafting of acrylic acid on the polypropylene nonwoven fabric using pulsed underwater diaphragm electrical discharge. Plasma Chem. Plasma Process. **44**(2), 983–1001 (2024). https://doi.org/10.1007/s11090-024-10454-y
98. D. Gao, H. Yang, G. Liu, C. Chen, J. Yao, C. Li, Effect of electrochemical oxidation degree of carbon fiber on the interfacial properties of carbon fiber–reinforced polyaryletherketone composites. J. Reinf. Plast. Compos. **42**(21–22), 1107–1118 (2022). https://doi.org/10.1177/07316844221145763
99. H. Li et al., Electrochemical oxidation of recycled carbon fibers for an improved interaction toward alkali-activated composites. J. Cleaner Prod. **368**, 133093 (2022). https://doi.org/10.1016/j.jclepro.2022.133093
100. Q. Li, X. Zhao, H. Quan, Y. Zhou, Establishing an energy-saving scouring/bleaching one-step process for cotton/spandex fabric using UVA-assisted irradiation. RSC Adv. **12**(15), 9404–9415 (2022). https://doi.org/10.1039/D2RA00659F
101. O. Sedaghat, N. Bahramifar, M. Nowrouzi, H. Younesi, Life cycle assessment of industrial wastewater treatment: evaluating the environmental impact of electrocoagulation technologies. J. Water Process Eng. **71**, 107257 (2025). https://doi.org/10.1016/j.jwpe.2025.107257
102. H. Afanga et al., Integrated electrochemical processes for textile industry wastewater treatment: system performances and sludge settling characteristics. Sustain. Environ. Res. **30**(1), 2 (2020). https://doi.org/10.1186/s42834-019-0043-2
103. R. Tanveer, A. Yasar, A.-S. Nizami, A.B. Tabinda, Integration of physical and advanced oxidation processes for treatment and reuse of textile dye-bath effluents with minimum area footprint. J. Cleaner Prod. **383**, 135366 (2023). https://doi.org/10.1016/j.jclepro.2022.135366
104. E. GilPavas, I. Dobrosz-Gómez, M.-Á. Gómez-García, Optimization and toxicity assessment of a combined electrocoagulation, H_2O_2/Fe^{2+}/UV and activated carbon adsorption for textile wastewater treatment. Sci Total Environ. **651**, 551–560 (2019). https://doi.org/10.1016/j.scitotenv.2018.09.125
105. I. Arslan-Alaton, I. Alaton, Degradation of xenobiotics originating from the textile preparation, dyeing, and finishing industry using ozonation and advanced oxidation. Ecotoxicol. Environ. Saf. **68**(1), 98–107 (2007) (in eng). https://doi.org/10.1016/j.ecoenv.2006.03.009
106. C.-w. Kan, H.-f. Cheung, Q. Chan, A study of plasma-induced ozone treatment on the colour fading of dyed cotton. J. Cleaner Prod. **112**, 3514–3524 (2016). https://doi.org/10.1016/j.jclepro.2015.10.100
107. A.N. Keobounnam, C. Lenert-Mondou, A. Kubik, M.J. Hawker, Evaluating hydrophobic recovery of N_2 and $H_2O(g)$ plasma modified silk fibroin films aged at ambient and elevated temperatures. J. Vac. Sci. Technol. A **41**(5), 050401 (2023). https://doi.org/10.1116/6.0002803
108. S. Watson, C. Pattyn, M. Mohseni Garakani, A. Ajji, S. Reuter, M.R. Wertheimer, Wetting and aging behaviors for several O_2 and NH_3 plasma treated nanofibrous and flat film polymer materials: a comparison. Plasma Process. Polym. **20**(6), e2200224 (2023). https://doi.org/10.1002/ppap.202200224
109. G. Primc, M. Mozetič, Hydrophobic recovery of plasma-hydrophilized polyethylene terephthalate polymers. Polymers **14**(12). https://doi.org/10.3390/polym14122496
110. J. Behnisch, A. Holländer, H. Zimmermann, Factors influencing the hydrophobic recovery of oxygen-plasma-treated polyethylene. Surf. Coat. Technol. **59**(1), 356–358 (1993). https://doi.org/10.1016/0257-8972(93)90112-2
111. M. Mozetič, Aging of plasma-activated polyethylene and hydrophobic recovery of polyethylene polymers. Polymers **15**(24). https://doi.org/10.3390/polym15244668
112. D. Essa, 04. Recent techniques of surface modification for textile fabrics. J. Meas. Sci. Appl. (JMSA) **4**(1), 66–87 (2024) (in en). https://doi.org/10.21608/jmsa.2024.360350

113. A. Zille, F.R. Oliveira, A.P. Souto, Plasma treatment in textile industry. Plasma Process. Polym. **12**(2), 98–131 (2015). https://doi.org/10.1002/ppap.201400052
114. M. Madhushree, P. Vairavel, G.T. Mahesha, K.S. Bhat, Oxidative modifications of cellulose: methods, mechanisms, and emerging applications. J. Nat. Fibers **22**(1), 2497910 (2025). https://doi.org/10.1080/15440478.2025.2497910
115. T. Huber, N. Graupner, J. Müssig, Regenerated cellulose fibres and their composites: From fundamental properties to advanced applications. Progress Mater. Sci. **156**, 101547 (2026). https://doi.org/10.1016/j.pmatsci.2025.101547
116. D. Hegemann, M.M. Hossain, D.J. Balazs, Nanostructured plasma coatings to obtain multi-functional textile surfaces. Progress Org. Coat. **58**(2), 237–240 (2007). https://doi.org/10.1016/j.porgcoat.2006.08.027
117. S. Shariful Islam, M. Alam, S. Akter, Identifying the values of whiteness index, strength and weight of cotton spandex woven fabric in peroxide bleaching of different concentration. Fibers Text. **26**(4), 96–109 (2019)

Chapter 5
Resource-Efficient Processes for Textile Functionalization

Mohammad Neaz Morshed, Sina Seipel, Sakil Mahmud, and Mohammad Raza Miah

Abstract The textile industry faces increasing pressure to minimize environmental impact through the adoption of sustainable and resource-conscious functionalization methods. This has driven the development of diverse advanced processing technologies aimed at improving material performance while reducing ecological footprints. Technologies such as inkjet printing, valve-jet printing, three-dimensional (3D) printing, supercritical carbon dioxide ($ScCO_2$) treatment, digital hydraulic spray atomization, plasma ecotechnology, electrospinning, and biofunctionalization—each offer both unique advantages as well as relevant challenges. While many of these approaches can reduce waste and lower chemical and water usage, their resource efficiency and production flexibility vary significantly depending on the specific application and scale. For instance, supercritical CO_2 treatment is notable for its minimal solvent use but may be limited to certain fabric types, whereas inkjet printing enhances pattern precision but requires optimization for industrial throughput. Moreover, some techniques, such as 3D printing and electrospinning, are still emerging and face scalability constraints. In this chapter, a general overview

M. N. Morshed (✉) · S. Seipel
Department of Textile Technology, Faculty of Textiles, Engineering and Business (Including the Swedish School of Textiles), University of Borås, Borås, Sweden
e-mail: mohammad_neaz.morshed@hb.se

S. Seipel
e-mail: sina.seipel@hb.se

S. Mahmud
Ivory V. Nelson Center for the Sciences, Department of Chemistry and Physics, Lincoln University, Oxford, PA, USA

Department of Textile Engineering, Faculty of Engineering, Daffodil International University, Dhaka, People's Republic of Bangladesh

S. Mahmud
e-mail: smahmud@lincoln.edu

M. R. Miah
Department of Textile Engineering, BGMEA University of Fashion and Technology, Dhaka, People's Republic of Bangladesh
e-mail: raza.miah@buft.edu.bd

M. N. Morshed and M. Shahid (eds.), *Advances in Surface Modification and Functionalization in Textiles*, Textile Science and Clothing Technology,
https://doi.org/10.1007/978-3-032-23377-6_5

of selected resource-efficient processes, along with their challenges and opportunities for textile functionalization will be presented.

Keyword Resource-efficient processes · Sustainability · Textile functionalization

5.1 Introduction

Traditional textile functionalization techniques, such as wet-chemical treatments, padding, and coating, have long been the backbone of textile processes [1]. However, these conventional processes are resource-intensive, consuming large volumes of water, energy, and chemicals, and often generating substantial waste and pollution [2–4]. The environmental burden and regulatory pressures and resource shortages are compelling the industry to rethink these conventional processes and transition toward more sustainable alternatives.

Recent advancements have increasingly focused on developing advanced, resource-efficient processes that maintain or enhance fabric performance while minimizing environmental impact, such as inkjet printing [5, 6], valve-jet printing [7, 8], 3D printing [9, 10], $ScCO_2$ treatment [11, 12], hydraulic spray atomization [13], plasma treatment [14, 15], electrospinning [16, 17], biofunctionalization [18], ultrasonic-assisted finishing [19], sol–gel coatings [20], enzyme-based treatments [21], microwave-assisted treatment [22], nanotechnology-enabled surface functionalization [23], and so on. These processes offer drastically reduce material waste, and significantly lower water, energy, and hazardous chemical consumption. The transition to these resource-efficient processes is not merely a response to environmental concerns but also a strategic move toward operational efficiency and cost-effectiveness. Often, these advanced methods offer enhanced production flexibility and scalability, allowing manufacturers to customize functionalities while reducing lead times and resource input [24]. In addition, their alignment with circular-economy principles (minimizing waste, enabling recyclability, and reducing emissions) makes them vital in steering the textile sector toward sustainability.

Despite the promise of these advanced techniques, their widespread adoption faces challenges related to technological maturity, integration into existing manufacturing lines, and economic viability [25]. Many of these resource-efficient processes require significant upfront investment in specialized equipment and skilled labor, which can be a barrier, especially for small and medium-sized enterprises [26]. Additionally, some technologies, such as $ScCO_2$ treatment and biofunctionalization, are still in the developmental stage or face limitations related to process scalability and consistency of results [27]. Another critical challenge lies in ensuring the durability and functionality of treated textiles under real-world conditions, which often require extensive testing and optimization [28]. The compatibility of these novel treatments with a wide range of textile substrates and end-use requirements also varies, necessitating tailored approaches for different fibers and applications. Furthermore, lifecycle assessments (LCAs) and comprehensive environmental impact studies are essential to fully validate the sustainability claims of these processes, as some may shift environmental burdens rather than eliminate them [29]. Therefore, the scientific community is focusing not only on improving the efficiency and effectiveness of these advanced and resource-efficient processes but also on developing integrated production models that can seamlessly incorporate them into existing manufacturing workflows, especially in dyeing and functional finishing [13].

Resource efficiency has been incorporated into textile functionalization approaches, linking raw material selection, process design, functional performance, and end-of-life considerations [18, 24]. However, their widespread adoption is still underway due to the challenges briefly mentioned earlier. Ongoing collaborative efforts between academic institutions and industry stakeholders are essential to accelerate the integration of these technologies into existing manufacturing infrastructures [30]. These efforts aim not only to address performance and cost barriers but also to meet evolving regulatory standards and consumer expectations for safer, more sustainable textile products. In parallel, the integration of digitalization and advanced technologies, such as artificial intelligence (AI), machine learning (ML), and data-driven process control, is opening new pathways to optimize efficiency, traceability, and product personalization [31, 32]. These advanced processes, when integrated with AI and ML systems, may enable predictive maintenance, real-time quality assurance, and adaptive process adjustments, thereby enhancing operational performance. Many of these innovations are being developed with the goal of ensuring long-term viability and adaptability for the future, ensuring that technologies remain adaptable, scalable, and resilient amid shifting environmental regulations, market dynamics, and technological advancements. By embedding flexibility, interoperability, and long-term sustainability into process design, the textile industry can transition toward a more circular, robust, and future-ready production ecosystem.

5.2 Overview of Sustainable Aspects in Textile Functionalization

The growing demand for textiles capable of offering functional properties such as flame retardancy, microbial resistance, water repellency, or UV protection has led to widespread demand for functionalization processes within the textile industry [30, 33–37]. While functionalization significantly enhances the versatility and value of textiles, it also introduces a complex set of sustainability challenges that span from environmental to human health, and end-of-life dimensions [38], as outlined in Fig. 5.1. The very attributes that enhance the appeal of functionalized textiles, including special properties, can often pose challenges to principles of circularity, safe chemical use, and ecological compatibility.

A key concern stems from reliance on synthetic, non-renewable chemicals, such as petroleum-derived binders, fluorinated compounds, and formaldehyde-based resins, which pose significant risks to human health and the environment throughout the

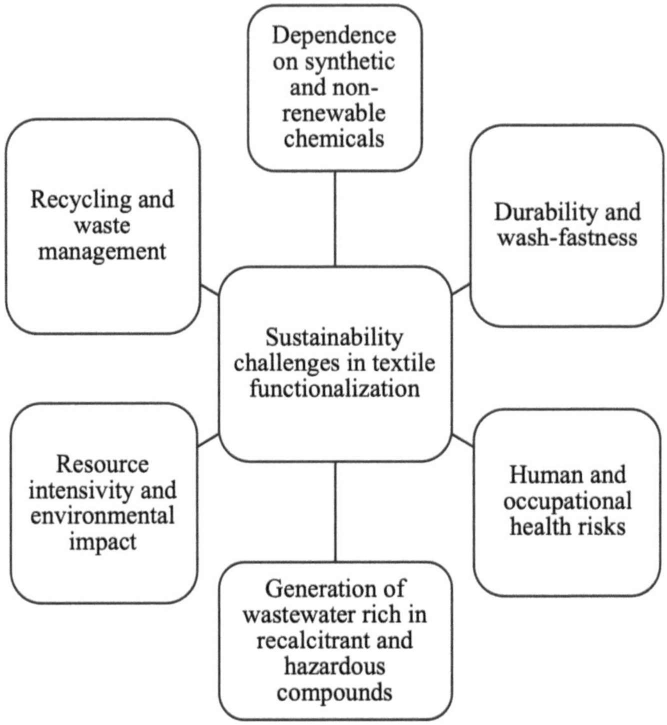

Fig. 5.1 Overview of complex sets of sustainability challenges in textile functionalization

product lifecycle [39–41]. In addition, the production of functional textiles is often associated with high energy consumption, greenhouse gas emissions, and hazardous by-products, while their poor biodegradability means they persist in the environment long after disposal, complicating end-of-life management of textile waste [42]. This is particularly problematic for fluorinated-based compounds, which can degrade into per- and polyfluoroalkyl substances (PFAS), a group of pollutants now recognized for their toxicity and extreme resistance to natural breakdown [39]. Another pressing issue is the generation of wastewater rich in recalcitrant and hazardous compounds during functionalization. Processes such as resin finishing, crease-resistant finish, water-repellent finish, antimicrobial finish, and specialty coatings often require auxiliary solvents, salts, and crosslinking agents, resulting in effluents with high chemical oxygen demand (COD) and complex chemical compositions [24]. Inadequate treatment of these effluents can lead to contamination of water bodies, eutrophication, and loss of aquatic biodiversity. Even advanced wastewater treatment systems struggle to remove emerging contaminants, such as nanoscale materials, which can accumulate in sediments and enter food webs [43].

Human and occupational health risks form an additional layer of concern. Workers in textile processing facilities may be exposed to formaldehyde from crease-resistant finishes, volatile organic compounds from solvents, or inhalable nanoparticles from advanced coatings [30, 44]. These exposures are often chronic and occur in poorly regulated environments, especially in production mills in low- and middle-income countries. The uneven distribution of risk raises ethical questions about global production systems, where the economic benefits of functional textiles in consumer markets may come at the expense of worker health in manufacturing regions. Subsequently, durability and wash-fastness, which are desirable qualities for consumer use, also carry unintended environmental consequences. Functional finishes that lack strong fixation often leach out during laundering, releasing chemicals and nanoparticles into domestic wastewater streams. This leaching contributes to diffuse pollution, with studies documenting nanosilver release that affects microbial ecosystems in wastewater treatment plants [45]. Similarly, specific hydrophobic treatments degrade into persistent pollutants that accumulate in soil and water, creating long-term ecological burdens. At the end-of-life stage, functionalized textiles present formidable challenges for recycling and waste management [38, 42]. The presence of coatings, crosslinked polymers, and composite finishes complicates both mechanical and chemical recycling pathways by reducing fiber recovery efficiency or contaminating recycled streams. As a result, incineration and landfilling remain common fates for such textiles, but these practices lead to toxic emissions or long-term persistence of non-degradable compounds. The durability that ensures performance in use simultaneously undermines biodegradability and recyclability, intensifying waste accumulation [46]. Finally, there is a methodological issue: current sustainability assessments often fail to capture the full scope of impacts associated with textile functionalization. While LCA is widely applied, most studies emphasize energy and water use, overlooking chemical toxicity, nanoparticle leaching, or recyclability barriers. This limited perspective risks underestimating the actual sustainability footprint of functionalized textiles and highlights the need for more comprehensive evaluation approaches.

Despite these concerns, functionalization can also bring indirect sustainability benefits if carefully managed. Durable finishes, for example, may extend the useful life of garments by providing wrinkle resistance, stain repellency, or antimicrobial properties that reduce laundering frequency, water consumption, and detergent use [30]. Enhanced UV-protective coatings can reduce material degradation, prolonging product lifespans and reducing replacement frequency [47]. Similarly, flame-retardant finishes contribute to user safety, potentially preventing fire-related waste and damage. In this sense, functionalization poses a paradox: while it complicates end-of-life management and raises toxicity concerns, it can also improve resource efficiency during use by lowering maintenance requirements and extending product longevity.

Overall, addressing the complex sustainability challenges outlined earlier necessitates the adoption of innovative, resource-efficient processes that minimize environmental and health impacts while maintaining or enhancing textile performance. Emerging technologies in textile functionalization focus on reducing chemical use, energy consumption, and waste generation throughout production, aiming to balance functional benefits with ecological responsibility. By integrating such approaches, the textile industry can mitigate the adverse effects of conventional treatments, promote safer chemical alternatives, and improve end-of-life options. The following section explores key resource-efficient functionalization methods, their potential advantages, and limitations in advancing sustainable textile manufacturing.

5.3 Resource-Efficient Processes for Textile Functionalization

The pursuit of sustainable, resource-efficient textile functionalization processes has driven the development of advanced processing technologies that minimize environmental impact while enhancing material performance. Approaches like digital inkjet and valve-jet printing allow precise deposition of functional materials with minimal waste, whereas 3D printing enables the fabrication of complex textile structures with tailored functionalities. Techniques like $ScCO_2$ processing offer solvent-free functionalization, reducing chemical usage and effluent generation, while plasma treatments modify surface properties without affecting bulk material. Additionally, innovative methods such as hydraulic spray atomization systems and electrospinning provide fine control over coating uniformity and nanofiber fabrication, opening new avenues for high-performance, sustainable textile products in a more resource-efficient manner.

5.3.1 Digital Inkjet Printing

Digital inkjet printing operates on the principle of precise, controlled ejection of tiny droplets of functional ink onto a textile substrate to create patterns or apply functional materials [48–50]. The key components include the ink reservoir, printhead, actuation system, and substrate transport mechanism. The printhead is the core element, responsible for generating and directing ink droplets. Two main types of actuation are commonly used for textile application: (a) Thermal (bubble) ink-jet (TIJ): localized heating creates a vapor bubble, forcing a droplet out of the nozzle (Fig. 5.2a). (b) Piezoelectric inkjet (PIJ): an electric pulse deforms a piezo element, pushing a droplet through the nozzle (Fig. 5.2b). Based on the method of generating droplets, printheads, the core component of printing systems, are generally divided into two primary categories: continuous inkjet (CIJ) and drop-on-demand (DOD). A printhead is composed of numerous ink channels, ranging from dozens to several thousand, each channel responsible for producing an individual droplet. The size of these droplets is inversely related to the printing resolution; narrower channels produce smaller droplets, which, in turn, enable higher printing resolution. Once ejected, droplets travel through the air and impact the textile surface. The substrate's absorbency, surface energy, and pre-treatment determine ink penetration, spreading, and adhesion. By selectively activating individual nozzles, complex patterns with

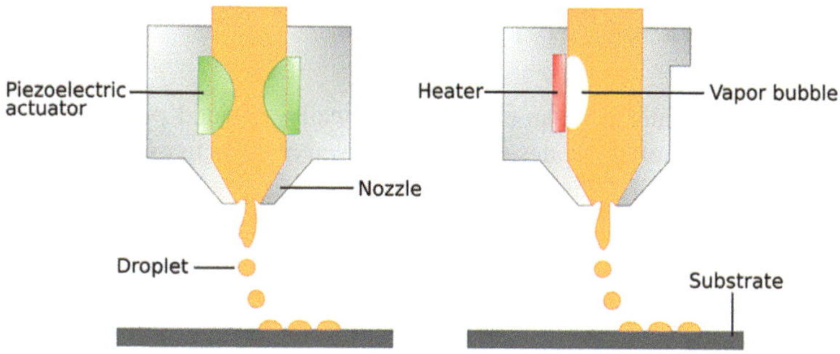

Fig. 5.2 Schematic illustration of **a** piezoelectric and **b** thermal (bubble-jet) inkjet printing mechanisms (Reused from Maleki and Bertola [52] under CC BY 4.0 license)

micrometer-scale resolution can be generated, enabling precise functionalization with minimal material and energy waste [51].

Digital inkjet printing has emerged as a resource-efficient method for textile functionalization, enabling precise deposition of functional materials directly onto fabric surfaces without the need for screens or masks. Unlike conventional printing techniques, inkjet printing enables high-resolution patterns, on-demand production, and a significant reduction in water, energy, and chemical consumption, aligning well with sustainable textile manufacturing goals. This method is particularly effective for applying dyes, pigments, conductive inks, enzymes, and other functional agents to natural and synthetic fibers. For instance, reactive dyes can be selectively printed onto cotton fabrics to achieve vibrant colors with minimal effluent generation [53]. Similarly, inkjet printing of antimicrobial nanoparticles such as silver or zinc oxide has been demonstrated to impart durable antimicrobial properties to textiles with controlled dosage and localized application [54, 55]. From a technological perspective, the formulation of functional inks is critical, as parameters like viscosity, surface tension, and particle size directly influence printing performance and fabric penetration. For digital inkjet printing, including valve-jet systems, the optimal surface tension typically lies in the range of 25–35 mN/m to ensure clean droplet formation, prevent satellite droplets, and facilitate precise deposition [56]. Studies have shown that optimization of ink rheology and pretreatment of textile substrates can significantly improve uniformity, adhesion, and functional efficacy [57]. Furthermore, digital inkjet printing supports multi-layered and multi-material printing, enabling complex functional designs such as gradient conductivity, selective hydrophobicity, or patterned enzyme immobilization [6]. Finally, it's worth noting the importance of proper maintenance. Adhering to a consistent maintenance routine is fundamental for ensuring the longevity and reliability of printheads, as the most prevalent cause of failure, clogged nozzles.

5.3.2 Digital Valve-Jet Printing

Digital valve-jet printers use an electromagnetically actuated printhead, in which an electromagnetic field activates a solenoid-driven plunger to precisely control the opening and closing of individual nozzles. These nozzles, arranged on the printhead, selectively eject ink droplets onto specific regions of the substrate based on a predefined digital design file. Ink delivery through the system is facilitated by pressurized air, which carries the ink through internal channels to the nozzles, enabling controlled, localized deposition. This mechanical actuation enables printing larger droplet volumes and supports a broader viscosity range than inkjet systems. Figure 5.3 shows the working principle of this electromagnetic type of DOD technology. As such, valve-jet printing is particularly well-suited for depositing high-solid-content inks, bio-inks, and formulations containing enzymes, nanoparticles, or other functional agents. The system's simplicity and robustness provide operational stability and compatibility with various ink chemistries, including shear-sensitive and biologically active materials [8]. Compared with other drop-on-demand technologies such as thermal and piezoelectric inkjet, valve-jet systems typically offer advantages such as higher production speeds, longer throw distances, greater durability in challenging environments, and lower total cost of ownership. However, TIJ and PIJ technologies tend to excel in applications demanding high-resolution output, such as fine barcodes or detailed logos.

Effective ink formulation is critical for reliable valve-jet printing performance. Key parameters include viscosity, surface tension, particle size, and all of which influence droplet formation, jet stability, and substrate wetting behavior. Since valve-jet printing accommodates higher viscosities (typically up to 500 mPa s), it enables the formulation of inks with greater functional loading. Particle size is another fundamental parameter in determining the suitability of an ink formulation for digital inkjet printing, particularly for valve-jet systems such as the Chromojet table-top printer. For reliable printing, the particle size should be at least 10 times smaller than the nozzle diameter to avoid nozzle blockage and ensure uniform droplet formation [49]. In digital valve-jet technology, particle size must be controlled to prevent nozzle

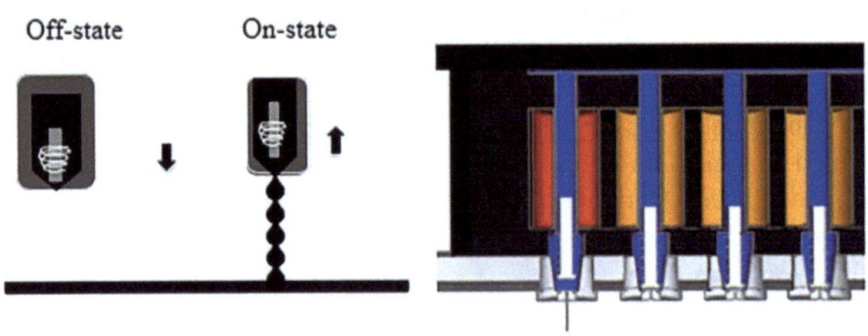

Fig. 5.3 Working principle of electromagnetic type (digital valve-jet) DOD technology (Reused from Chromojet Manual by Zimmer Austria under CC BY 4.0 license)

clogging, ensure consistent droplet formation, and maintain printhead functionality during extended printing operations. Surface tension is another crucial parameter that determines the jetting performance and droplet behavior of ink formulations in digital inkjet printing. It affects not only the formation, break-off, and stability of ink droplets at the nozzle but also governs how droplets spread and adhere to the substrate upon impact.

Valve-jet print technology has achieved widespread success in industrial textile applications due to a combination of key advantages that make it both highly reliable and cost-effective. It stands out as one of the most affordable printing methods available today, offering precise control over ink consumption to suit specific application needs. Its versatility allows for a wide range of mark sizes, from under a quarter inch to several feet in height, achievable by stitching multiple printheads together. Unlike many other technologies, valve-jet systems are adaptable to the production line, not the other way around. They can be mounted over, under, beside, or even on robotic arms to fit seamlessly into diverse industrial setups. Additionally, they support long throw distances, enabling printing even on uneven or irregular surfaces (like carpets). The technology is compatible with a broad spectrum of substrates, both porous and non-porous, and can operate effectively with virtually any industrial-grade ink, benefiting from decades of development and an ever-expanding catalog of formulations. Valve-jet printers also offer high-speed performance suitable for fast-moving production lines and are supported by bulk ink supply systems that enhance usability and extend production runtime.

5.3.3 *Three-Dimensional (3D) Printing*

3D printing, also known as additive manufacturing, represents a paradigm shift in textile functionalization by constructing or modifying fabrics layer by layer from digital models. Unlike conventional methods that treat existing fabrics, 3D printing introduces functionality through direct material deposition, enabling the creation of complex, 3D structures that are impossible to achieve with traditional techniques [58]. The principle hinges on the precise, computer-controlled placement of functional materials onto or integrated within textile substrates. Various 3D printing technologies have been developed based on the specific characteristics of the materials used and the desired properties or geometry of the final product. Among these, only a limited number have reached a level of technological maturity suitable for widespread application [59]. The most prominent and widely adopted techniques include fused deposition modeling (FDM), stereolithography (SLA), digital light processing (DLP), selective laser sintering (SLS), and polymer jetting (commonly referred to as PolyJet) [60].

As illustrated in Fig. 5.4a, FDM is an extrusion-based 3D printing method where a thermoplastic filament is heated and deposited layer by layer through a nozzle to build the object. It is one of the most common and cost-effective techniques used for prototyping and functional parts. SLA is a vat photopolymerization process that uses a UV laser to selectively cure liquid photopolymer resin layer by layer. It offers high-resolution prints with fine details and smooth surface finishes, making it suitable for applications in dentistry, jewelry, and prototyping (Fig. 5.4b). DLP is similar to SLA but uses a digital light projector to cure entire layers of photopolymer resin simultaneously. This results in faster print speeds than SLA while maintaining high resolution and accuracy (Fig. 5.4c). SLS is a powder bed fusion technique that uses a laser to sinter powdered materials, typically nylon or other polymers, layer by layer to form a solid structure. It does not require support structures and is well-suited for producing durable, complex parts. PolyJet is an inkjet-based 3D printing technology that works by jetting tiny droplets of liquid photopolymer onto a build platform, which are then instantly cured using UV light. This process is repeated layer by layer to build the final object (Fig. 5.4d). PolyJet stands out for its ability to produce parts with excellent resolution, smooth surface finishes, and multi-material or multi-color capabilities within a single print (Fig. 5.4e) [60].

The applications of 3D printing in textile functionalization span diverse fields, leveraging its capacity for customization and structural innovation. In wearable technology, it enables the seamless integration of flexible sensors, antennas, and energy-harvesting components directly into garments, facilitating real-time health monitoring and communication without compromising comfort [61, 62]. In medical textiles, 3D printing enables the creation of customized orthotic supports, smart bandages with embedded drug-delivery systems, and biocompatible tissue-engineering scaffolds that conform precisely to anatomical contours. In protective clothing, it enhances performance through strategically printed reinforcement pads

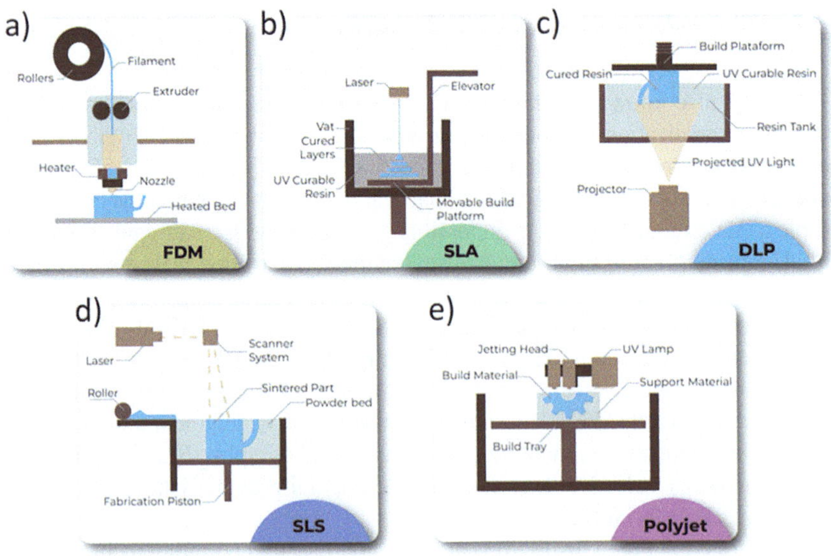

Fig. 5.4 Most prominent and widely adopted techniques for 3D printing (Reused with permission from Martins et al. [60])

for impact resistance, patterned flame-retardant barriers that maintain fabric breathability, and graded stiffness for ergonomic support in athletic or industrial gear. Additionally, 4D printing, where 3D-printed textiles dynamically respond to environmental stimuli like heat or moisture, opens new avenues for adaptive apparel and smart architectural fabrics [63]. By merging digital design freedom with advanced materials, 3D printing not only redefines textile functionality but also paves the way for sustainable production through minimal waste and on-demand manufacturing.

5.3.4 Supercritical CO_2 Technology

A supercritical fluid (SCF) is a substance maintained at conditions where both its temperature and pressure exceed its critical point values. This unique state enables a wide array of industrial applications, including extraction, dyeing, impregnation, polymer synthesis, and the production of powdered polymers. Among various substances used as SCFs, carbon dioxide (CO_2) is the most widely adopted due to its favorable properties. CO_2 is non-toxic, readily available, inexpensive, non-flammable, and considered environmentally benign. $ScCO_2$ technology leverages the unique physical properties of carbon dioxide above its critical point (31.1 °C and 7.39 MPa) [12]. In this supercritical state, CO_2 exhibits a hybrid behavior, possessing the penetrating power of a gas and the solvating power of a liquid. This dual character is the core fundamental behind its mechanism. Furthermore, the absence of surface tension in $ScCO_2$ allows for thorough infiltration without damaging or distorting delicate materials. A key benefit is that the process yields a dehydrated and solvent-free final product, a critical requirement in sensitive industries like pharmaceuticals and food packaging [64].

The process functions as an advanced solvent system in which $ScCO_2$ serves as a carrier for dyes and functional agents, as illustrated in Fig. 5.5. Thermodynamic principles govern the mechanism and involve three key stages: impregnation, equilibration/diffusion, and deposition. First, the dry textile is placed in a high-pressure vessel, and $ScCO_2$, loaded with functional compounds, is pumped in. Due to its gas-like properties, the $ScCO_2$-fluid mixture rapidly penetrates the textile's microstructure. The system is then maintained at specific pressure and temperature to reach thermodynamic equilibrium, where the concentration of the functional solute is balanced between the $ScCO_2$ phase and the fibers' polymer matrix. This equilibrium drives the second stage, as the dissolved compounds spontaneously diffuse from the $ScCO_2$ into the fiber polymer to equalize their chemical potential. Finally, a controlled depressurization disrupts this equilibrium; the solubility of the functional agents in the CO_2 plummets, causing them to nucleate and deposit. The CO_2 itself reverts to a gas and evaporates, leaving the functional agents permanently trapped within the fiber structure. This "dry" process eliminates the need for water and the subsequent energy-intensive drying required in conventional aqueous processing.

The practical implementation of $ScCO_2$ technology has demonstrated remarkable versatility across multiple textile sectors. In synthetic fiber dyeing, particularly

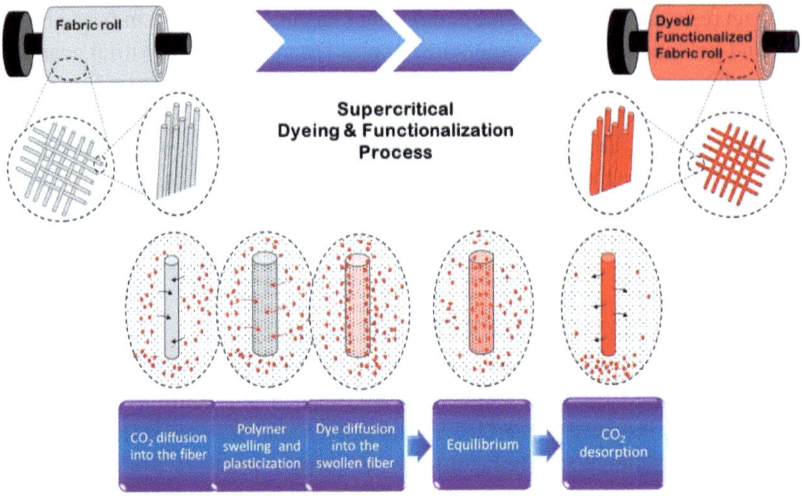

Fig. 5.5 Schematic illustration of supercritical dyeing and functionalization process (Reused with permission from Goni et al. [65])

for polyester, the technology has revolutionized coloration processes by eliminating water use while achieving superior dye penetration and fastness. Beyond simple coloration, researchers have successfully developed multi-functional textiles through simultaneous processing, as evidenced by Molla et al.'s work creating polyester fabrics with integrated coloration, antimicrobial protection, and water repellency in a single processing step [12]. The technology has shown particular promise in medical textiles in achieving controlled drug release profiles by impregnating anti-inflammatory compounds into surgical meshes [66]. Similarly, flame retardancy applications have benefited from $ScCO_2$ processing, with studies demonstrating deep penetration of phosphorus-based retardants into polypropylene fabrics while maintaining the material's essential tactile properties and breathability [67].

Despite its significant potential, $ScCO_2$ technology faces substantial barriers that limit its widespread adoption in industry. The most immediate challenge is economic: the initial capital investment for high-pressure processing equipment far exceeds that of conventional textile machinery, creating a formidable entry barrier for many manufacturers. Operational complexities compound this financial hurdle, as the technology requires specialized expertise in thermodynamics and high-pressure system management. Technical limitations continue to constrain the widespread implementation of supercritical CO_2 ($ScCO_2$) treatment. The process demonstrates variable efficacy across different textile substrates, performing particularly well with hydrophobic synthetic fibers like polyester but showing reduced effectiveness with natural fibers such as cotton and wool due to their hydrophilic nature and porous morphology. Additionally, many functional compounds exhibit limited solubility in $ScCO_2$, a nonpolar medium, necessitating the use of specialized chemical formulations or co-solvents to enhance solubility. The prevalent use of batch and semi-continuous processing systems restricts production throughput, while the lack of standardized processing parameters for diverse textiles requires extensive customization for each application.

5.3.5 *Plasma Treatment*

Plasma treatment represents a versatile and eco-friendly approach to textile surface modification, utilizing partially ionized gas containing reactive species to alter material properties while preserving bulk characteristics [14, 68]. Often called the "fourth state of matter," plasma is created by applying electromagnetic energy to a gas, generating a complex mixture of ions, electrons, free radicals, photons, and excited molecules that collectively interact with textile surfaces. As illustrated in Fig. 5.6. The fundamental theory involves multiple simultaneous interaction mechanisms. Surface etching occurs through physical sputtering and chemical ablation, where high-energy ions remove surface contaminants and create micro-roughness. Active reactive species, particularly oxygen and nitrogen radicals, initiate surface activation by forming new polar functional groups (hydroxyl, carbonyl, carboxyl) via free-radical reactions. Additionally, plasma-induced deposition enables the polymerization of monomer precursors into functional nano-coatings.

Plasma systems are broadly classified by operational pressure and thermal characteristics, with each category employing distinct generation techniques. Low-pressure plasma, also known as cold plasma, operates under vacuum conditions typically ranging from 0.01 to 10 mbar. In this environment, electrons achieve high energies while the gas temperature remains low, making it particularly suitable for processing heat-sensitive textiles. Common excitation methods for low-pressure plasma include direct current (DC) glow discharge, radio frequency (13.56 MHz), and microwave (2.45 GHz) energy sources. In contrast, atmospheric-pressure plasma eliminates the need for vacuum systems, enabling continuous processing that is better

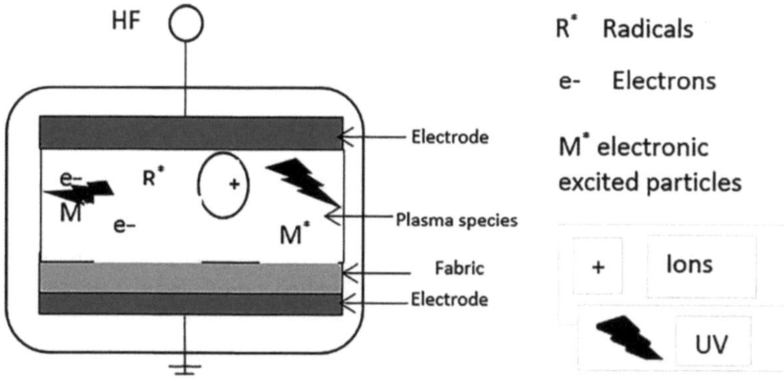

Fig. 5.6 Illustration of a mixture of ions, electrons, free radicals, and photons during plasma treatment of textiles (Reused with permission from Saleem et al. [69])

suited to industrial production lines. This category encompasses several configurations, such as corona discharge, dielectric barrier discharge (DBD), and plasma jet systems. Despite their enhanced industrial feasibility, atmospheric-pressure systems require precise control of operating parameters to prevent electrical arcing and ensure uniform treatment across the textile substrate.

The efficacy of plasma treatment is governed by several interconnected factors that collectively determine the surface modification outcome. Gas composition is a primary influence: oxygen plasma introduces polar functional groups that enhance hydrophilicity via surface oxidation; argon plasma primarily induces physical etching, increasing surface roughness; and fluorine-containing gases deposit hydrophobic fluorinated groups. Power density and exposure time directly affect the generation of reactive species and treatment depth, with optimal conditions varying widely among different textile substrates. Excessive power or prolonged exposure may cause fiber damage. System pressure critically influences the mean free path and collision frequency of plasma species, thereby affecting treatment uniformity and penetration depth, with low-pressure systems typically enabling deeper modifications. Additionally, substrate properties, including fiber chemical composition, crystallinity, and surface morphology, significantly impact modification effectiveness, as materials differ in their susceptibility to plasma-induced chemical and structural changes. Other factors, such as plasma source type and electrode configuration, also play important roles in achieving uniform and controlled treatments [14].

The sophisticated interaction between plasma species and textile surfaces enables precisely tailored functionalization for diverse applications. For synthetic fibers such as polypropylene, oxygen plasma treatment creates permanent hydrophilic surfaces by forming polar functional groups, dramatically improving dye uptake and printability when conventional processing proves ineffective [70]. Alternatively, hexafluorocarbon plasma deposition provides durable water- and oil-repellency by forming low-surface-energy CF_3-rich layers on the fiber surface. Antimicrobial functionality is achieved through strategic incorporation of silver nanoparticles or grafting of quaternary ammonium compounds via plasma-induced surface reactions, while enhanced adhesion in technical textiles results from the synergistic combination of improved surface roughness and optimized chemical compatibility between fibers and matrix materials [71].

5.3.6 Hydraulic Spray Atomizing System

Hydraulic spray atomizing systems represent a digital application technology for textile dyeing and finishing that utilizes controlled spray deposition to achieve uniform distribution. The fundamental principle involves atomizing liquid finishing formulations into fine droplets through hydraulic pressure and directing them onto moving textile substrates [13, 72]. In these systems, chemical solutions are pressurized and forced through specially designed nozzles that break the lamella flow of a liquid into micron-sized droplets ranging from 10 to 100 μm. The atomized droplets are then transported to the fabric surface via controlled streams, resulting in a uniform deposition pattern across the entire textile width. A current spray dyeing machine available on the market is the rotary disk atomizer, which uses rotating nozzle atomizers to spray solutions onto the fabric with low wet pick-up. The atomizer gets in contact with the front and back sides of the fabric twice, based on the positioning of the guided rollers [72]. Unlike conventional immersion methods, this contact-free application allows precise control over chemical application without saturating the fabric, significantly reducing both chemical consumption and the energy required for subsequent drying.

Hydraulic spray systems demonstrate exceptional versatility in applying a wide range of functional finishes to textiles. For water-repellent treatments, these systems enable uniform distribution of fluorocarbon- or silicone-based finishes, with a significant reduction in chemical usage compared to padding techniques. For example, studies on hydraulic spray atomizing methods for wool dyeing and finishing report substantial reductions in water use up to about 88% compared to conventional processes and markedly lower chemical consumption due to precise application and minimized bath preparation [13, 72]. In antimicrobial functionalization, the technology ensures optimal biocide distribution while minimizing active ingredient consumption. The systems excel at applying moisture-management finishes, enabling precise patterning of hydrophilic and hydrophobic compounds within designated fabric zones. Furthermore, sequential spray stations enable multi-functional finishing in a single pass, applying different chemical formulations without cross-contamination. The minimal application moisture significantly reduces drying energy requirements and often eliminates the need for subsequent washing steps, making the process both economically and environmentally advantageous.

The advantages of hydraulic spray atomizing systems include dramatic reductions in chemical consumption (typically 30–50%), water usage (60–80%), and energy requirements for drying. The technology enables rapid recipe changes with minimal waste and provides exceptional control over application parameters across diverse fabric constructions [13]. However, limitations include potential challenges with spray penetration on dense or piled fabrics, requiring specialized nozzle configurations and optimized spray angles. Additionally, the technology demands careful formulation of chemical solutions to prevent nozzle clogging and ensure consistent atomization. Regular maintenance of nozzles and filtration systems is essential to maintain application quality. Despite these limitations, hydraulic spray atomization technology represents a significant advancement in sustainable textile processing, offering manufacturers a viable solution to reduce environmental impact while maintaining high-quality functional finishes.

5.3.7 Electrospinning

Electrospinning represents an advanced nanofiber production technique that utilizes electrostatic forces to create continuous polymer fibers with diameters ranging from nanometers to several micrometers [16]. The fundamental principle is to apply a high voltage (typically 5–30 kV) to a polymer solution or melt, creating charged jets that undergo stretching and whipping motions before depositing onto a grounded collector, as illustrated in Fig. 5.7a. As the polymer solution is pumped through a capillary nozzle or spinneret, the electrostatic forces overcome the solution's surface tension, forming a Taylor cone from which a continuous jet is ejected. This jet undergoes vigorous bending instability and stretching during its trajectory to the collector, leading to dramatic fiber thinning and solvent evaporation. The resulting nanofibers form non-woven mats with high surface area-to-volume ratios, interconnected porous structures, and tunable morphology, making them ideal for advanced functional applications [73].

Modern electrospinning systems comprise several critical components, including high-voltage power supplies, precision syringe pumps, spinneret assemblies, and grounded collectors. The process parameters significantly influencing fiber morphology can be categorized into three main groups: solution parameters (polymer concentration, viscosity, conductivity, and surface tension), process parameters (applied voltage, flow rate, needle-to-collector distance, and collection speed), and environmental parameters (temperature, humidity, and atmospheric composition). Advanced systems incorporate multiple spinneret configurations, rotating collectors for aligned fiber production, and environmental control chambers for consistent results. Precise control of these parameters enables the production of fibers with tailored diameters, surface morphologies, and alignment characteristics for specific functional applications [74].

Electrospun nanofiber mats demonstrate exceptional capabilities in textile functionalization across multiple domains, as highlighted in Fig. 5.7b [75, 76]. For protective applications, electrospun membranes provide effective barrier properties against particulate matter, biological agents, and chemical hazards while maintaining breathability. In biomedical textiles, the technology enables the incorporation of therapeutic agents, growth factors, and antimicrobial compounds for controlled-release applications in wound dressings and tissue-engineering scaffolds. The high surface area enhances filtration efficiency in technical textiles, while the tunable porosity enables sophisticated moisture management in sportswear. Smart textiles benefit from incorporating conductive nanofibers for sensing applications, phase-change materials for thermal regulation, and responsive polymers for adaptive functionality. The technology also enables surface functionalization of conventional textiles through direct electrospinning onto existing fabrics, creating composite structures with enhanced properties [77].

The significant advantages of electrospinning include the ability to produce ultrafine fibers with controlled diameters and architectures, high surface area for enhanced functionality, and versatile material compatibility with natural polymers, synthetic polymers, and composites [17]. The process enables precise control over fiber alignment, porosity, and surface morphology, allowing for tailored performance characteristics. However, limitations include relatively low production rates compared to conventional textile processes, challenges in scaling up for industrial production, and potential issues with the mechanical strength of fiber mats. Additional considerations

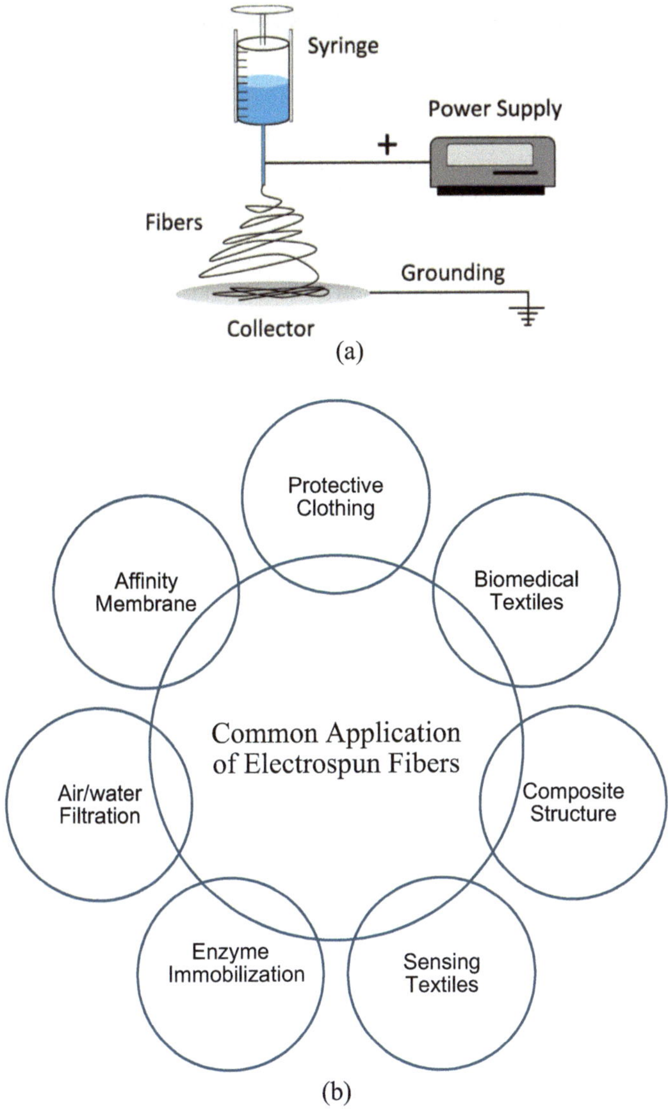

Fig. 5.7 **a** Schematic illustration of electrospinning process; **b** common application of electrospun fibers

include solvent management in solution electrospinning and the need for specialized expertise in process optimization. Despite these limitations, electrospinning remains a powerful technology for creating advanced functional textiles with capabilities beyond conventional manufacturing methods, particularly in high-value applications where nano-scale features and precise morphological control are essential [78].

5.4 Biofunctionalization of Textiles for Sustainable Development

The emerging field of textile biofunctionalization represents a paradigm shift toward sustainable textile engineering, utilizing biological agents and biomimetic principles to create advanced multifunctional fabrics. This approach harnesses nature's own toolkit to impart sophisticated properties to textiles while minimizing environmental impact [79]. One significant application involves creating water-repellent surfaces through biomimicry, using biowax esters and fungal proteins that replicate the self-cleaning properties of natural surfaces like lotus leaves, providing an eco-friendly alternative to conventional fluorocarbon treatments. In the area of protective textiles, bio-based flame retardancy is achieved using protein-rich compounds derived from milk and phytate sources from plants, which form protective char layers when exposed to heat [80]. For personal comfort and hygiene, biological odor management systems employ molecular encapsulation techniques using starch-derived compounds and natural antimicrobial agents that prevent bacterial growth responsible for unpleasant smells. Protective functionality is further enhanced through natural antimicrobial treatments utilizing crustacean shell extracts and plant-based nanoparticles that provide adequate protection against pathogens while being biodegradable [81].

Innovative applications extend to environmental remediation, where textiles are engineered with enzyme-based systems and functionalized biopolymers that capture atmospheric carbon dioxide [82, 83]. Additionally, integrating biological sensing elements creates smart diagnostic textiles that can detect and respond to specific biomarkers via visible color changes, enabling continuous health monitoring [84]. These diverse biofunctionalization strategies collectively contribute to sustainable development by replacing hazardous chemicals with biodegradable alternatives, reducing energy requirements for processing, and supporting circular-economy principles in which textile waste can serve as feedstock for new bio-based materials, ultimately transforming textiles from passive materials into active, environmentally responsible systems.

5.5 Challenges and Opportunities of Resource-Efficient Processes for Textile Functionalization

The transition toward resource-efficient textile functionalization represents a critical evolution in sustainable manufacturing, yet this shift presents a complex landscape of technical and economic challenges alongside significant opportunities for innovation. While emerging technologies offer substantial reductions in water consumption, chemical use, and energy requirements, their implementation faces significant barriers, including high initial capital investment, technical complexity requiring specialized expertise, and limitations on scalability for mass production. Many resource-efficient processes also encounter material compatibility issues, particularly with natural fibers, and struggle to achieve the consistent treatment uniformity required, along with durability requirements for industrial applications.

Despite these challenges, the opportunities for advancement are considerable. The growing market demand for sustainable textiles, coupled with increasingly stringent environmental regulations, creates a compelling economic case for adopting resource-efficient technologies. The integration of digital technologies with advanced functionalization processes enables precision in chemical applications, which has high potential for reducing material consumption while enhancing performance. The development of multi-functional treatments that combine multiple properties in a single processing step offers both economic and environmental advantages. Furthermore, the convergence of nanotechnology with traditional textile processes creates opportunities to develop smart textiles with responsive properties while minimizing resource inputs. The circular economy paradigm offers additional opportunities to create closed-loop systems in which waste streams serve as inputs for new functionalization processes.

Looking ahead, the field of resource-efficient textile functionalization is poised for transformative developments driven by several converging trends. The integration of artificial intelligence and machine learning with process control systems will enable real-time optimization of functionalization parameters, dramatically improve efficiency, and reduce waste. We anticipate the emergence of bio-intelligent textiles that combine biological systems with advanced materials, creating fabrics capable of self-regulation and adaptive functionality. The development of modular, scalable systems will lower entry barriers for smaller manufacturers, while advances in green chemistry will yield new biodegradable functional agents with enhanced performance characteristics. However, significant sustainability challenges must be addressed, particularly concerning recycling and end-of-life management. The integration of functional finishes often complicates textile recycling processes, as many functional agents contaminate recycling streams and hinder fiber-to-fiber recycling. Current functionalization methods can compromise the biodegradability of natural textiles and create challenges in mechanical and chemical recycling systems. Future research must focus on developing reversible functionalization systems that can be deactivated or removed during recycling, and on designing functional agents compatible with circular economy principles. The development of standardized protocols for assessing the environmental impact of functionalized textiles throughout their lifecycle will be crucial.

The growing emphasis on circularity will drive innovation in recyclable functional finishes and disassembly-compatible treatments, supporting the creation of truly circular textile systems. Furthermore, the integration of digital fabrication techniques, such as inkjet or 3D printing, with resource-efficient textile functionalization processes has the potential to enable highly customized, on-demand treatments, reducing overproduction and material waste in conventional supply chains. As these methods continue to mature, they may move from niche applications toward broader industrial adoption, facilitating the production of textiles with targeted, multifunctional properties. When coupled with sustainable material choices and attention to end-of-life recyclability, such approaches can transform textiles from passive substrates into active, intelligent systems that deliver enhanced performance while minimizing environmental impacts throughout their lifecycle.

Acknowledgements The authors would like to acknowledge the institutional support from the Department of Textile Technology, Faculty of Textiles, Engineering and Business (including the Swedish School of Textiles), University of Borås, Sweden, during the preparation of this chapter.

AI-Usage Declaration During the preparation of this chapter, the authors used artificial intelligence to improve the language of the manuscript. After using this tool/service, the authors reviewed and edited the content as needed and take full responsibility for the content of the publication.

References

1. A. Nadi et al., Evolution in the surface modification of textiles: a review. Text. Prog. **50**(2), 67–108 (2018)
2. P.S. Kumar, G.J. Joshiba, Sustainability in dyeing and finishing, in *Sustainability in the Textile and Apparel Industries: Production Process Sustainability*. (Springer, 2020), pp.165–177
3. M. Chowdhury et al., Optimizing textile dyeing and finishing for improved energy efficiency and sustainability in fleece knitted fabrics. Cleaner Energy Syst. **9**, 100154 (2024)
4. H. Zheng et al., An ecofriendly dyeing of wool with supercritical carbon dioxide fluid. J. Clean. Prod. **143**, 269–277 (2017)
5. J. Yu, S. Seipel, V.A. Nierstrasz, Digital inkjet functionalization of water-repellent textile for smart textile application. J. Mater. Sci. **53**(18), 13216–13229 (2018)
6. T. Biswas, Enzyme Printed Fabrics: Bio-functionalisation of Synthetic Textiles by Digital Inkjet Printing (Högskolan i Borås, 2022)
7. M.N. Morshed, J. Yu, V. Nierstrasz, Fabrication of antibacterial textiles through digital printing of functional chitosan ink, in *12th International Conference on Fiber and Polymer Biotechnology* (2024)
8. J. Akhter, Printing of redox enzyme on surface activated polyester fabric by using digital valve jet printing technology, in *A Resource Efficient Enzyme Immobilization Approach* (2025)
9. R.H. Sanatgar, C. Campagne, V. Nierstrasz, Investigation of the adhesion properties of direct 3D printing of polymers and nanocomposites on textiles: effect of FDM printing process parameters. Appl. Surf. Sci. **403**, 551–563 (2017)
10. K. Chatterjee, T.K. Ghosh, 3D Printing of textiles: potential roadmap to printing with fibers. Adv. Mater. **32**(4), 1902086 (2020)
11. A.G. Hassabo et al., Applications of supercritical carbon dioxide in textile finishing: a review. J. Text. Color. Polym. Sci. **19**(2), 179–187 (2022)
12. M. Tadesse Abate, Supercritical CO_2 technology in resource-effectiveproduction of functional and smart textiles (Högskolan i Borås, 2020)
13. R. Mulder et al., Study on hydraulic spray atomizing system as a new resource-efficient dyeing-finishing method for wool fabric. Sci. Rep. **12**(1), 21814 (2022)
14. R. Shishoo, *Plasma Technologies for Textiles* (Elsevier, 2007)
15. M.N. Morshed et al., Surface modification of polyester fabric using plasma-dendrimer for robust immobilization of glucose oxidase enzyme. Sci. Rep. **9**(1), 15730 (2019)
16. W.E. Teo, S. Ramakrishna, A review on electrospinning design and nanofibre assemblies. Nanotechnology **17**(14), R89 (2006)
17. S. Agarwal, A. Greiner, J.H. Wendorff, Functional materials by electrospinning of polymers. Prog. Polym. Sci. **38**(6), 963–991 (2013)

18. L.J. Rather et al., Biofunctionalization of various textile materials using enzyme biotechnology as a green chemistry alternative, in *Enzymes in Food Technology: Improvements and Innovations*. (Springer, 2018), pp.263–276
19. A. Körlü, M.İ. Bahtiyari, Ultrasound-based wet processes in textile industry, in *Innovative and Emerging Technologies for Textile Dyeing and Finishing* (2021), pp. 265–299
20. B. Mahltig, H. Haufe, H. Böttcher, Functionalisation of textiles by inorganic sol–gel coatings. J. Mater. Chem. **15**(41), 4385–4398 (2005)
21. A. Cavaco-Paulo, G. Gubitz, *Textile Processing With Enzymes*, vol. 29 (Woodhead publishing, 2003)
22. S. Adeel et al., Microwave-assisted sustainable dyeing of wool fabric using cochineal-based carminic acid as natural colorant. J. Nat. Fibers (2019)
23. A.K. Yetisen et al., Nanotechnology in textiles. ACS Nano **10**(3), 3042–3068 (2016)
24. M. Shahid, M. Yusuf, Sustainability in textile finishing: an introduction to current challenges and innovations, in *Sustainable Finishing Techniques in Textiles*. (Springer, 2025), pp.283–293
25. E. Ozturk et al., Sustainable textile production: cleaner production assessment/eco-efficiency analysis study in a textile mill. J. Clean. Prod. **138**, 248–263 (2016)
26. N. Astha, A. Madhu, Supercritical fluid dyeing technology: innovations and applications, in *Sustainable Coloration Techniques in Textiles* (Springer, 2025), pp. 237–253
27. L.S. Abd-Elaal et al., Supercritical carbon dioxide as an impregnation medium for producing functional materials in textile finishing. J. Text. Coloration Polym. Sci. **22**(1), 187–195 (2025)
28. M.S. Hossain et al., Coloration and functionalization of organic cotton fabric using Chinese fringe leaf extraction: a greener approach. Results Eng. **25**, 104054 (2025)
29. A. Vade, A. Athalye, Environmental impact of digital printing versus traditional textile printing, in *Digital Printing: Sustainable Solution to Fashion Industry*. (Springer, 2026), pp.295–333
30. M.K. Singh, *Textiles Functionalization-A Review of Materials, Processes, and Assessment* (IntechOpen, 2021)
31. A. Ghosh, S. Das, B. Saha, *Artificial Intelligence in Textile Engineering: Basic Concepts and Applications* (Elsevier, 2024)
32. J. Qin et al., AI-assisted dyeing optimization of polyimide fibers via backpropagation neural networks. Prog. Org. Coat. **209**, 109611 (2025)
33. S. Mahmud et al., UV protection and antibacterial treatment of wool using green silver nanoparticles. Asian J. Chem. **30**(1), 116–122 (2018)
34. M.S. Sk et al., Antimicrobial performance of silver–copper–zeolite microparticle-treated organic cotton fabric using versatile methods. Surf. Innov. **11**(4), 223–230 (2022)
35. K.M.F. Hasan et al., Wool functionalization through AgNPs: coloration, antibacterial and wastewater treatment. Surf. Innov. **9**(1), 25–36 (2020)
36. S. Mahmud et al., In situ synthesis of green AgNPs on ramie fabric with functional and catalytic properties. Emerg. Mater. Res. **8**(4), 623–633 (2019)
37. S. Mahmud et al., Surface functionalization of "Rajshahi silk" using green silver nanoparticles. Fibers **5**(3), 35 (2017)
38. R.B. Baloyi et al., Recent advances in recycling technologies for waste textile fabrics: a review. Text. Res. J. **94**(3–4), 508–529 (2024)
39. S.M. Tari, A.R. Athalye, Biodegradability and sustainability in textile finishing—current challenges and future perspectives, in *Sustainable Finishing Techniques in Textiles*. (Springer, 2025), pp.17–39
40. S. Schellenberger et al., An outdoor aging study to investigate the release of per-and polyfluoroalkyl substances (PFAS) from functional textiles. Environ. Sci. Technol. **56**(6), 3471–3479 (2022)
41. S. Mahmud et al., Organic silicone based poly-acrylate binder synthesis for textile pigment printing. Am. J. Polym. Sci. Eng. **3**(1), 119–128 (2015)
42. C. Ribeiro et al., Textile waste pre-treatments for enhanced chemical recycling. Discov. Appl. Sci. **7**(6), 534 (2025)
43. G.A. Kallawar, B.A. Bhanvase, A review on existing and emerging approaches for textile wastewater treatments: challenges and future perspectives. Environ. Sci. Pollut. Res. **31**(2), 1748–1789 (2024)

44. K.-H. Kim, S.A. Jahan, J.-T. Lee, Exposure to formaldehyde and its potential human health hazards. J. Environ. Sci. Health C **29**(4), 277–299 (2011)
45. D. Voelker et al., Approach on environmental risk assessment of nanosilver released from textiles. Environ. Res. **140**, 661–672 (2015)
46. I. Wojnowska-Baryła et al., The growing problem of textile waste generation—the current state of textile waste management. Energies **17**(7), 1528 (2024)
47. S. Mahmud et al., Multifunctional organic cotton fabric based on silver nanoparticles green synthesized from sodium alginate. Text. Res. J. **90**(11–12), 1224–1236 (2020)
48. I.M. Hutchings, G.D. Martin, *Inkjet Technology for Digital Fabrication* (Wiley Online Library, 2013)
49. S.D. Hoath, *Fundamentals of Inkjet Printing: The Science of Inkjet and Droplets* (Wiley, 2016)
50. S. Seipel, *Ink Jetting of Photochromic Ink: Towards the Design of a Smart Textile Sensor* (Högskolan i Borås, 2020)
51. G.D. Martin, I.M. Hutchings, Fundamentals of inkjet technology, in *Inkjet Technology for Digital Fabrication* (2012), pp. 21–44
52. H. Maleki, V. Bertola, Recent advances and prospects of inkjet printing in heterogeneous catalysis. Catal. Sci. Technol. **10**(10), 3140–3159 (2020)
53. L. Wang, C. Hu, K. Yan, A one-step inkjet printing technology with reactive dye ink and cationic compound ink for cotton fabrics. Carbohyd. Polym. **197**, 490–496 (2018)
54. A. Soleimani-Gorgani, O. Avinc, R. Alborz, Sustainable antibacterial cotton fabrics with in situ formed silver nanoparticles by bio-inkjet printing. J. Clean. Prod. **386**, 135796 (2023)
55. H. Shahariar et al., Inkjet printing of reactive silver ink on textiles. ACS Appl. Mater. Interfaces **11**(6), 6208–6216 (2019)
56. P.J. Smith, D.H. Shin, *Inkjet-Based Micromanufacturing* (Wiley, 2012)
57. J. Yu et al., Uniform-colored synthetic textiles by digital inkjet printing: effect of vacuum plasma treatment and multiple printing passes. Fibers Polym. 1–14 (2025)
58. R.H. Sanatgar, *FDM 3D Printing of Conductive Polymer Nanocomposites: A Novel Process for Functional and Smart Textiles L'Impression 3D de Nanocomposites Polymères Conducteurs: un Nouveau Procédé Pour la Fonctionnalisation de Surface De Smart Textiles*
59. J. Izdebska-Podsiadły, Classification of 3D printing methods, in *Polymers for 3D Printing*. (Elsevier, 2022), pp.23–34
60. P. Martins et al., Advances in printing and electronics: from engagement to commitment. Adv. Func. Mater. **33**(16), 2213744 (2023)
61. P. Jiang et al., Surface functionalization–a new functional dimension added to 3D printing. J. Mater. Chem. C **8**(36), 12380–12411 (2020)
62. P.A. Eutionnat-Diffo et al., Development of flexible and conductive immiscible thermoplastic/elastomer monofilament for smart textiles applications using 3D printing. Polymers **12**(10), 2300 (2020)
63. M.C. Biswas et al., 4D printing of shape memory materials for textiles: mechanism, mathematical modeling, and challenges. Adv. Func. Mater. **31**(19), 2100257 (2021)
64. M. Tadesse Abate, V. Nierstrasz, *Impregnation of materials in supercritical $CO_2$to impart various functionalities* (2019)
65. M.L. Goni, N.A. Ganan, R.E. Martini, Supercritical CO_2-assisted dyeing and functionalization of polymeric materials: a review of recent advances (2015–2020). J. CO_2 Utilization **54**, 101760 (2021)
66. O.R. Davies et al., Applications of supercritical CO_2 in the fabrication of polymer systems for drug delivery and tissue engineering. Adv. Drug Deliv. Rev. **60**(3), 373–387 (2008)
67. Y. Liu et al., Preparation of durable and flame retardant polyester fabric using eco-friendly supercritical carbon dioxide. Colloids Surf. A Physicochem. Eng. Aspects 139534 (2026)
68. R. Morent et al., Non-thermal plasma treatment of textiles. Surf. Coat. Technol. **202**(14), 3427–3449 (2008)
69. M. Saleem et al., Functionality and applications of non-thermal plasma activated textiles: a review. Mater. Today Proc. **47**, S74–S82 (2021)

70. M.S. Kang, B. Chun, S.S. Kim, Surface modification of polypropylene membrane by low-temperature plasma treatment. J. Appl. Polym. Sci. **81**(6), 1555–1566 (2001)
71. H. Ssekasamba et al., Review of plasma surface engineering technology toward sustainable textile materials. Mater. Circular Econ. **6**(1), 27 (2024)
72. K. Mae Amandoron, *Establishment of a Resource-efficient Spray Dyeing Process for Polyester Fabrics: A study on the key process steps of the hydraulic spray atomising system—pre-treatment, dyeing, predrying, and fixation* (2024)
73. J. Doshi, D.H. Reneker, Electrospinning process and applications of electrospun fibers. J. Electrostat. **35**(2–3), 151–160 (1995)
74. A. Shabani et al., Electrospinning technology, machine learning, and control approaches: a review. Adv. Eng. Mater. **27**(7), 2401353 (2025)
75. G.B. Medeiros et al., Modification and functionalization of fibers formed by electrospinning: a review. Membranes **12**(9), 861 (2022)
76. M.T. Hossain et al., Progress and prospects of chemical functionalization of textiles via nanotechnology. ACS Appl. Eng. Mater. **3**(1), 1–20 (2025)
77. D. Demir, A. Vaseashta, N. Bölgen, Recent advances of electrospinning and multifunctional electrospun textile materials for chemical and biological protection, in *Nanoscience and Nanotechnology in Security and Protection against CBRN Threats* (2020), 275–289
78. D. Ji et al., Electrospinning of nanofibres. Nat. Rev. Methods Primers **4**(1), 1 (2024)
79. D. Weerasinghe, S. Perera, D. Dissanayake, Application of biomimicry for sustainable functionalization of textiles: review of current status and prospectus. Text. Res. J. **89**(19–20), 4282–4294 (2019)
80. M. Chen et al., Recent advancements of bio-derived flame retardants for polymeric materials. Polymers **17**(2), 249 (2025)
81. R. Broadhead, L. Craeye, C. Callewaert, Odour control or inhibition using antimicrobial finishing, in *Odour in Textiles*. (CRC Press, 2022), pp.95–116
82. J. Shen, Y. Yuan, S. Salmon, Carbonic anhydrase immobilized on textile structured packing using chitosan entrapment for CO2 capture. ACS Sustain. Chem. Eng. **10**(23), 7772–7785 (2022)
83. Y.R. Maghraby et al., Enzyme immobilization technologies and industrial applications. ACS Omega **8**(6), 5184–5196 (2023)
84. E. Battista et al., Enzymatic sensing with laccase-functionalized textile organic biosensors. Org. Electron. **40**, 51–57 (2017)

Chapter 6
Supercritical CO_2 Process for Surface Modification and Functionalization of Textiles

Molla Tadesse Abate, Michael Rauch, and Abdulrhman Hamdy Mustafa Aly

Abstract Textile processing using supercritical carbon dioxide ($scCO_2$) medium is gaining prominence due to benefits such as being a non-toxic, eco-friendly solvent with zero-liquid-discharge and facilitating easy dye recovery and recycling. Its status as an environmentally benign solvent offers significant advantages over traditional aqueous and solvent-based methods, as proven in polyester dyeing. Applications of $scCO_2$ impregnation are diverse, encompassing the incorporation of pharmaceuticals, functional finishing agents, dyes and other active compounds into polymeric matrices and surface modifications. Though still limited, recent studies have expanded the potential of $scCO_2$ in textile surface modification and functionalization. This has enabled the development of materials with properties such as antimicrobial activity, UV protection, antioxidant capability, drug-loading and improved dyeing efficiency while reducing water and chemical consumption. This chapter reviews key studies that utilize $scCO_2$ for the development of functional textiles through surface modification and impregnation. Recent advances and developments in textile surface modification and functionalization using $scCO_2$ are presented. The review aims to encourage further research into the use of $scCO_2$ technology for developing advanced functional textile materials and promoting its application on an industrial scale.

M. T. Abate (✉) · M. Rauch · A. H. M. Aly
Engineering Department, Division Sustainable Textiles, Hof University of Applied Sciences, Hof, Germany
e-mail: mollat2008@gmail.com

M. Rauch
e-mail: michael.rauch@hof-university.de

A. H. M. Aly
e-mail: abdulrhman.aly@hof-university.de

M. T. Abate
Ethiopian Institute of Textile and Fashion Technology, Bahir Dar University, P.O. Box 1037, Bahir Dar, Ethiopia

M. N. Morshed and M. Shahid (eds.), *Advances in Surface Modification and Functionalization in Textiles*, Textile Science and Clothing Technology,
https://doi.org/10.1007/978-3-032-23377-6_6

Keywords Supercritical CO_2 ($scCO_2$) · Impregnation · Textile functionalization · Sustainable processing

6.1 Introduction

The textile industry is a major contributor to global water pollution, primarily due to its high-water consumption and the discharge of chemically laden wastewater. Traditional textile processing methods consume large amounts of water and chemicals, leading to substantial environmental impacts, including pollution and resource depletion [1]. For example, to produce 1 kg of finished textile, it requires 100–150 L of water, with wastewater outputs reaching up to 300 m^3 per tonne containing dyes, salts, acids and other pollutants [2, 3]. Notably, 5–10% of dyes are lost in conventional dyeing processes, and textile dyeing and finishing operations account for 17–20% of all industrial water pollution globally. As a result, it is facing growing scrutiny for its significant environmental impact. In response, the textile industry is increasingly adopting eco-friendly practices to minimize waste and reduce reliance on toxic chemicals. There is a strong focus on cleaner, more cost-effective and sustainable production methods that tackle both environmental and health challenges. Technological innovation and green manufacturing are now essential not just in the final product, but across the entire production process [4]. Within this shift, supercritical CO_2 technology stands out as a promising green alternative capable of driving a transformative move towards sustainable textile processing. Supercritical CO_2 is increasingly utilized in the textile industry, particularly for dyeing synthetic fibres like polyester. This method offers several advantages over traditional aqueous dyeing processes, including reduced water consumption, lower energy usage and minimized environmental impact due to the absence of wastewater generation [5]. The $scCO_2$ dyeing process is characterized by its ability to dissolve hydrophobic dyes effectively, leading to shorter dyeing times [6] and potentially lower operational costs by up to 50% compared to conventional methods [7, 8]. In addition to dyeing, recent advances in supercritical carbon dioxide ($scCO_2$) technology have demonstrated its effectiveness in other textile applications such as scouring, bleaching, surface modification and the incorporation of functional agents to textiles. Initially successful in polyester dyeing, $scCO_2$ has enabled the integration of functionalities such as antimicrobial, antioxidant and UV-blocking properties [9]. To expand these advantages, current research is actively exploring its application to dyeing natural fibres like cotton and their blends. Studies have shown that $scCO_2$ can be used to dye these materials, enhancing both colour strength and fastness while supporting environmental sustainability [10]. Beyond dyeing, the technology is also being investigated for scouring [11], bleaching [12] and functional finishing [13], broadening its industrial relevance. In addition, its surface modification capability enhances the interaction between the fibre and dyes, nanoparticles or functional compounds [14]. This chapter briefly reviews the use of $scCO_2$ technology for textile surface modification and the functional finishing of textiles. It explores the fundamentals of $scCO_2$ processing, its mechanisms, surface modification techniques, applications across various textile functionalization processes, recent advancements and potential for transforming sustainable textile processing. Finally, it highlights the limitations and future prospects of the technology. Through this exploration, the chapter aims to establish $scCO_2$ processing as a viable and superior alternative to conventional textile modification methods, contributing to the development of environmentally sustainable and economically viable textile manufacturing practices that meet the evolving demands of modern industry and society.

6.2 Fundamentals of Supercritical CO_2

6.2.1 Definition, Properties and Phase Behaviour

A supercritical fluid (SCF) represents a unique state of matter that exists when pure substances are subjected to conditions exceeding their critical temperature and pressure simultaneously. In this supercritical state, the distinction between liquid and gas phases disappears, creating a single-phase fluid that exhibits extraordinary properties combining characteristics of both phases. SCF possesses liquid-like density, enabling effective solvation capabilities, while maintaining gas-like viscosity and diffusivity, facilitating rapid mass transfer and penetration into porous materials [15]. In this supercritical region, increasing the temperature at constant pressure (isobaric) above the critical temperature, or increasing the pressure at constant temperature (isothermal) above the critical pressure, keeps the fluid in a supercritical state without undergoing any phase transition [16]. Among SCFs, carbon dioxide (CO_2) is the most popular as it offers several advantages, including low toxicity, ready availability, low cost, non-flammability, environmental sustainability and being chemically inert under many conditions. In addition, CO_2 exhibits a critical temperature (Tc) of 31.1 °C and a critical pressure (Pc) of 73.8 bar, representing relatively mild conditions compared to other supercritical fluids. This enhances its appeal as a solvent in industrial processes [17, 18]. The critical density is 0.468 g/cm^3, establishing the threshold conditions necessary for achieving the supercritical state. The phase behaviour of CO_2 is characterized by distinct regions separated by the critical point, where the liquid–vapour equilibrium curve terminates, and the supercritical region begins. Figure 6.1 shows the phase diagram of CO_2 emphasizing its critical temperature (304.13 K or 31.0 °C) and critical pressure (7.38 MPa).

The phase diagram of CO_2 in Fig. 6.1 illustrates the relationship between pressure, temperature and density, revealing the unique characteristics of the supercritical region. Below the critical point, CO_2 exists as either a gas or a liquid with distinct phase boundaries. However, above the critical point, the fluid properties change continuously without phase transitions, allowing for precise control over

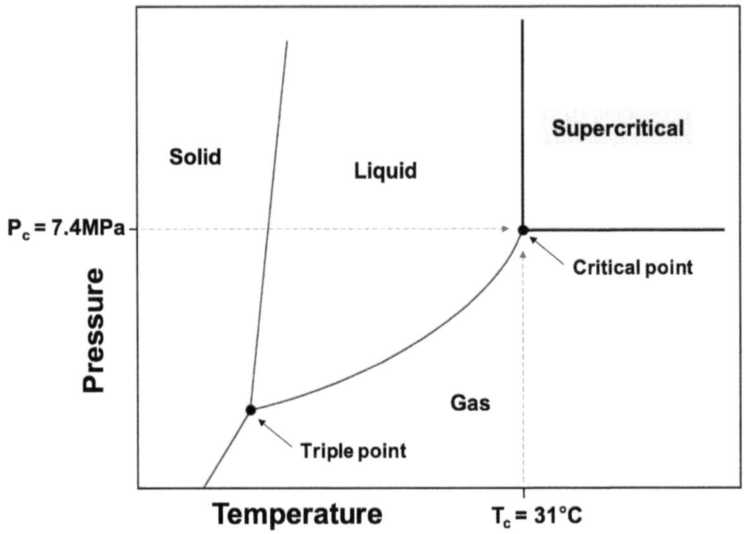

Fig. 6.1 Phase diagram of CO_2 [19]

solvent characteristics through manipulation of temperature and pressure parameters. This continuous variation in properties enables the tuning of $scCO_2$ characteristics to optimize specific textile processing applications. The fundamental properties of supercritical CO_2 that make it particularly attractive for textile processing are due to its unique properties that lie between those of gases and liquids. It combines liquid-like density with gas-like viscosity and diffusivity. As shown in Table 6.1, CO_2 has density values closer to liquids, while its viscosity and diffusivity resemble those of gases. This combination results in high compressibility, low viscosity, high diffusivity and tunable density depending on operating conditions. These properties enable $scCO_2$ to penetrate textile structures more effectively than conventional solvents while providing excellent mass transfer characteristics [20–22]. The absence of surface tension in the supercritical state allows for superior wetting and penetration into textile fibres, eliminating the limitations associated with conventional liquid processing, where surface tension can impede uniform treatment distribution [23, 24].

Table 6.1 Comparison of some physical properties of CO_2 at different states [25]

Fluide properties	Gas (313 K, 1 bar)	SCF (313 K, 100 bar)	Liquid (300 K, 500 bar)
Density (Kg m^{-3})	2	632	1029
Viscosity (μPa s)	16	17	133
Diffusion Coeff. (m^2 s^{-1})	$5.1*10^{-6}$	$1.4*10^{-8}$	$8.7*10^{-9}$

6.2.2 Solvent Characteristics and Tunability

Supercritical CO_2 offers tunable solvent properties ideal for textile applications such as dyeing, finishing and functionalization. Its solvating power depends on density, which can be adjusted through pressure and temperature changes [20]. The solvating power of $scCO_2$ is primarily determined by its density, which varies continuously with operating conditions. At low densities (near-critical conditions), $scCO_2$ exhibits poor solvating ability, while at higher densities (achieved through increased pressure), its solvating power approaches that of conventional organic solvents. The polarity of supercritical CO_2 is relatively low, making it an excellent solvent for nonpolar and weakly polar compounds. However, the addition of co-solvents or modifiers can significantly alter its polarity and solvating characteristics, expanding its applicability to a broader range of chemical species [5]. Common co-solvents include ethanol, methanol and water, which can be added in small quantities (typically 1–10% by volume) to enhance the solubility of polar compounds without significantly compromising the environmental benefits of the process. The tunability of $scCO_2$ extends beyond simple density variations to include temperature-dependent effects on molecular interactions and mass transfer properties. For example, higher temperatures generally increase diffusivity and reduce viscosity, improving penetration into textile structures, while simultaneously reducing density and solvating power. This temperature–pressure relationship provides multiple pathways for optimizing processing conditions to achieve specific textile functionalization objectives while maintaining operational efficiency [26].

6.2.3 Environmental, Safety and Operational Aspects

Supercritical CO_2 technology offers significant environmental and operational advantages over conventional water- or solvent-based textile finishing methods. Its non-toxic, recyclable and zero-liquid-discharge characteristics make it a truly sustainable alternative. As a non-hazardous and naturally occurring gas, CO_2 eliminates the health and safety risks associated with traditional solvents [5]. A major benefit lies in its recyclability, in which CO_2 can be recovered and reused after processing through simple depressurization, achieving recovery efficiencies above 95%. As the CO_2 fully evaporates and is collected after processing, there are no solvent residues on textiles, ensuring high product safety and quality. This closed-loop operation minimizes material losses, reduces operational costs and maintains consistent process performance without solvent degradation or purification [27]. Furthermore, unlike conventional methods, $scCO_2$ produces no wastewater, eliminating the need for costly treatment systems and making it especially valuable in water-scarce regions or under strict environmental regulations [26, 28]. Additional environmental and operational advantages include low energy consumption (no drying and operating typically at lower temperature), complete elimination of solvent residues and reduced greenhouse gas emissions. The superior mass transfer and diffusivity of $scCO_2$ enhance chemical utilization, shorten processing times and increase production efficiency. Moreover, the process generates no volatile organic compound (VOC) emissions, improving air quality and regulatory compliance. From an operational perspective, $scCO_2$ systems offer simplified process control, lower maintenance requirements and improved equipment longevity due to the absence of corrosive solvents. Together, these factors make supercritical CO_2 processing a clean, efficient and economically viable technology for sustainable textile manufacturing [5, 8, 26, 27].

6.3 Functionalization and Surface Modification Techniques in $scCO_2$

The unique physicochemical properties of $scCO_2$ discussed above render it an exceptional medium for functionalizing and modifying material surfaces for a diverse set of surface engineering techniques. These techniques leverage the superior transport properties of $scCO_2$ to achieve uniform, conformal and deep-penetrating modifications that are often unattainable with conventional liquid or gas-phase processes. This section delves into the primary methodologies employed for functionalization and surface modification in $scCO_2$, categorizing them into chemical and physical approaches (Fig. 6.2) and highlighting their mechanisms, advantages and applications.

6.3.1 Chemical Functionalisation Techniques

Chemical functionalization involves the covalent attachment of molecular species or polymers to a substrate surface, permanently altering its chemical composition and properties. $scCO_2$ acts as both a solvent and a transport vehicle for reactants, facilitating reactions at the fluid-substrate interface.

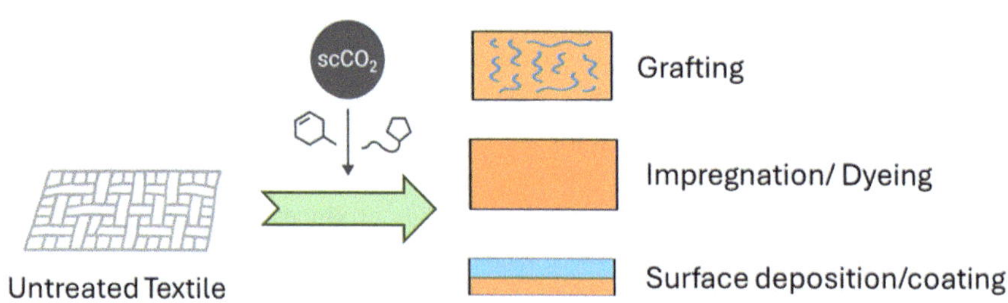

Fig. 6.2 Schematic illustrating the process of functionalization and surface modification techniques in $scCO_2$

6.3.1.1 Grafting and Polymerization Reactions

Grafting polymers onto solid substrates is one of the most established $scCO_2$-based surface modification strategies. The low viscosity, high diffusivity and tunable solvent strength of $scCO_2$ allow monomers to penetrate polymer matrices while enabling controlled heterogeneous reactions [29]. This makes $scCO_2$ an excellent platform for producing high-performance polymer and textile surfaces with tailored functionality [30]. The most common and effective approach in $scCO_2$-based surface modification is surface-initiated polymerisation. In this technique, immobilised initiators generate active centres that polymerise monomers dissolved in $scCO_2$. Efficient mass transport enables dense, well-defined polymer brushes [31]. Various polymerisation mechanisms can be employed within this framework. Some of them are Free-radical [32], ring-opening [33] and cationic [34] polymerisation mechanisms, which benefit from the high diffusivity and anhydrous nature of the medium, supporting controlled architectures and high-molecular weight grafts. The approach is especially advantageous for modifying high-aspect-ratio or porous materials such as membranes, fibres and monoliths. This grafting strategy has been applied to a wide range of polymers and substrates. Examples include grafting maleic anhydride [35], acrylonitrile [36], 2-hydroxyethyl methacrylate (HEMA) [37], glycidyl methacrylate (GMA) [38], 1,4-dichlorobutane [39] and methyl methacrylate (MMA) [40] onto polypropylene and other polymers, yielding improved hydrophilicity, interfacial adhesion, mechanical properties and surface energy. $scCO_2$-assisted solid-phase grafting has also demonstrated higher grafting yields and better control over reaction kinetics compared with melt or solution methods.

6.3.1.2 Chemical Deposition and Functionalization

Chemical deposition in $scCO_2$ involves reactions that form covalent bonds or chemically generate functional species directly on the substrate surface. A key example is the silanization and chemical functionalization of metal oxides (e.g. SiO_2, TiO_2) and porous materials, where organosilane molecules undergo controlled silanol condensation reactions in the dry, anhydrous $scCO_2$ environment. The deposition of organosilane molecules from $scCO_2$ results in highly uniform, monolayer coatings without the aggregation and multilayer formation often encountered in liquid-phase processes due to solvent retention [41]. The dry, anhydrous nature of $scCO_2$ is critical for controlled silanol condensation reactions. This approach is especially effective for natural or high-performance fibres like cellulosics and aramids, which possess accessible reactive sites suitable for bonding [30]. Successful grafting of silane coupling agents onto nanocellulose materials, such as cellulose nanofibril (CNF) nanopapers, using supercritical CO_2 has been reported [42]. This approach is crucial for developing high-performance bio-nanocomposites by improving the dispersion and interfacial compatibility of nanocellulose in hydrophobic polymer composites. A wide range of polymers have been modified using these grafting strategies including isotactic polypropylene (iPP) [32, 35, 37, 43, 44], Styrene–butadiene–styrene (SBS) triblock copolymers [45, 46], Poly(vinyl chloride) (PVC) [47, 48], Polystyrene [48], Chitosan membrane [49, 50] and so on.

Furthermore, $scCO_2$ is an ideal medium for the in situ synthesis and deposition of metal nanoparticles (e.g. Ag, Au, Pt) onto various supports. Metal–organic precursors dissolved in $scCO_2$ can be reduced (e.g. chemically with H_2 or thermally) to form well-dispersed nanoparticles within the pores or on the surface of substrates, a process pivotal for creating advanced catalysts and antimicrobial surfaces [51]. For instance, metal nanoparticles such as platinum can be chemically generated through in situ reduction of precursor complexes dissolved in $scCO_2$, resulting in highly dispersed catalytic coatings [52]. Similarly, silver nanoparticles have been synthesized on wool using a polyol reduction mechanism, producing uniform coatings formed by chemical transformation within the fibre environment [53–55]. Another example is the synthesis and application of a CO_2-philic pyridinium silicone coating, where the functional layer is chemically created and subsequently deposited onto cotton using $scCO_2$ to generate an effective biocidal surface [56]. These approaches illustrate how $scCO_2$ can facilitate both the formation and deposition of reactive species, enabling functionalization pathways not achievable through purely physical processes.

6.3.2 Physical Deposition and Impregnation Techniques

In contrast to chemical methods, physical techniques rely on the solubility and deposition of solutes from the $scCO_2$ phase onto a substrate without forming covalent bonds. This technique depends primarily on diffusion, swelling and physical interaction rather than covalent bonding. Through its high diffusivity, near-zero surface tension and tunable solvent strength, $scCO_2$ enables uniform and binder-free deposition of microspheres, polymers and nanoparticles.

6.3.2.1 $scCO_2$ as a Solvent for Deposition (Precipitation/Casting)

This technique exploits the principle of the anti-solvent precipitation approach in which functional compounds are deposited onto fibre surfaces by exploiting the controlled precipitation that occurs during depressurization of $scCO_2$. A solute (e.g. a polymer, drug, or dye) is first dissolved in $scCO_2$ under high pressure. The substrate is then placed in contact with this solution, allowing the solute to adsorb onto its surface. Upon rapid depressurization, the solvent power of CO_2 drops dramatically, causing the solute to precipitate in situ onto the substrate, forming a thin, contiguous film or a distribution of particles [14, 57]. This method, often referred to as Supercritical Anti-Solvent (SAS) deposition or $scCO_2$-assisted casting, is highly effective for creating polymer coatings on heat-sensitive materials and for the impregnation of active compounds (e.g. pharmaceuticals, antioxidants) into polymeric matrices for controlled release applications [58]. This technique is particularly useful for applying hydrophobic and omniphobic coatings such as Polyphosphazene-based microspheres

[59], silicone-based water repellents [60] and fluoropolymer or copolymer finishes [61–63]. In addition, nanoparticles including silver, TiO_2, hematite, gold and platinum can be dispersed in $scCO_2$ and deposited onto fibres with excellent uniformity [64]. Examples include the permanent impregnation of gold nanoparticles into various silk fibres [65] and silver nanoparticle coatings on cotton [66].

6.3.2.2 $scCO_2$ as Impregnation Media

The use of supercritical CO_2 for the impregnation of dyes and functional agents in textiles presents a sustainable alternative to traditional wet processing. This approach is considered resource-efficient and eco-friendly, leading to lower operational costs due to reduced energy and chemical consumption [8]. This is a simple diffusion of functional agents (dyes, hydrophobic polymers, drugs, nanoparticles, antimicrobials) into the textile fibres, followed by fixing either by chemical bonds, crosslinkers or physical entrapment as CO_2 is released. During impregnation processes, $scCO_2$ functions as both a solvent for dyes and active substances and as a transport medium [67]. Its unique properties, such as high diffusivity, low viscosity, high diffusion coefficients and zero surface tension, enable it to deliver dissolved substances deep into the solid matrix of a material [68]. This facilitates a homogeneous distribution of the active substance throughout the entire volume of the solid phase, a significant advantage over conventional impregnation methods. Furthermore, $scCO_2$ possesses the crucial ability to swell and plasticize polymers. This plasticization increases chain mobility and free volume within the polymer, which is vital for facilitating the penetration and incorporation of active compounds into the material structure [69, 70]. The impregnation of textiles in $scCO_2$ is a multi-step process involving solute dissolution and transport, sorption into the polymer matrix and depressurization. These stages are functionally integrated and can occur concurrently, particularly the dissolution and sorption steps [5, 71]. The extent of solute dissolution in $scCO_2$ depends on the chemical nature of the solute and the specific processing conditions, such as pressure and temperature. Supercritical CO_2 is a good solvent for nonpolar and low-molecular-weight molecules [72, 73]. The solubility of various compounds in $scCO_2$ has been measured extensively, and many studies that report the solubility of dyes have been published [74].

Sorption of the solute-laden $scCO_2$ into the polymer is largely driven by the extent of fibre swelling and plasticization induced by $scCO_2$. Polymers like polyester swell significantly under supercritical conditions, increasing free volume and enhancing the diffusion of active agents. The level of swelling is closely tied to the polymer's molecular structure and the specific impregnation parameters, with greater swelling generally leading to higher solute uptake [69, 71]. The gas-like diffusion properties of $scCO_2$ enable it to disperse the dye evenly into the small pores and crevices of the fibre [5]. The final step, depressurization, serves to remove CO_2 from the system and promotes the deposition or entrapment of the solute within the polymer matrix. The rate of depressurization is a critical process parameter. A rapid depressurization leads to a sudden drop in the solvating power of $scCO_2$, which can enhance solute

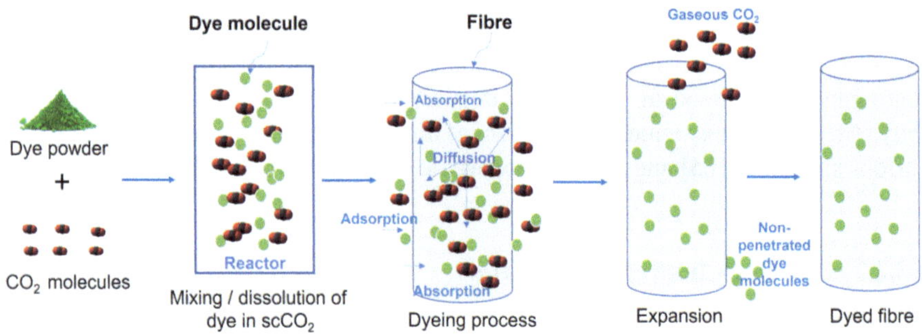

Fig. 6.3 Schematic of dyeing or impregnation mechanism in $scCO_2$ [19]

retention within the polymer. However, excessively fast depressurization can induce undesirable effects such as polymer foaming, cracking or degradation [5]. Therefore, the rate of depressurization must be carefully optimized based on the nature of the polymer and the functional agent to ensure both structural integrity and effective impregnation. A schematic overview of this process is illustrated in Fig. 6.3.

6.3.3 Methods for Enhancing Fixation and Performance

The durability and performance of textiles treated with functional agents via $scCO_2$ are strongly determined by the degree of fixation achieved. To enhance fixation efficiency, a variety of strategies have been developed, including chemical bonding, physical entrapment and the use of binding agents. The optimal approach largely depends on the chemical nature of the functional agent, the type of textile substrate and the intended end-use of the material [68]. Chemical bonding represents one of the most durable and effective mechanisms for anchoring functional agents onto textile fibres. It involves the formation of covalent linkages between the agent and reactive sites on the fibre surface. This process can be significantly enhanced using crosslinking or coupling agents, as well as surface activation techniques that generate additional reactive sites for bonding. For example, epoxy-based crosslinking agents have been applied concurrently with functional agents in $scCO_2$ environments to improve fixation and long-term durability [75]. Similarly, reactive dyes such as fluorotriazines exhibit excellent fixation on cotton substrates under acidic $scCO_2$ conditions [10]. In addition, vinylsulphone reactive disperse dyes have been shown to chemically bond with wool fibres when applied through $scCO_2$ processing [76].

6.4 Some Applications of $scCO_2$ in Textile Functionalization

The application of functional finishes using $scCO_2$ has gained significant attention as a sustainable alternative to conventional textile finishing processes. Supercritical CO_2 enables transformative functionalization of textiles by depositing or impregnating bioactive compounds, nanoparticles and polymeric coatings directly into fibre matrices [9]. During $scCO_2$ processing, finishing agents are either dissolved or dispersed within the supercritical medium, ensuring their even distribution throughout the textile structure. The absence of surface tension ensures a uniform coating, while rapid depressurization deposits the active agents onto the fibre surfaces [8]. Functional properties achieved through this method include hydrophobic and oleophobic treatments, antimicrobial activity, thermal regulation, flame-retardant finishes, UV-protective finishes, conductive and other advanced functional finishes and some of them are described in this section.

6.4.1 Hydrophobic and Oleophobic Finishes

Among its most successful applications of $scCO_2$ medium are hydrophobic and oleophobic treatments, which provide water and oil repellency for protective and high-performance fabrics. It is a highly effective medium due to its strong solvency for silicone- and fluoropolymer-based compounds, which are often poorly compatible with aqueous systems. Low-molecular-weight PTFE has been uniformly deposited onto polyester, producing consistently high water repellency [77]. Ultradisperse polytetrafluoroethylene (PTFE) coatings were applied to polyethylene terephthalate (PET) materials in $scCO_2$ with aliphatic alcohol co-solvents, resulting in the formation of a surface layer with reduced surface energy and improved stability under dry friction conditions [62]. Nylon finished with perfluoroalkyl methacrylate and hydroxyalkyl methacrylate copolymers in the presence of diisocyanate crosslinkers exhibited outstanding water and oil repellency (Fig. 6.4) [63]. $scCO_2$ has also been used to introduce fluorine-containing solutions onto polyester, resulting in improved repellency while maintaining air permeability and mechanical strength [78]. While fluorinated compounds (PFAS) have traditionally dominated this field, environmental and health concerns have driven research towards PFAS-free alternatives, such as silicone- and hydrocarbon-based repellents [79, 80]. Silicone-based water-repellent functionalization was performed in $scCO_2$ combined with dyeing, resulting in good colour strength and improved hydrophobicity [81]. Silicone-based finishes offer high water repellency, breathability and good wash durability, whereas hydrocarbon-based finishes provide sufficient water resistance with improved environmental compatibility, though typically without oleophobic performance. Innovative approaches, such as incorporating hematite nanoparticles into cellulosic fibres to achieve water resistance, further demonstrate the adaptability of $scCO_2$ systems for developing

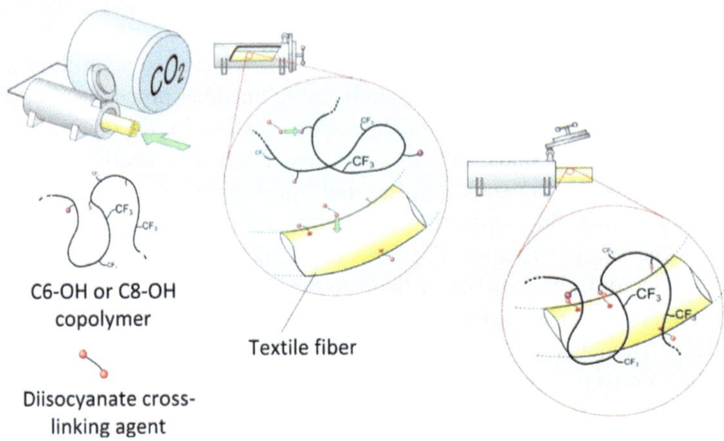

Fig. 6.4 Deposition perfluoroalkyl methacrylate/hydroxyalkyl methacrylate copolymers from $scCO_2$ medium [63]

high-performance protective materials [82]. Impregnation with $scCO_2$ often showed enhanced uniformity and durability compared to conventional application methods [21].

6.4.2 Antimicrobial Treatments

Antimicrobial functionalization is one of the most intensively explored applications of $scCO_2$ due to the medium's ability to uniformly deliver active compounds into textile matrices while preserving agent stability. Supercritical CO_2 technology also offers a sustainable, water-free method for imparting durable antibacterial and antifungal properties to textiles such as polyester, cotton and blends. Several antimicrobial agents have been tested in this system with promising results [83]. Organometallic systems, particularly silver-based materials, represent the most widely applied antimicrobial agents in $scCO_2$ [8]. Wool treated with silver nanoparticles demonstrated excellent catalytic, antistatic and antibacterial behaviour [53–55], while cotton impregnated with silver precursors exhibited strong bactericidal activity after in situ reduction [66]. These studies highlight the ability of $scCO_2$ to distribute metal-based agents homogeneously while avoiding agglomeration issues common in aqueous treatments. Silicon-containing antimicrobial agents have also been incorporated through $scCO_2$. CO_2-philic quaternary ammonium siloxanes applied to cotton produced durable antimicrobial activity resistant to UV exposure and washing [84]. Polyethylene fibres treated with silicone-based 2,2,6,6-tetramethyl4-piperidinol (TMP)-based N-chloramine using $scCO_2$ showed uniform

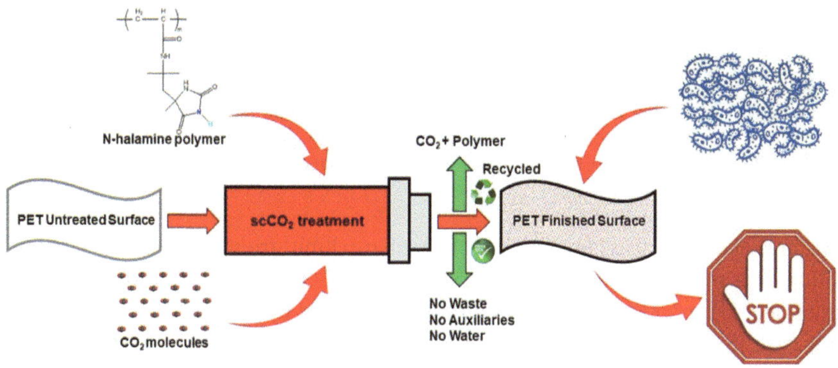

Fig. 6.5 Antibacterial surface modification of polyester with N-halamine, in $scCO_2$ medium [85]

nanometre-scale coatings and long-lasting biocidal action. Fluorinated pyridinium silicon-based compounds further strengthened the efficacy of silicon-driven antimicrobial finishes by forming well-distributed surface layers with strong antibacterial activity [56]. Non-metal-based systems, including N-halamines, have been applied to synthetic fibres via $scCO_2$, yielding high chlorine loading and substantial antibacterial effects on polyester (Fig. 6.5) [85]. Natural antimicrobial substances such as thymol, carvacrol and chitosan derivatives have also been successfully impregnated into cotton, polyester, sutures and biodegradable polymer composites. Thymol-loaded substrates consistently demonstrated broad-spectrum antibacterial activity, and chitosan lactate and low-molecular-weight chitosan penetrated polyester effectively in $scCO_2$, producing materials with strong and durable antimicrobial behaviour [86–88]. The combination of synthetic, natural and hybrid antimicrobial agents highlights $scCO_2$'s versatility for generating hygiene- and healthcare-relevant functional textiles.

6.4.3 UV Protection and Antioxidant Functionality

UV-protective functionality has been achieved through the deposition of both inorganic and organic agents using $scCO_2$. A notable approach involves the pretreatment of silk fibres with silane in $scCO_2$, followed by microwave-assisted deposition of silver nanoparticles. Silane pretreatment substantially improves nanoparticle adhesion and distribution, resulting in significantly reduced UV–Vis transmittance and enhanced UV-shielding performance compared with conventional methods [54]. Silane pretreatment followed by nanoparticle deposition in $scCO_2$ produces silk and polyester with reduced UV–Vis transmittance, leading to superior UV-blocking capability. These finishes also improve durability and fabric compatibility [21]. Curcumin and very low-molecular-weight chitosan derivatives have been used to impart antioxidant properties via $scCO_2$ impregnation [89]. Similarly, textiles coloured with natural

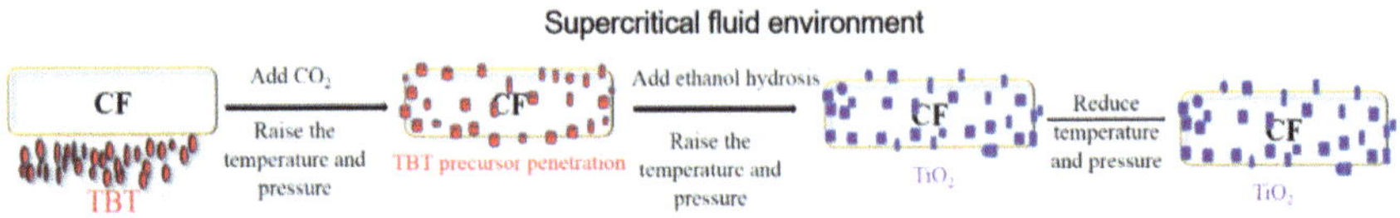

Fig. 6.6 Process illustration of TiO_2 synthesis on the surface of cotton fibre (CF) in $scCO_2$ [90]

bioactive agents such as curcumin in $scCO_2$ also exhibit strong intrinsic UV-blocking antioxidant properties, demonstrating the medium's potential for producing multi-functional UV-protective materials through combined dyeing and functionalization processes [89]. $scCO_2$ has also been used to deposit TiO_2 particles onto cotton fibres, resulting in enhanced UV resistance (Fig. 6.6) [90].

6.4.4 Other Functionalities and Applications

Supercritical CO_2 has also been explored for imparting other functionalities such as flame retardant, Electrical conductivity, catalytic performance and others. Flame retardancy is another functional domain where $scCO_2$ demonstrates distinct advantages, particularly in delivering phosphorus- and silicon-based flame-retardant compounds. Cotton fabrics impregnated with 9,10-dihydro9-oxa-10-phosphaphenanthrene-10-oxide (DOPO), 5060, SiO_2 and their combinations showed substantial improvements in flame resistance, with the most significant effect observed in DOPO-treated samples [91]. Cotton fabrics were treated in supercritical CO_2 with novel phosphorus–nitrogen and sulphur–nitrogen piperazine derivatives, producing durable flame-retardant coatings. The treated fabrics showed self-extinguishing behaviour and higher char yield, confirming improved flame resistance [92]. Cotton fabrics developed pleat-like surface wrinkles after treatment with supercritical CO_2 and rapid depressurization under specific conditions (40 °C, 20 MPa, 60 min). The wrinkled surfaces effectively anchored TiO_2 nanoparticles without binders, demonstrating a simple method for functionalizing cotton fabrics [93]. Cotton treated with palladium(II) hexafluoroacetylacetonate achieved conductive properties due to uniform deposition of palladium complexes throughout the fibre structure [94]. Wool fabrics coated with silver nanoparticles in $scCO_2$ showed a combination of antistatic, catalytic and antibacterial functions, illustrating the ability of $scCO_2$ processing to produce multifunctional advanced textiles [53–55]. Additional studies on silicone-based dimethylsiloxane polymers further demonstrate how $scCO_2$ can introduce complex hybrid materials that modify both the surface chemistry and physical properties of textile substrates [95]. Silk fabrics have been pretreated with 3-aminopropyltrimethoxysilane in supercritical CO_2, followed by silver nanoparticle synthesis under microwave irradiation. The resulting coatings showed improved adhesion, softness and UV protection compared to untreated fabrics [54].

Supercritical CO_2 was also used for metallizing polyester fibres with organometallic compounds to impart electrical, antibacterial and catalytic properties. This water-free process enables multifunctional textiles for applications such as antistatic fabrics, flexible heaters and antibacterial or catalytic materials [96]. Polyarylate fibres were impregnated with palladium acetate using $scCO_2$ to act as a catalyst for electroless plating, improving plating adhesion [97]. Furthermore, supercritical CO has emerged as an effective alternative for pharmaceutical manufacturing processes, enabling the design and control of particle size and drug loading for drug delivery systems [98, 99]. It has also been used for cleaning and pretreating textiles, removing residues or activating surfaces without water, preparing them for subsequent functionalization steps [100].

6.5 Challenges and Limitations

The application of $scCO_2$ in textile surface modification faces several challenges and limitations, primarily concerning fibre compatibility, additive solubility, equipment costs and process control.

6.5.1 Fibre Compatibility: Natural, Synthetic and Blended Fibres

The effectiveness of $scCO_2$ treatments varies significantly across different fibre types due to their distinct chemical compositions, morphologies and polarities [101]. Synthetic fibres, particularly polyester, exhibit high compatibility with $scCO_2$ due to their nonpolar structure, which aligns well with the low polarity of supercritical CO. This compatibility facilitates efficient transport and diffusion of functional additives [5, 8]. Polyester is a well-studied and industrially viable fibre for $scCO_2$ applications [27]. In contrast, natural fibres such as cotton, wool and silk present greater challenges. These fibres are hydrophilic, possess polar functional groups and often retain moisture, which can limit the solubility and penetration of nonpolar compounds in the $scCO_2$ medium [102]. While some natural fibres like wool and cotton can be dyed in $scCO_2$, specific dye structures or conditions are often required [76, 103]. High temperatures (e.g. 160 °C) in $scCO_2$ can also lead to damage in natural fibres like cotton, viscose, silk and wool [104]. Blended fabrics, such as wool/polyester, introduce complexity, as optimal conditions for one component may not suit the other, though the nonpolar synthetic component can sometimes improve colour strength in $scCO_2$ dyeing [76]. Despite these challenges, research shows opportunities for natural fibres, including $scCO_2$-assisted impregnation of nanoparticles into silk [65] and flax fibres [105], and the preparation of TiO_2 particles on cotton for UV resistance

[90, 106]. Pretreatment with $scCO_2$ can also improve properties of natural fibres like flax [107] and Lyocell for metal plating [108].

6.5.2 Compatibility with Functional Additives

The success of $scCO_2$ processing also depends on the nature of the functional additives. $scCO_2$ is effective for dissolving and transporting nonpolar and low-molecular-weight compounds [5]. However, its solubility parameters are less suited for polar or high-molecular-weight substances, including many hydrophilic pharmaceuticals, antimicrobial peptides and enzymes, which exhibit limited solubility in pure $scCO_2$ [109]. This often necessitates the use of co-solvents or chemical modification [110]. Incorporating nanoparticles, such as titanium dioxide, zinc oxide and silver, into textile fibres using $scCO_2$ remains challenging due to issues like particle aggregation, poor dispersion and weak binding [65]. However, $scCO_2$ has been successfully used for micronization and encapsulation of functional pigments, with solubility in $scCO_2$ being a critical factor [110]. Research is exploring ways to enhance the solubility of polymers in $scCO_2$, for instance, by introducing functional groups like ether, acetate and ester, or through fluorination, which can increase intermolecular interactions and improve solubility [111, 112].

6.5.3 Equipment Cost and Scalability

The widespread adoption of $scCO_2$ processing in the textile industry is hindered by the high capital investment required for equipment [113]. Unlike conventional machinery, $scCO_2$ systems demand robust, high-pressure vessels, precision pumps, compressors, heat exchangers and integrated safety controls [5]. While $scCO_2$ dyeing offers operational cost benefits such as reduced water and energy consumption and elimination of wastewater, the initial investment remains a significant barrier [27, 113]. Scalability for different applications also poses a challenge. Maintaining uniform temperature, pressure and fluid dynamics in large systems requires sophisticated engineering and control. Furthermore, the efficient recovery and reuse of CO_2, crucial for sustainability and cost-effectiveness, becomes more complex at larger volumes [5].

6.5.4 Process Optimization and Control

Achieving reliable and reproducible results in $scCO_2$-based textile treatments necessitates precise process optimization and control. Critical parameters such as temperature, pressure, treatment time and additive concentration must be carefully adjusted to suit both the fibre type and the desired functional outcome. These parameters directly influence the solubility and transport of additives in the supercritical phase, as well as their interaction with the fibre surface [109]. Flow dynamics, residence time and system configuration also impact treatment uniformity, particularly in continuous or semi-continuous processes. However, real-time monitoring of treatment conditions and in situ analysis of functionalization effects are currently limited [26]. The complex interactions between CO_2, fibre substrates and additives often require extensive trial-and-error experimentation to determine optimal settings. Developing advanced control systems, including model-predictive controllers, machine learning algorithms and in-line sensors, could enhance the reliability and scalability of the process [114].

6.6 Recent Developments and Future Directions

6.6.1 Recent Technological Developments

Significant advancements have been made in $scCO_2$ dyeing technology, particularly for synthetic fibres such as polyester, polyamide and their blends. Modern systems can effectively dye polyester and cellulose diacetate at lower temperatures and shorter dyeing times compared to conventional aqueous methods, reducing energy consumption and the need for water. For instance, polyamide 6.6 fabrics have been successfully dyed in $scCO_2$ at 100 °C and 220 bar, achieving high colour strength and good fastness properties [115, 116]. Similarly, wool/polyester and wool/nylon blended fabrics have been dyed in $scCO_2$ using synthesized reactive disperse dyes, showing improved colour strength and saturated colours compared to aqueous dyeing [76]. Innovations in dye chemistry have expanded the applicability of $scCO_2$ dyeing. Researchers have developed novel dyes, such as an anthraquinonoid disperse reactive dye for cotton, silk and wool [117], and a vinylsulphone (VS) reactive disperse dye that acts as a disperse dye for synthetic fibres and a reactive dye for protein fibres in $scCO_2$ [76]. Modified polyacrylate adhesives containing silicon and fluorinated silicon have also been developed for wash-free digital inkjet dyeing of polyester, demonstrating enhanced colour yield and fastness properties [118]. Furthermore, the use of co-solvents and carriers like salicylic acid and p-Vanillin can reduce $scCO_2$ dyeing time and temperature for polyester and cellulose acetate [115]. For cellulosic fibres like cotton, pretreatments such as polyethylene glycol (PEG) and mordants like aluminium acetate have been shown to improve the dyeability with natural dyes in $scCO_2$ [119]. Cationic modification of cotton is also being explored to enhance the utilization of reactive dyes and mitigate environmental concerns associated with conventional dyeing.

Beyond colouration, $scCO_2$ technology is increasingly applied in functional finishing of textiles. The application of $scCO_2$ in textile functionalization has seen considerable evolution, driven by the increasing demand for sustainable processing methods, advancements in green chemistry and progress in high-pressure technologies [120, 121]. The $scCO_2$ medium eliminates the need for water, thereby avoiding energy-intensive drying and preventing chemical hydrolysis, which can lead to more durable finishes [26]. This includes the deposition of various functional agents, such as flame retardants, antimicrobial agents and UV-protective substances [122–124]. Moreover, $scCO_2$ has been used in scouring, desizing and various finishing applications [8]. For synthetic fibres like polyester, $scCO_2$ has been shown to achieve over 99% oil removal during scouring and superior fastness properties in dyeing, which are enhanced by optimizing pressure and temperature [11]. The refinement of process parameters, including pressure and temperature, is crucial for enhancing treatment uniformity and control over surface modification [11]. As the understanding of its physicochemical behaviour advances, $scCO_2$ process becomes more predictable and reproducible, facilitating its expansion into technical and high-performance textile sectors applications [8].

6.6.2 Future Perspectives and Research Directions

Current research focuses on extending $scCO_2$ processing capabilities to cotton, wool and other natural fibres. The primary challenge lies in the limited solubility of conventional reactive and acid dyes in $scCO_2$ [103, 125]. Promising approaches include development of CO_2-soluble reactive dye systems, use of modified supercritical fluids with co-solvents and exploration of alternative dyeing mechanisms compatible with supercritical conditions [126]. For example, wool/polyester and wool/nylon blended fabrics have been dyed in $scCO_2$ using synthesized reactive disperse dyes, showing improved colour strength and saturated colours compared to aqueous dyeing. Most importantly, the integration of $scCO_2$ technology into circular textile manufacturing represents a promising research and development trajectory. Future work will likely focus on developing recyclable and renewable fibre systems that are inherently compatible with $scCO_2$-based treatments, aligning with circular economy principles [113, 115, 127]. This includes the use of agro-waste-derived biopolymers and recycled materials [128]. Research will also likely continue in the direction of closed-loop systems where CO_2 is continuously recovered and reused, aligning with the principles of circularity and resource efficiency [5, 113]. Additionally, digital process monitoring and AI-driven optimization are essential to ensure consistency and traceability across treatment batches, with AI techniques like artificial neural networks (ANNs) and adaptive neuro fuzzy interface systems (ANFIS) already being employed to predict colour strength in $scCO_2$ dyeing [129]. Emphasis will be placed on life-cycle assessments to quantify environmental benefits and economic feasibility. As regulatory pressures on water use, chemical discharge and carbon emissions increase, $scCO_2$-based functionalization stands as a transformative solution for the textile sector. Realizing its full potential will require interdisciplinary collaboration between chemists, materials scientists, textile engineers and industrial designers to overcome current limitations and fully hintegrate the technology into sustainable production ecosystems [113].

6.7 Conclusion

This review highlights the growing significance of supercritical CO_2 technology as a sustainable and versatile platform for textile functionalisation and surface modification. Across diverse studies, scCO_2 has proven effective in enabling dyeing, functional finishing through impregnation, grafting and surface coating processes that impart functional properties such as colour, antimicrobial activity, UV protection, hydrophobicity and other functionalities. By eliminating water and reducing chemical waste, scCO_2 offers a cleaner and more energy-efficient alternative to conventional wet-processing methods. Its ability to uniformly distribute functional agents and combine multiple finishing steps into a single process further underscores its technological and environmental advantages. However, despite its demonstrated potential at the laboratory scale, industrial implementation remains limited due to high equipment costs, the need for process optimization for various fibre types and challenges related to the solubility and compatibility of certain finishing agents. Future research should therefore focus on developing cost-effective reactor designs, identifying renewable functional agents and integrating scCO_2 with complementary green technologies such as plasma and enzymatic treatments. Additionally, techno-economic analyses and life-cycle assessments are essential to quantify the true sustainability benefits and identify pathways towards commercialization. Overall, scCO_2 represents a transformative approach that aligns with the textile industry's transition towards circularity, resource efficiency and environmental responsibility.

AI-Usage Declaration

During the preparation of this chapter, the authors used ChatGPT to improve the language of the manuscript. After using this tool/service, the authors reviewed and edited the content as needed and take full responsibility for the content of the publication.

References

1. G. Mujtaba, N. Muhammad, Q. Khan, M. Shah, N.K. Abbood, Water pollution hazards and toxicity caused by textile industries effluent. J. Ict Des. Eng. Technol. Sci. **7**(2), 7785 (2023)
2. U. Rott, Multiple use of water in industry: the textile industry case. J. Environ. Sci. Health A **38**(8), 1629–1639 (2003)
3. Z. Wang, M. Xue, K. Huang, Z. Liu, Textile dyeing wastewater treatment. Adv. Treat. Text. Effluent **64**, 91–116 (2011)

4. J. Sawant, R. Guru, D. Grewal, S.C. Talekar, S.P. Kulkarni, Sustainability in textiles: a critical review of eco–friendly practices and materials. Shodhkosh J. Vis. Perform. Arts **5**(2), 121 (2024)
5. B.J. Agrawal, (2015). Supercritical carbon-dioxide assisted dyeing of textiles: an environmental benign waterless dyeing process (2015)
6. M. Van der Kraan, M.V.F. Cid, G.F. Woertee, W.J.T. Veugeters, G.J. Witkamp, Equilibrium study on the disperse dyeing of polyester textile in supercritical carbon dioxide. Text. Res. J. **77**(8), 549–558 (2007)
7. T.A. Elmaaty, K. Sayed-Ahmed, M. Magdi, H. Elsisi, An eco-friendly method of extracting alizarin from rubia tinctorum roots under supercritical carbon dioxide and its application to wool dyeing. Sci. Rep. **13**(1), 124 (2023)
8. H.A. Eren, O. Avinç, S. Eren, Supercritical carbon dioxide for textile applications and recent developments. IOP Confer. Ser. Mater. Sci. Eng. **254**(8), 082011 (2017)
9. M.T. Abate, A. Ferri, J. Guan, G. Chen, V. Nierstrasz, Supercritical CO_2 technology in resource-effective production of functional and smart textiles. J. Fiber Bioeng. Inform. **15**(1), 35–44 (2022)
10. M.V.F. Cid, J.V. Spronsen, M.V.D. Kraan, W.J.T. Veugelers, G.F. Woerlee, G.J. Witkamp, Excellent dye fixation on cotton dyed in supercritical carbon dioxide using fluorotriazine reactive dyes. Green Chem. **7**(8), 609 (2005)
11. C.T. Wang, W.-F. Lin, Scouring and dyeing of polyester fibers in supercritical carbon dioxide. J. Chem. Eng. Jpn. **34**(2), 244–248 (2001)
12. S. Eren, O. Avinc, Z. Saka, H.A. Eren, Waterless bleaching of knitted cotton fabric using supercritical carbon dioxide fluid technology. Cellulose **25**(10), 6247–6267 (2018)
13. M.T. Abate, S. Seipel, J. Yu, M. Viková, M. Vik, A. Ferri, J. Guan, G. Chen, V. Nierstrasz, Supercritical CO_2 dyeing of polyester fabric with photochromic dyes to fabricate UV sensing smart textiles. Dyes Pigm. **183**, 108671 (2020)
14. J. Patadiya, P.V. Chougale, M. Naebe, B. Kandasubramanian, P. Mahajan-Tatpate, Recent advancement in surface modification of aramid fiber assisted by supercritical carbon dioxide. Polym. Adv. Technol. **34**(12), 3748–3758 (2023)
15. J.M.H.L. Sengers, *Supercritical Fluids: Their Properties and Applications* (2000), pp. 1–29
16. S.P. Nalawade, F. Picchioni, L. Janssen, Supercritical carbon dioxide as a green solvent for processing polymer melts: processing aspects and applications. Progr. Polym. Sci. **31**(1), 19–43 (2006)
17. S.L. Wells, J. Desimone, CO_2 technology platform: an important tool for environmental problem solving. Angew. Chem. **40**(3), 518–527 (2001)
18. X. Zhang, S. Heinonen, E. Levänen, Applications of supercritical carbon dioxide in materials processing and synthesis. RSC Adv. **4**(105), 61137–61152 (2014)
19. M.T. Abate, Supercritical CO_2 technology in resource-effective production of functional and smart textiles, Högskolan i Borås (2020)
20. S.L. Draper, G.A. Montero, B. Smith, K. Beck, Solubility relationships for disperse dyes in supercritical carbon dioxide. Dyes Pigm. **45**(3), 177–183 (2000)
21. A.G. Hassabo, M. Zayed, M. Bakr, E.A. El-Aziz, H. Othman, Applications of supercritical carbon dioxide in textile finishing: a review. J. Text. Colorat. Polym. Sci. **19**(2), 179–187 (2022)
22. H. Zheng, J. Zhang, L. Zheng, Optimization of an ecofriendly dyeing process in an industrialized supercritical carbon dioxide unit for acrylic fibers. Text. Res. J. **87**(15), 1818–1827 (2016)
23. T.A. Elmaaty, E.A. El-Aziz, Supercritical carbon dioxide as a green media in textile dyeing: a review. Text. Res. J. **88**(10), 1184–1212 (2017)
24. L. Lara, I. Cabral, J. Cunha, Ecological approaches to textile dyeing: a review. Sustainability **14**(14), 8353 (2022)
25. A.A. Clifford, J.R. Williams, Introduction to supercritical fluids and their applications, in *Supercritical Fluid Methods and Protocols*, eds. by J.R. Williams, A.A. Clifford (Humana Press, Totowa, 2000), pp. 1–16

26. G.A. Montero, C.B. Smith, W.A. Hendrix, D.L. Butcher, Supercritical fluid technology in textile processing: an overview. Ind. Eng. Chem. Res. **39**(12), 4806–4812 (2000)
27. W.A. Hendrix, Progress in supercritical CO_2 dyeing. J. Ind. Text. **31**(1), 43–56 (2001)
28. W. Saus, D. Knittel, E. Schollmeyer, Dyeing of textiles in supercritical carbon dioxide. Text. Res. J. **63**(3), 135–142 (1993)
29. A.I. Cooper, Polymer synthesis and processing using supercritical carbon dioxide. J. Mater. Chem. **10**(2), 207–234 (2000)
30. W. Yi-ming, Bulk modification and grafting of polymer in supercritical CO_2. Polym. Bull. **98**, 7454 (2011)
31. P.B. Zetterlund, F. Aldabbagh, M. Okubo, Controlled/living heterogeneous radical polymerization in supercritical carbon dioxide. J. Polym. Sci. A Polym. Chem. **47**(15), 3711–3728 (2009)
32. T. Liu, G.H. Hu, G. Tong, L. Zhao, G.P. Cao, W. Yuan, Supercritical carbon dioxide assisted solid-state grafting process of maleic anhydride onto polypropylene. Ind. Eng. Chem. Res. **44**(12), 4292–4299 (2005)
33. H.S. Ganapathy, H.S. Hwang, Y.T. Jeong, W.-K. Lee, K.T. Lim, Ring-opening polymerization of l-lactide in supercritical carbon dioxide using PDMS based stabilizers. Eur. Polym. J. **43**(1), 119–126 (2007)
34. M.R. Clark, J.M. DeSimone, Cationic polymerization of vinyl and cyclic ethers in supercritical and liquid carbon dioxide. Macromolecules **28**(8), 3002–3004 (1995)
35. A. Galia, R.D. Gregorio, G. Spadaro, O. Scialdone, G. Filardo, Grafting of maleic anhydride onto isotactic polypropylene in the presence of supercritical carbon dioxide as a solvent and swelling fluid. Macromolecules **37**(12), 4580–4589 (2004)
36. Y.Z. Wang, D. Ma, C. Yue, W.N. Jia, Y. Li, Study on the graft copolymerization of AN onto PP by using supercritical carbon dioxide. Adv. Mater. Res. **815**, 717–721 (2013)
37. D. Li, B. Han, Z. Liu, Grafting of 2-hydroxyethyl methacrylate onto isotactic poly(propylene) using supercritical CO_2 as a solvent and swelling agent. Macromol. Chem. Phys. **202**(11), 2187–2194 (2001)
38. M.H. Kunita, A.W. Rinaldi, E.M. Girotto, E. Radovanovic, E.C. Muniz, A.F. Rubira, Grafting of glycidyl methacrylate onto polypropylene using supercritical carbon dioxide. Eur. Polym. J. **41**(9), 2176–2182 (2005)
39. C. Jia, C. Yuan, Z. Ma, Y. Du, L. Liu, Y. Huang, Improving the mechanical and surface properties of aramid fiber by grafting with 1,4-dichlorobutane under supercritical carbon dioxide. Materials **12**(22), 3766 (2019)
40. C.S. Li, A.N. Zheng, S.X. Zhang, H. Ren, Grafting methyl methacrylate onto styrene-butadiene-styrene tri-block copolymer in supercritical carbon dioxide. Huadong Ligong Daxue Xuebao J. East China Univ. Sci. Technol. **32**(6), 724–727 (2006)
41. J.J. Watkins, T.J. McCarthy, Polymerization of styrene in supercritical CO_2-swollen poly (chlorotrifluoroethylene). Macromolecules **28**(12), 4067–4074 (1995)
42. C. Darpentigny, C. Sillard, M. Menneteau, E. Martínez, P.R. Marcoux, J. Bras, B. Jean, G. Nonglaton, Antibacterial cellulose nanopapers via aminosilane grafting in supercritical carbon dioxide. ACS Appl. Bio-Mater. **3**(12), 8402–8413 (2020)
43. Z. Dong, Z. Liu, B. Han, J. He, T. Jiang, G. Yang, Modification of isotactic polypropylene film by grafting of acrylic acid using supercritical CO_2 as a swelling agent. J. Mater. Chem. **12**(12), 3565–3569 (2002)
44. H. Rao, Z. Zhang, Graft copolymerization of maleic anhydride/styrene onto isotactic polypropylene using supercritical CO_2. Polym. Adv. Technol. **19**(7), 770–774 (2007)
45. L. Hui, Grafting of SBS and maleic anhydride in supercritical carbon dioxide and surface properties. Polym. Mater. Sci. Eng. **12**, 6556 (2010)
46. C. Li, S. Zhang, A. Zheng, X.Z. Kong, B. Geng, Grafting of maleic anhydride onto styrene–butadiene–styrene triblock copolymer using supercritical CO_2 as a swelling agent. J. Appl. Polym. Sci. **102**(5), 4425–4429 (2006)
47. D. Li, B. Han, Phase behavior of supercritical CO_2/styrene/poly(vinyl chloride) system and synthesis of polystyrene/poly(vinyl chloride) composites. Macromolecules **33**(12), 4555–4560 (2000)

48. A.H. Trivedi, S. Kwak, S. Lee, Grafting of poly(vinyl chloride) and polypropylene with styrene in a supercritical CO_2 solvent medium: synthesis and characterization. Polym. Eng. Sci. **41**(11), 1923–1937 (2001)
49. L. Cao, X. Wang, G. Wang, J. Wang, A pH-sensitive porous chitosan membrane prepared via surface grafting copolymerization in supercritical carbon dioxide. Polym. Int. **64**(3), 383–388 (2014)
50. O. García-Valdez, D.G. Ramírez-Wong, E. Saldívar-Guerra, G. Luna-Bárcenas, Grafting of chitosan with styrene and maleic anhydride via nitroxide-mediated radical polymerization in supercritical carbon dioxide. Macromol. Chem. Phys. **214**(12), 1396–1404 (2013)
51. F. Cansell, C. Aymonier, Design of functional nanostructured materials using supercritical fluids. J. Supercrit. Fluids **47**(3), 508–516 (2009)
52. Y.-L. Chen, C.-H. Tsai, M.-Y. Chen, Y.-C. Lai, Green fabrication of supported platinum nanoparticles by supercritical CO_2 deposition. Materials **11**(12), 2587 (2018)
53. L. Peng, R. Guo, J. Lan, S. Jiang, C. Li, Z. Zhang, Synthesis of silver nanoparticles on wool fabric in supercritical carbon dioxide. Mater. Exp. **7**(5), 405–410 (2017)
54. L. Peng, R. Guo, J. Lan, S. Jiang, X. Wang, S. Lin, C. Li, Silver nanoparticles coating on silk fabric with pretreatment of 3-aminopropyltrimethoxysilane in supercritical carbon dioxide. J. Ind. Text. **47**(5), 883–896 (2016)
55. L.H. Peng, R.H. Guo, J.W. Lan, S.X. Jiang, C. Li, Z.Y. Zhang, Synthesis of silver nanoparticles on wool fabric in supercritical carbon dioxide. Mater. Express **7**(5), 405–410 (2017)
56. Y. Chen, Q. Zhang, Y. Ma, Q. Han, Surface-oriented fluorinated pyridinium silicone with enhanced antibacterial activity on cotton via supercritical impregnation. Cellulose **25**(2), 1499–1511 (2018)
57. N. Singh, M. Vinjamur, M. Mukhopadhyay, Insights into adsorptive drug loading on silica aerogels from supercritical CO_2. Langmuir **38**(43), 13075–13083 (2022)
58. E. Reverchon, R. Adami, Nanomaterials and supercritical fluids. J. Supercrit. Fluids **37**(1), 1–22 (2006)
59. N. Liubimtsev, A. Deniz, I.V. Elmanovich, M.O. Gallyamov, A. Pich, Morphology and properties of flame-retardant superhydrophobic polymer coatings deposited on cotton fabrics from supercritical CO_2. ACS Appl. Polym. Mater. **2**(7), 2919–2926 (2020)
60. A.I. Stakhanov, I.V. Elmanovich, E.I. Kravchenko, E.A. Khakina, A.A. Pavlov, E.P. Kharitonova, БВ Локшин, M.O. Gallyamov, New fluorinated cyclophosphazenes: synthesis, properties, applications. Phosp. Sulf. Silicon Relat. Elements **198**(4), 321–329 (2022)
61. P.S. Kazaryan, A.A. Tyutyunov, M.S. Kondratenko, I.V. Elmanovich, A.I. Stakhanov, V.V. Zefirov, M.O. Gallyamov, Н.В Благодатских, АР Хохлов, Superhydrophobic coatings on textiles based on novel poly(perfluoro-tert-hexylbutyl methacrylate-co-hydroxyethyl methacrylate) copolymer deposited from solutions in supercritical carbon dioxide. J. Supercrit. Fluids **149**, 34–41 (2019)
62. T.Y. Kumeeva, N.P. Prorokova, Ultrathin hydrophobic coatings obtained on polyethylene terephthalate materials in supercritical carbon dioxide with co-solvents. Russian J. Phys. Chem. A **92**(2), 346–351 (2018)
63. V.V. Zefirov, N.A. Lubimtsev, A.I. Stakhanov, I.V. Elmanovich, M.S. Kondratenko, B.V. Lokshin, M.O. Gallyamov, A.R. Khokhlov, Durable crosslinked omniphobic coatings on textiles via supercritical carbon dioxide deposition. J. Supercrit. Fluids **133**, 30–37 (2018)
64. J.S. Wang, C.M. Wai, G.J. Brown, S.D. Apt, Two-dimensional nanoparticle cluster formation in supercritical fluid CO_2. Langmuir **32**(18), 4635–4642 (2016)
65. M. Singh, E.S. Dey, S. Bhand, C. Dicko, Supercritical carbon dioxide impregnation of gold nanoparticles demonstrates a new route for the fabrication of hybrid silk materials. Insects **13**(1), 18 (2022)
66. S.D. Gittard, D. Hojo, G.K. Hyde, G. Scarel, R.J. Narayan, G.N. Parsons, Antifungal textiles formed using silver deposition in supercritical carbon dioxide. J. Mater. Eng. Perform. **19**(3), 368–373 (2009)
67. I. Žižović, Potential of supercritical solvent impregnation for development of materials with antibacterial properties. Int. Arch. Med. Microbiol. **1**(1), 1–6 (2017)

68. M.T. Abate, V. Nierstrasz, Impregnation of materials in supercritical CO_2 to impart various functionalities (2019)
69. O.S. Fleming, S.G. Kazarian, Polymer processing with supercritical fluids. Supercrit. Carbon Dioxide **32**, 205–238 (2005)
70. S.M. Gross, R.D. Givens, M. Jikei, J.R. Royer, S.A. Khan, J.M. DeSimone, P.G. Odell, G.K. Hamer, Synthesis and swelling of poly(bisphenol a carbonate) using supercritical CO_2. Macromolecules **31**(25), 9090–9092 (1998)
71. M. Banchero, Supercritical fluid dyeing of synthetic and natural textiles—a review. Colorat. Technol. **129**(1), 2–17 (2013)
72. E. Bach, E. Cleve, E. Schollmeyer, Past, present and future of supercritical fluid dyeing technology—an overview. Rev. Progr. Colorat. Relat. Top. **32**(1), 88–102 (2002)
73. T. Kim, G. Kim, J.Y. Park, J.S. Lim, K.P. Yoo, Solubility measurement and dyeing performance evaluation of Aramid NOMEX yarn by dispersed dyes in supercritical carbon dioxide. Ind. Eng. Chem. Res. **45**(10), 3425–3433 (2006)
74. R.B. Gupta, J.J. Shim, *Solubility in Supercritical Carbon Dioxide* (CRC Press, Florida, 2006)
75. W.X. Ma, C.S. Zhao, S. Okubayashi, I. Tabata, K. Hisada, T. Hori, A novel method of modifying poly(ethylene terephthalate) fabric using supercritical carbon dioxide. J. Appl. Polym. Sci. **117**(4), 1897–1907 (2010)
76. A.A. Mousa, F.A. Mohamed, S.A.A. El-Megied, Y.A. Youssef, Dyeing of synthetic fiber-based wool blended fabrics in supercritical carbon dioxide. Sci. Rep. **14**(1), 542 (2024)
77. N.P. Prorokova, T.Y. Kumeeva, A.V. Khorev, V.M Buznik, ЛН Никитин, Ensuring a high degree of water repellency of polyester textile materials by treating them with supercritical carbon dioxide. Fibre Chem. **42**(2), 109–113 (2010)
78. Y.-Y. Xu, L.-J. Zheng, F. Ye, Y.-F. Qian, J. Yan, X.-Q. Xiong, Water/oil repellent property of polyester fabrics after supercritical carbon dioxide finishing. Therm. Sci. **19**(4), 1273–1277 (2015)
79. S. Chen, J. Lai, Y. Du, Z. Hong, Review on non-fluorinated durable water repellent alternatives for textile and fabric coatings. Sci. Technol. Eng. Chem. Environ. Protect. **1**(3), 1252 (2025)
80. A. Verbič, B. Golja, B. Stres, B. Likozar, U. Novak, Unveiling PFAS-free solutions for hydrophobic and oleophobic textile coatings, in *Proceedings of Socratic Lectures* (2024)
81. A. Schulz, Supercritical carbon dioxide as a green media for simultaneous dyeing and functionalisation: a study on disperse dyeing and silicone functionalisation for water repellency of polyester fabric (2019)
82. S. Xu, D. Shen, P. Wu, Fabrication of water-repellent cellulose fiber coated with magnetic nanoparticles under supercritical carbon dioxide. J. Nanopart. Res. **15**(4), 1577 (2013)
83. I. Žižović, Supercritical fluid applications in the design of novel antimicrobial materials. Molecules **25**(11), 2491 (2020)
84. Y. Chen, M.Q. Niu, S. Yuan, H.N. Teng, Durable antimicrobial finishing of cellulose with QSA silicone by supercritical adsorption. Appl. Surf. Sci. **264**, 171–175 (2013)
85. M. Orhan, F. Demirci, H.B. Kocer, V. Nierstrasz, Supercritical carbon dioxide application using hydantoin acrylamide for biocidal functionalization of polyester. J. Supercrit. Fluids **165**, 104986 (2020)
86. M.T. Abate, A. Ferri, J. Guan, G. Chen, J.A. Ferreira, V. Nierstrasz, Single-step disperse dyeing and antimicrobial functionalization of polyester fabric with chitosan and derivative in supercritical carbon dioxide. J. Supercrit. Fluids **147**, 231–240 (2019)
87. T. Baba, K. Hirogaki, I. Tabata, S. Okubayashi, K. Hisada, T. Hori, Impregnation of chitin/chitosan into polyester fabric using supercritical carbon dioxide. Sen'i Gakkaishi **66**(3), 63–69 (2010)
88. S. Milovanovic, M. Stamenic, D. Markovic, M. Radetic, I. Zizovic, Solubility of thymol in supercritical carbon dioxide and its impregnation on cotton gauze. J. Supercrit. Fluids **84**, 173–181 (2013)
89. M.T. Abate, Y. Zhou, J. Guan, G. Chen, A. Ferri, V. Nierstrasz, Colouration and bio-activation of polyester fabric with curcumin in supercritical CO_2: Part II—effect of dye concentration on the colour and functional properties. J. Supercrit. Fluids **157**, 104703 (2020)

90. S. Ye, H. Sun, J. Wu, L. Wan, Y. Ni, R. Wang, Z. Xiang, X. Deng, Supercritical CO_2 assisted TiO_2 preparation to improve the UV resistance properties of cotton fiber. Polymers **14**(24), 5513 (2022)
91. H. Ke, Q. Fang, Y. Han, H. Zheng, L. Zheng, Y. Dong, Facile flame-retardant finishing for cotton in supercritical CO_2. Text. Res. J. **92**(13–14), 2306–2316 (2022)
92. S. Chang, B. Condon, J. Smith, M. Easson, Preparation of flame retardant cotton fabric using environmental friendly supercritical carbon dioxide. Int. J. Mater. Sci. Appl. **6**(6), 269 (2017)
93. S. Katayama, L. Zhao, S. Yonezawa, Y. Iwai, Modification of the surface of cotton with supercritical carbon dioxide and water to support nanoparticles. J. Supercrit. Fluids **61**, 199–205 (2012)
94. Y. Iwai, S. Sameshima, S. Yonezawa, S. Katayama, Fabrication of conductive cotton by electroless plating method with supercritical carbon dioxide. J. Supercrit. Fluids **100**, 46–51 (2015)
95. A.L. Mohamed, M. Er-Rafik, M. Moller, Supercritical carbon dioxide assisted silicon based finishing of cellulosic fabric: a novel approach. Carbohydr. Polym. **98**(1), 1095–1107 (2013)
96. K. Opwis, D. Plohl, J. Schneider, J. Seithuemmer, D.S. Wunschik, J.S. Gutmann, Metallization of PET fibers in supercritical carbon dioxide and potential applications in the textile sector. J. Supercrit. Fluids **191**, 105722 (2022)
97. T. Sakai, I. Tabata, T. Hori, K. Hirogaki, Improved plating adhesion of polyarylate fibers via supercritical CO_2 and impregnation of slightly soluble oil on the fiber surface. Fibers Polym. **25**(5), 1737–1750 (2024)
98. R.K. Kankala, Y.S. Zhang, S.B. Wang, C.H. Lee, A.Z. Chen, Supercritical fluid technology: an emphasis on drug delivery and related biomedical applications. Adv. Healthcare Mater. **6**(16), 6528 (2017)
99. S.H. Soh, L.Y. Lee, Microencapsulation and nanoencapsulation using supercritical fluid (SCF) techniques. Pharmaceutics **11**(1), 21 (2019)
100. D. Aslanidou, C. Tsioptsias, C. Panayiotou, A novel approach for textile cleaning based on supercritical CO_2 and pickering emulsions. J. Supercrit. Fluids **76**, 83–93 (2013)
101. M.P. Gashti, J.A. Willoughby, P. Agrawal, Surface and bulk modification of synthetic textiles to improve dyeability (2011)
102. L. Bao-mei, The theory and research advances of supercritical CO_2 dyeing. Dyestuffs Colorat. **65**, 715 (2010)
103. Y. Fan, Y.Q. Zhang, K. Yan, J. Long, Synthesis of a novel disperse reactive dye involving a versatile bridge group for the sustainable coloration of natural fibers in supercritical carbon dioxide. Adv. Sci. **6**(1), 6539 (2018)
104. A. Schmidt, E. Bach, E. Schollmeyer, Damage to natural and synthetic fibers treated in supercritical carbon dioxide at 300 bar and temperatures up to 160 °C. Text. Res. J. **72**(11), 1023–1032 (2002)
105. A. Langhorst, A. Singhal, D.F. Mielewski, M. Banu, A. Taub, Nanoparticle modification of natural fibers for structural composites (2021)
106. Q. Yu, P. Wu, P. Xu, L. Li, T. Liu, L. Zhao, Synthesis of cellulose/titanium dioxide hybrids in supercritical carbon dioxide. Green Chem. **10**(10), 1061 (2008)
107. A. Langhorst, D. Zhang, J. Berman, X. Biraku, J.M. Rieland, M. Yu, B.J. Love, M. Banu, A. Taub, Mechanical property enhancement of flax fibers via supercritical fluid treatment. Sci. Rep. **14**(1), 542 (2024)
108. H. Cho, I. Tabata, K. Hisada, K. Hirogaki, T. Hori, Characterization of copper-plated conductive fibers after pretreatment with supercritical carbon dioxide and surface modification using lyocell fiber. Text. Res. J. **83**(8), 780–793 (2012)
109. Q. Xu, Y.N. Chang, Complex interactions among additive/supercritical CO_2/polymer ternary systems and factors governing the impregnation efficiency. J. Appl. Polym. Sci. **93**(2), 742–748 (2004)
110. D.T. Santos, M.Â.A. Meireles, Micronization and encapsulation of functional pigments using supercritical carbon dioxide. J. Food Process Eng **36**(1), 36–49 (2011)

111. Z. Dong, T. Hou, S. Hu, Z. Liu, C. Guo, W. Li, K. Lin, Screening and evaluation of CO_2 thickening agents based on molecular dynamics simulation technology. ACS Omega **10**(16), 16384–16395 (2025)
112. M.A. McHugh, A. Garach-Domech, I.-H. Park, D. Li, E. Barbu, P. Graham, J. Tsibouklis, Impact of fluorination and side-chain length on poly(methylpropenoxyalkylsiloxane) and poly(alkyl methacrylate) solubility in supercritical carbon dioxide. Macromolecules **35**(17), 6479–6482 (2002)
113. M.L. Catarino, F. Sampaio, A.L. Gonçalves, Sustainable wet processing technologies for the textile industry: a comprehensive review. Sustainability **17**(7), 3041 (2025)
114. T. Kim, G. Park, W.B. Kong, Y.-W. Lee, Supercritical dyeing technology. Clean Technol. **24**, 1–8 (2018)
115. P.J. Broadbent, C. Carr, D.M. Lewis, M. Rigout, E.J. Siewers, N.S. Kaveh, P. Desai, Investigation into the aqueous and supercritical carbon dioxide dyeing of polyester and cellulose acetate: the influence of temperature and carriers on fabric coloration and dye levelling. Rev. Progr. Colorat. Relat. Top. **35**, 652 (2025)
116. S. Eren, İ Özyurt, Waterless dyeing of polyamide 6.6. Polymers **16**(11), 1472 (2024)
117. R. Penthala, S.H. Park, H. Oh, I.Y. Lee, E.H. Ko, Y.A. Son, An ecofriendly dyeing of nylon and cotton fabrics in supercritical CO_2 with novel tricyanopyrrolidone reactive disperse dye. J. CO2 Util. **60**, 542 (2022)
118. L. Wang, H. Dong, K. Zheng, N. Zeng, M. Wu, X. Wang, H. Li, Organofluorosilicon modified polyacrylate with the unidirectional migration promotion of disperse dyes toward polyester fabric for wash-free digital inkjet dyeing. ACS Appl. Mater. Interf. **16**(30), 40030–40045 (2024)
119. C. Agbo, S. Okubayashi, Exploring the dyeability and sustainability of lac dye on cotton: a comparative analysis between water and supercritical CO_2 dyeing. Int. J. Cloth. Sci. Technol. **37**(3), 480–494 (2025)
120. T. Abou-Elmaaty, A. Mousa, H. Gaafar, A. Hebeish, H. Sorour, Development of textile dyeing using the green supercritical fluid technology: a review. Mater. Int. **2**, 373–390 (2020)
121. İ Yiğit, A.A. Özenç, S. Eren, The application of $scCO_2$ medium in cellulosic fibers: a sustainable approach in textile production. Clean Technol. Environ. Policy **27**(9), 3959–3968 (2025)
122. M. Fernandes, J. Padrão, A.I. Ribeiro, R.D.V. Fernandes, L. Melro, T. Nicolau, B. Mehravani, C. Alves, R. Rodrigues, A. Zille, Polysaccharides and metal nanoparticles for functional textiles: a review. Nanomaterials **12**(6), 1006 (2022)
123. K.R. Fontenot, M.W. Easson, J. Smith, C. Madison, S. Nam, T.-M.D. Nguyen, S. Chang, B.D. Condon, Application of a phosphazene derivative as a flame retardant for cotton fabric using conventional method and supercritical CO_2. AATCC J. Res. **1**(6), 16–26 (2014)
124. J. Ghosh, N.S. Rupanty, T. Noor, T.R. Asif, T. Islam, V. Reukov, Functional coatings for textiles: advancements in flame resistance, antimicrobial defense, and self-cleaning performance. RSC Adv. **15**(14), 10984–11022 (2025)
125. W.W. Ju, F. Wen-xiu, Dyeing performance of cotton fabrics with natural curcuminoid in supercritical carbon dioxide. J. Text. Eng. Fash. Technol. **5**(2), 652 (2019)
126. H.-K. Bae, J.-H. Jeon, H. Lee, Influence of co-solvent on dye solubility in supercritical carbon dioxide. Fluid Phase Equil. **222–223**, 119–125 (2004)
127. C.-H.H. Lo, Degumming silk by CO_2 supercritical fluid and their dyeing ability with plant indigo. Int. J. Cloth. Sci. Technol. **33**(3), 465–476 (2020)
128. B. Niu, Z. Xu, J. Xiao, Y. Qin, Recycling hazardous and valuable electrolyte in spent lithium-ion batteries: urgency, progress, challenge, and viable approach. Chem. Rev. **123**(13), 8718–8735 (2023)
129. A. Hajı, M. Vadood, M. Öztürk, İ Yiğit, S. Eren, H.A. Eren, Prediction of colour strength in environmentally-friendly dyeing of polyester fabric with madder using supercritical carbon dioxide. Rev. Prog. Colorat. Relat. Top. **141**(1), 44–53 (2024)

Chapter 7
2D and 3D Printing Technologies for Textile Surface Functionalization

A. M. Ehsanul Haque, Mustakimul Hossain Saeef, Md. Motakabbir Hasan, and Md. Rashedul Islam

Abstract The evolving demands of consumers have transformed textiles from conventional clothing materials into multifunctional systems. Surface functionalization plays a central role in enabling these tailored properties, and the emergence of two-dimensional (2D) printing and three-dimensional (3D) printing has greatly expanded the design space for such modifications. This chapter examines how contemporary 2D and 3D printing techniques are reshaping textile surface functionalization, reviewing major 2D methods such as screen, gravure, inkjet, and flexographic printing, as well as key 3D processes including vat photopolymerization, material jetting, binder jetting, and material extrusion. It discusses critical factors such as ink formulation, adhesion, material selection, and process optimization. The chapter illustrates how 2D printing can impart conductivity, protection, sensing capabilities, and integrated electronic features while preserving fabric flexibility, whereas 3D printing enables fully or partially printed textile structures and direct polymer deposition for applications in healthcare, sportswear, defense, automotive components, and filtration media. In addition, it addresses sustainability aspects, including the use of biopolymers, waste-reduction strategies, and circular design concepts, and outlines current challenges and future research opportunities for advancing the functionalization of printed textiles.

A. M. Ehsanul Haque
Department of Textile Machinery Design and Maintenance, Bangladesh University of Textiles, Dhaka, Bangladesh
e-mail: ehsan@tmdm.butex.edu.bd

M. H. Saeef
Department of Textile Materials Engineering, Bangladesh University of Textiles, Dhaka, Bangladesh
e-mail: mustakim.saeef@tme.butex.edu.bd

Md. M. Hasan · Md. R. Islam (✉)
Department of Wet Process Engineering, Bangladesh University of Textiles, Dhaka, Bangladesh
e-mail: rashedul.islam@wpe.butex.edu.bd

Md. M. Hasan
e-mail: mmhasan@wpe.butex.edu.bd

M. N. Morshed and M. Shahid (eds.), *Advances in Surface Modification and Functionalization in Textiles*, Textile Science and Clothing Technology,
https://doi.org/10.1007/978-3-032-23377-6_7

Keywords Textile printing · Surface modification · 2D printing · 3D printing · Printed textiles

7.1 Introduction

Surface functionalization of textiles has become an important area of focus in research and industrial practice due to changing consumer demand and technological progress. Textiles are no longer seen solely as traditional protective and aesthetic materials; modern applications require fabrics that are multifunctional, high-performing, and sustainable. Properties including hydrophobicity, antimicrobial activity, ultraviolet (UV) resistance, conductivity, flame retardancy, and enhanced dyeability are now regarded as essential functional attributes for innovative approaches across several sectors such as technical wear, healthcare, military, and smart fashion [1]. The achievement of these attributes mainly relies on the effectiveness of surface functionalization methods.

For generations, textile functionalization depended mostly on traditional wet chemical finishing techniques, which involved hazardous chemicals, energy and water-intensive approaches, and noxious emissions [2]. Moreover, conventional methodologies often do not provide accurate, controlled material deposition, thereby limiting the ability to achieve highly localized or customizable functionalities [3]. These constraints have inspired further investigations into more sophisticated alternatives to offer sustainable and efficient solutions. Printing technologies, in this regard, are becoming the paradigms of a new textile engineering era. Such techniques are used to selectively add materials onto the surface of fibers or fabrics, enabling decorative effects as well as more advanced functionalities. 2D printing, including spray, screen, gravure, and inkjet printing, facilitates by providing opportunities for high-resolution surface patterning as well as integration of conductive pathways and application of functional inks or nanoparticles. In contrast, 3D printing further widens the scope by manufacturing three-dimensional complex architectures with increased surface area that can additionally provide mechanical reinforcement, stiffening properties, and smart interactions [4].

The primary objective of this chapter is to delineate two key technologies: 2D and 3D printing techniques dedicated to textile surface functionalization. This begins with the basics of surface functionalization, along with the role of printing for surface functionalization of textiles. We further categorize various printing technologies, their basic mechanisms, and materials utilized in both 2D and 3D textile printing techniques. We also explore the applications of such printing techniques in various functional applications. Finally, we conclude with the sustainability challenges and future research directions, indicating how these technologies could transform the textile industry toward greater customizability, versatility, and sustainability.

7.2 Surface Functionalization of Textiles Through Printing

Textiles are present in almost every aspect of daily life and are now expected to provide protection, responsiveness, and even connectivity, rather than only aesthetics. Many of these advanced functions can be introduced without redesigning the fiber bulk; instead, they arise from modifying the outermost layer where the material interacts with the environment, a process known as surface functionalization [5]. Although this surface represents only a small fraction of the textile volume, tailored modification can markedly alter fabric behavior, enabling water repellency, antimicrobial activity, electrical conductivity, or responsiveness to heat and light while maintaining softness, breathability, and comfort [5–7].

Originally used mainly for color and patterns, textile printing has evolved into a versatile surface engineering tool that can deposit functional entities such as conductive particles, antimicrobial agents, UV blocking compounds, shape-changing inks, and soft electronic components [8, 9]. A key advantage of such printing-based functionalization is spatial and quantitative control: instead of treating the entire fabric with bulk chemicals, printing enables selective modification of defined regions with controlled dosage and high edge definition [10, 11]. As textiles become more technical and multifunctional, printing offers micro-scale accuracy, flexibility, and compatibility with a wide range of functional inks, making it a strategic route for implementing advanced surface functionalities in textile products.

7.3 Printing Technologies and Their Classification

As surface functionalization (SF) has become a key approach for turning everyday textiles into high performance, smart, and responsive materials, both 2D and 3D printing play pivotal roles in transforming conventional textiles into such materials. These processes provide digital control over the location, amount, and architecture of deposited materials, enabling customized structures, integrated responsive features, and fabrics that adapt to user or environmental demands [12]. They also support rapid pattern updates, on-demand or small batch production, and reduced material waste, which are advantageous for sustainable smart textile manufacturing [11, 13].

Though textile printing has continuously evolved from conventional methods [14], based on the dimensionality of material deposition and structure formation, textile printing can be broadly categorized into 2D and 3D approaches [15]. Two-dimensional printing includes conventional contact-based methods such as block, roller, screen (flat bed and rotary), and gravure printing, as well as digital, non-contact methods such as inkjet (continuous and drop on demand), flexographic, and offset (lithographic) printing [16, 17]. These techniques enable planar deposition of functional inks, coatings, and pigments, allowing, for example, conductive tracks, antimicrobial or UV blocking layers, and thermochromic motifs to be placed only where needed while preserving fabric drape and breathability [11, 18].

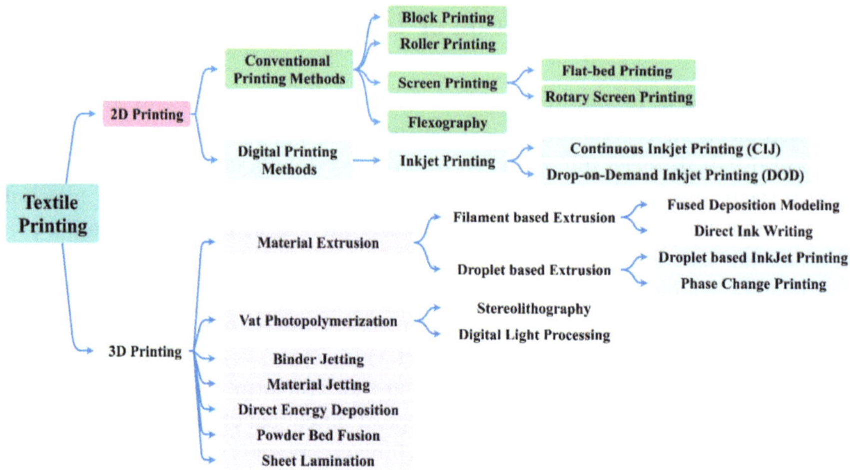

Fig. 7.1 Classification of printing technologies

Three-dimensional printing or additive manufacturing encompasses processes such as fused deposition modeling (FDM), stereolithography (SLA), selective laser sintering (SLS), powder liquid 3D printing (PLP), digital light processing (DLP), and electron beam melting (EBM) [19]. When applied to textiles, these methods build layer-by-layer structures directly on or within fabrics, forming cushioning lattices, protective ribs, tactile sensors, flexible meshes, and hybrid surfaces that enhance mechanical performance and enable interactive or adaptive functionality [20]. Together, 2D and 3D printing thus provide a complementary toolbox for precise, multifunctional textile surface engineering [11, 21–23] (Fig. 7.1).

7.4 2D Printing Technology for Surface Functionalization

7.4.1 *Introduction to 2D Printing*

2D printing is considered one of the modern era's most dynamic and transformative surface modification technologies in the textile sector. Its goal is to apply inks, pastes, or functional materials over textile substrates in a precisely controlled planar manner [24]. Unlike most of the conventional methods of finishing or coating, which primarily rely on techniques like immersion, padding, or exhaustion that treat the entire bulk of the textile, 2D printing is highly localized to target specific regions or patterns on the substrate [25]. Such high precision and localized application allow the fabrication of detailed designs, conductive traces, and functional coatings without compromising the textile's natural flexibility or breathability. As a result, 2D printing

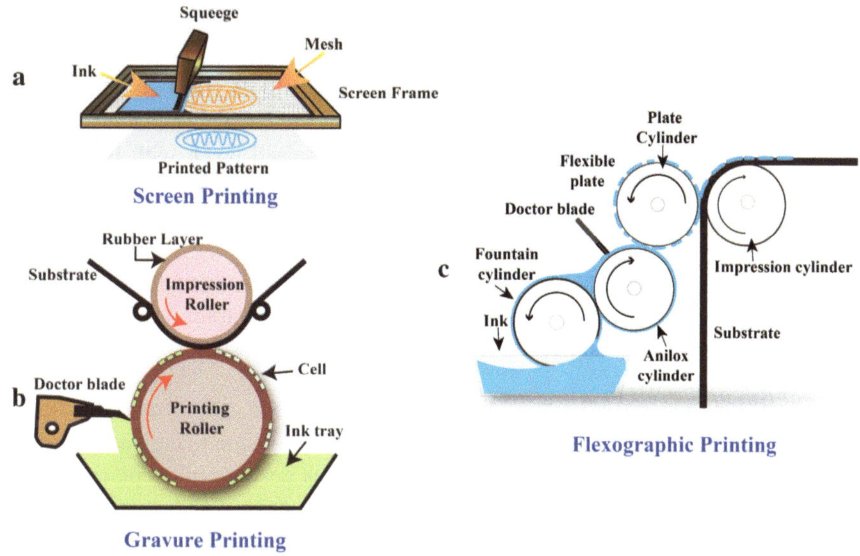

Fig. 7.2 Printing techniques **a** screen printing, **b** gravure printing, **c** flexographic printing [26]

has become a major facilitator of smart and functional textile development, especially for uses like wearable electronics, sensors, and responsive surfaces [27, 28]. 2D printing can be further classified into two major types (Fig. 7.2):

(a) Conventional Printing Methods: which include the traditional, contact-based printing techniques such as block printing, roller printing, screen printing (flat bed and rotary), gravure printing, flexography, and offset printing (also known as lithography).
(b) Digital Printing Methods: These represent the digital and non-contact printing mechanisms like inkjet printing (both continuous and drop-on-demand modes) [29, 30].

7.4.2 *Methods of 2D Printing*

The field of 2D printing consists of several established and emerging methods. These methods differ from each other in their mechanism, material compatibility, and production capacity. It is essential to comprehend these methods or techniques in order to pick the optimal procedure for the intended functional result and application.

7.4.2.1 Screen Printing: Thick Film Functionalization

Screen printing (also known as silk screen printing) is still the most established and widely used method for thick film functionalization in textiles. This contact-based process presses high-viscosity pastes through a patterned mesh with the help of a squeegee. It can apply larger material volumes, often from tens to hundreds of micrometers thick layers in a single pass. This level of mass loading can hardly be achieved by any other digital printing methods [31, 32]. So, this particular ability makes it more suitable for high electrical conductivity and applications where substantial volumes of active material are needed. For example: printed batteries, wearable heating elements, and other high-performance e-textile components.

Mechanism of Screen Printing

In screen printing, a polyester mesh is coated with an emulsion to create a stencil in which only the image areas remain open [33]. A thick paste or ink is placed on the screen, and a rubber or polyurethane squeegee is drawn across it with controlled pressure and speed, forcing ink through the open mesh onto the textile. As the squeegee moves along, the tensioned mesh briefly contacts the substrate and then "snaps off," leaving a defined layer of ink on the fabric surface.

Control Parameters and Rheological Mechanism

The quality and thickness of the deposited ink layer are primarily governed by the mesh count (threads per inch) and thread diameter, alongside squeegee parameters such as angle, hardness, pressure, and snap off distance. This precise control over deposition parameters highlights the primary scientific advantage of screen printing over digital methods like inkjet printing: the ability to achieve high mass loading through exceptional rheological versatility.

From a functionalization perspective, the process relies on the non-Newtonian, thixotropic behavior of the print pastes. Unlike low viscosity inkjet fluids (<20 mPa·s), screen printing pastes exhibit high initial viscosities ranging from 500 to over 50,000 mPa·s [34]. The process utilizes pseudoplastic (shear thinning) behavior: when the squeegee applies shear stress, the viscosity temporarily drops, allowing the paste to flow through the mesh openings. Crucially, the paste then exhibits thixotropy (structural recovery), where viscosity instantly recovers once the shear stress is removed. This rapid recovery prevents the paste from bleeding into the porous yarn structure, ensuring sharp design edges and maintaining the functional layer on the fabric surface rather than losing it within the fiber matrix (Fig. 7.3).

Functional Capabilities: Thickness and Durability

This rheological adaptability facilitates the deposition of thick layers (typically 80–100 μm) in a single pass, significantly exceeding the sub-micrometer (sub-μm) deposits typical of inkjet printing. This capacity for thick-layer deposition allows for the use of high-solid-content pastes containing metal powders, carbon materials, ceramic fillers, or conductive polymers [35]. Such high mass loading is essential for attaining specific functional goals, including high-current-conducting

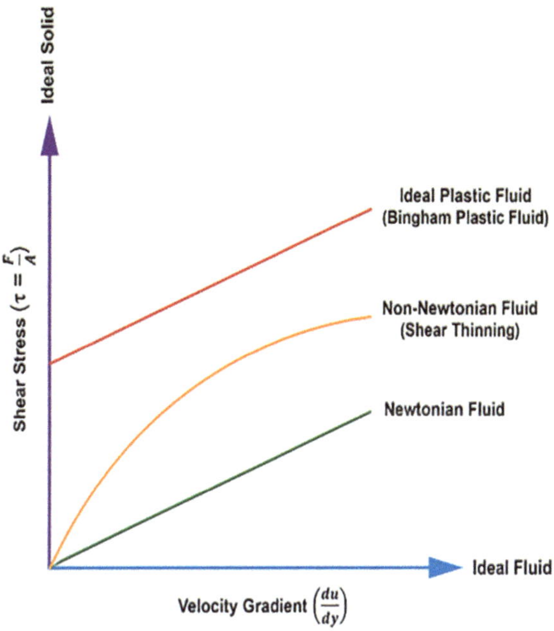

Fig. 7.3 Rheological flow curves illustrating the relationship between shear stress (τ) and velocity gradient (du/dy) for Newtonian, non-Newtonian (shear thinning) and Bingham plastic fluids compared to ideal solids and fluids [36, 37]

tracks, energy-storage electrodes, and robust heating elements [32, 38]. Furthermore, these thick layers provide the mechanical robustness, excellent adhesion, and physical strength required for wearable e-textiles to withstand repeated bending and laundering [26, 39].

7.4.2.2 Gravure Printing: High-Speed Roll to Roll Surface Functionalization

Gravure printing (intaglio or rotogravure) is a renowned, high-speed roll-to-roll (R2R) printing method that is famous for its accuracy and reliability in continuous large-scale production. The process uses engraved cells, which meter inks with remarkable consistency. This makes gravure printing essential for applying coatings and designs over extensive areas of flexible textiles or foil materials [40].

Mechanism: Volumetric Metering and Ink Transfer

Gravure printing is a direct contact process that uses an engraved cylinder with recessed image areas to transfer ink at high speed. The metal gravure cylinder is covered with millions of etched or laser-engraved cells of defined depth and size, each acting as a miniature ink reservoir. As the cylinder rotates through an ink fountain, these cells fill with ink, after which a flexible doctor blade wipes excess ink from the land areas, leaving ink only inside the recesses. An impression roller then presses the ink-filled cylinder against the textile (or an intermediate blanket), drawing ink out of the cells and onto the fabric surface through pressure and capillary action. Because each cell has a fixed volume, gravure can deliver highly consistent ink transfer with every rotation, which underpins its precision and suitability for high-speed production [41].

7.4.2.3 Flexographic Printing: Resilient Relief Transfer for Textured Surfaces

Flexographic printing combines the speed of rotary printing processing with the flexibility or adaptability of relief printing. Unlike gravure printing, which uses rigid metallic cylinders, flexography uses flexible photopolymer plates [42, 43]. This basic difference makes it suitable for textile surface functionalization. The mechanical flexibility of the printing plate is crucial, since the plate can easily follow the natural uneven surfaces of woven or knitted fabrics. This ensures continuous film formation where rigid methods might struggle.

Mechanism: Anilox Metering Coupled with Plate Deformation

The effectiveness of flexographic printing mainly relies on the interaction between the anilox roller and the flexible relief plate. The anilox roller is a ceramic cylinder engraved with millions of microscopic cells, each holding a defined volume of ink; these cells pick up a precise amount of functional ink and deliver it to the raised areas of the printing plate, ensuring consistent ink transfer even at high speeds [44]. The photopolymer relief plate then contacts the textile and undergoes slight elastic deformation, allowing ink to reach both the "peaks" and part of the "valleys" of the fabric structure under low "kiss" pressure, which minimizes the risk of crushing high-loft fabrics or damaging sensitive textile products [45, 46].

7.4.2.4 Inkjet Printing

Inkjet printing technology has evolved from being a simple graphic reproduction tool to a sophisticated digital micro-dispensing system for textile surface functionalization. Unlike traditional wet processing methods like padding or coating, it allows for precise dispensing of functional ink droplets that range from picoliter level 10^{-12} to nanoliter level 10^{-9} onto textile substrates. Inkjet printing also supports higher resolution patterning, usually exceeding 600 DPI (dots per inch), with some systems achieving up to 1200 DPI [47, 48]. These capabilities make it possible to apply functional fluids in specific areas, including conductive nanoparticle suspensions, biological agents, or hydrophobic coatings.

Principles of Inkjet Printing

The inkjet printing process operates mainly on two principles: continuous inkjet (CIJ) and drop on demand (DOD). In continuous inkjet (CIJ) printing, a stream of ink continuously breaks up into droplets due to the Rayleigh Plateau instability in

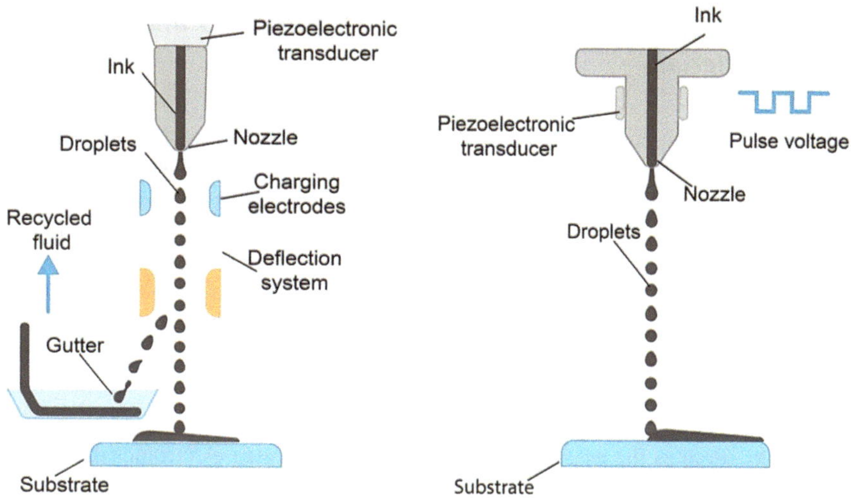

Fig. 7.4 Schematic showing the mechanism of continuous and drop-on-demand inkjet printing [26]

the jet. These small droplets are sorted, where a small fraction is steered toward the substrate, and the remaining are recycled. In the case of drop-on-demand (DOD) printing, droplets are generated only when required, thus optimizing material usage and enabling dynamic pattern control. The DOD mechanism functions through the fast and exact transformation of electrical signals into a mechanical pulse that produces a single droplet ejection from the nozzle [49, 50] (Fig. 7.4).

Generally, this is achieved by two primary actuation systems: piezoelectric actuation and thermal (or bubble jet) actuation. The piezoelectric actuation system consists of piezoelectric materials (such as lead zirconate titanate or PZT) that are placed either around the ink reservoir or embedded within the diaphragm of the inkjet printhead. When a voltage pulse is applied to such materials, due to the inverse piezoelectric effect, they undergo rapid expansion or contraction in response to the applied electric field. This produces a pressure pulse to push out a specific amount of ink through the nozzle orifice. The advantage of piezoelectric systems is their versatility with a wide range of solvent and nanoparticle-based inks, as they do not require heating the fluid [51–53].

Alternatively, thermal actuation functions through the operation of a high-current pulse which activates a small heating element (resistor) located within the ink channel. The ink vehicle produces an instant vapor bubble through the intense heat, which reaches up to 300 °C in a brief time span. The bubble grows quickly, which produces a sudden high-pressure force to release a droplet. The bubble then breaks down as the channel fills up again through capillary action. Thermal systems that operate at lower costs and with less complexity function only with solvents that have low boiling points, yet they damage the structure of delicate functional materials [49, 54].

7.4.3 Formulation and Performance of Printable Functional Materials

The success of 2D-printed textiles really hinges on the design and performance of the specialized inks used. These formulations need to strike a balance between two often conflicting goals: they need to be printable (which means they have to have the right flow properties for the printing method like low viscosity for inkjet printing or high viscosity for screen printing) and they also need to be effective in their function (delivering the desired electrical, chemical, or biological properties).

7.4.3.1 Conductive Inks: The Foundation of Smart Textiles

These inks serve as the essential building blocks for printed electronics on fabrics, allowing for the development of conductive pathways that facilitate circuits, power transmission, and sensing elements. One of the main hurdles is achieving excellent electrical conductivity while also ensuring flexibility and washability of the textiles. To overcome this, researchers have used several compositions. Metallic nanoparticles such as silver (Ag) and copper (Cu) nanoparticles [55], along with carbon-based materials like carbon nanotubes (CNTs) and graphene [56], have shown promise in achieving excellent conductivity without sacrificing the flexibility and washability of textiles [57–59]. Conductive inks are mainly created as stable mixtures, featuring highly conductive fillers that are evenly distributed within a binder system. The binder helps them stick to the fabric, while the solvent or vehicle manages the flow properties. After printing, a curing or sintering process is usually necessary to eliminate the solvent and binder, solidifying the conductive network.

7.4.3.2 Nanoparticles and Hybrid Composites: Multifunctional Performance

Multifunctional performance has become a highly desired quality parameter in recent years. The introduction of various types of nanostructures and hybrid composites into ink formulations is helping turn this desire into reality. **Inorganic Nanoparticles**: These materials exploit their high surface area and unique quantum effects at the nanoscale [60, 61]. **Titanium Dioxide (TiO_2) and Zinc Oxide (ZnO)**: These are two semiconductor materials that shine when it comes to broad-spectrum UV protection, thanks to their strong absorption in the UV range. Additionally, they exhibit photocatalytic properties when exposed to UV light, which can help with self-cleaning and break down air pollution when applied to fabrics [62]. **Silver (Ag) and Copper (Cu) Nanoparticles**: Beyond their electrical applications, these metals have a remarkably high surface area-to-volume ratio, allowing for the gradual release of metal ions (Ag^+, Cu^{2+}). This feature gives them powerful antimicrobial and antifungal properties, making them ideal for medical, hygiene and military textiles [63]. **Hybrid**

Composites: These advanced formulations intentionally blend multiple materials to achieve multifunctionality within a single printed layer. For instance, combining a conductive filler like silver nanowires with a flame-retardant inorganic particle such as nano clay or a thermochromic pigment can create a circuit that is both antimicrobial and fire-resistant. This approach maximizes functional integration while minimizing the number of printing steps [64, 65].

7.4.3.3 Sustainable and Biodegradable Materials: Eco-friendly Functionalization

There's a growing movement to reduce the environmental footprint of textile functionalization, encouraging the use of green, biodegradable printable materials.

- **Biopolymer Based Inks**: These inks come from natural, renewable sources like cellulose, chitin, and starch. They act as binders or functional matrices that encapsulate active components while reducing our dependence on synthetic polymers [66].
- **Natural Pigments and Dyes**: By using pigments derived from plants, algae, or microbes, the use of synthetic colorants can be reduced [67].
- **Bio-sourced Binders and Additives**: Protein-based polymers and natural gums enhance wash fastness and durability while still being biodegradable, tackling the issue of microplastic pollution that comes from traditional textile finishes.

7.4.4 Process Parameters and Optimization of Functional Performance

To achieve greater functional efficiency and long-term durability in 2D printed textiles, a complex set of process parameters must be optimized. These parameters are very important for determining the final properties, such as how well it conducts electricity, how well it washes, and how strong it is. Getting the right parameters is crucial to ensuring that the printed layers can withstand real-world stresses without losing functionality. This results in measurable metrics (Table 7.1):

- **Durability and Wash Cycles**: This is assessed by how much of the original functionality (e.g., electrical conductivity) is retained after multiple washing cycles, according to standards such as AATCC or ISO. This presents a significant challenge for bringing wearable electronics to market.
- **Mechanical Resilience**: This refers to how well the printed track can endure bending, stretching and abrasion without failing either electrically or physically. Performance is typically evaluated by the $\left(\frac{R}{R_0}\right)$ (the change in resistance compared to the initial resistance) when under strain [68].

Table 7.1 Process parameters and performance metrics

Process parameter	Typical range/condition	Impact on functional performance	References
Ink viscosity	1–20 or typically lower than 100 mPa·s (inkjet), higher ranges up to 1,000–100,000 mPa·s reported in some studies (screen)	Affects droplet formation, prints uniformity and prevents bleeding	[69, 70]
Surface tension	Common inks have 25–50 mN/m	Influences wetting, adhesion, and spreading	[71]
Substrate wettability	Contact angle <90°	Ensures even ink spreading and strong adhesion	[72]
Curing temperature	60–200 °C (depending on substrate)	Solvent removal, sintering, cross linking	[73]
Post treatment	Lamination, encapsulation, and chemical fixation	Improves wash fastness, mechanical resilience, durability	[74, 75]

- **Functional Efficiency**: This is the output measured from the printed property, such as the sheet resistance (Ω/sq) for conductive tracks, the UV protection factor (UPF) or the antimicrobial kill rate (percentage reduction).

Overall, 2D printing technologies offer a strong, scalable and flexible platform for the next wave of textile functionalization. Their ability to integrate with digital production processes and a variety of material chemistries makes them vital for creating sustainable, highly customizable, and multifunctional smart fabrics.

7.4.5 Applications of 2D-Printed Functional Textiles

2D printing (including screen printing, rotary printing, gravure, flexographic, and inkjet) supports a variety of applications, including fashion, healthcare, industrial systems, and on-demand manufacturing.

7.4.5.1 Smart and Functional Textiles for Wearables and Healthcare

2D printing is an indispensable method for integrating electronics and responsive materials into flexible textile substrates. Accurate deposition of conductive, sensing, or semi-functional layers enables scalable, cost-effective production of wearable devices.

7.4.5.1.1 Printed Flexible Electronics and Sensors

Functional inks such as silver nanoparticles, carbon nanotubes (CNTs), graphene, and conductive polymers (e.g., PEDOT: PSS) can be printed via screen printing, inkjet printing, and gravure printing. Also, the deposition of nanomaterials and sensitive bio-functional inks assists in the creation of small sensors for real-time monitoring of mechanical strain, temperature, humidity, biomedical markers (e.g., enzyme-based glucose detection) or antibody-doped inks (e.g., disease diagnosis) [76–78]. The flexibility of textile fabrics, along with the mechanical compliance of printed conductive patterns, makes them suitable for motion tracking, athletic monitoring, and rehabilitation systems [79].

7.4.5.1.2 Chromic and Responsive Surfaces

Functional 2D printing can also utilize a range of chromic materials that change color in response to external stimuli (e.g., temperature, light, electric current, mechanical force) [80]. Thermochromic inks react to heat, while photochromic inks respond to UV light. These chromic colorants are printed on garments to create visual effects or provide practical feedback, like signaling changes in body temperature or exposure to sunlight. Hydrochromic inks become transparent when they come into contact with water [81]. They are applied in swimwear and umbrella designs to reveal hidden patterns.

7.4.5.1.3 Healthcare and Biomedical Monitoring

Moreover, 2D printing is widely used to create sturdy, comfortable wearable electrodes for electrocardiography (ECG) and electromyography (EMG) by printing with carbon or silver/silver chloride (Ag/AgCl) inks [32, 82]. This technology also extends to biosensing layers that detect sweat metabolites, pH, or electrolytes, as well as moisture-responsive printing for wound healing management and smart bandages.

7.4.5.1.4 Printed Drug Delivery and Therapeutic Patches

Methods like screen or gravure printing are capable of depositing microdoses of therapeutic molecules or polymer drug mixes onto nonwovens or medical textiles, which allow controlled transdermal release [83].

7.4.5.2 Industrial Applications: Filtration, Composites, Energy, and Automotive

In the case of industrial applications, 2D printing provides a way to apply high performance finishes to the exact locations or zones of the textile substrate.

7.4.5.2.1 Super Hydrophobicity and Self-Cleaning

The principles of "Lotus Effect" are used in spraying printing methods to deposit hydrophobic nanoparticles, like silica or fluorocarbons, to create rough water repellent surfaces. This rough surface lets the water droplets to roll off rather than spreading. As a result, it develops a self-cleaning property where water droplets carry dirt particles away [84].

7.4.5.2.2 Camouflage and Defense

The defense industry depends on rotary and digital printing (inkjet) to produce advanced multispectral camouflage [85, 86]. Furthermore, these fabrics are printed with dyes and finishes that change near infrared (NIR) reflectance, making soldiers harder to see with night vision gear. Such digital printing helps to quickly change camouflage patterns to match specific terrains, such as urban or jungle, without the delay of engraving new screens [87].

7.4.5.2.3 Automotive and Aerospace

Screen printing is used in vehicle interiors to apply anti-slip dots, abrasion-resistant coatings, and localized flame retardants to seat covers and carpets. Besides, phase change materials (PCMs) can be printed onto headliners to help in regulating cabin temperature [88].

7.5 3D Printing Technology

7.5.1 Introduction of 3D Printing in Textiles

3D printing, also known as additive manufacturing, is changing the field of textile engineering by making it possible to make complicated, customized and multifunctional textile structures that are hard to make with traditional methods [89–91]. 3D printing, on the other hand, makes materials layer by layer directly from digital designs. This is different from traditional methods like weaving or knitting, which rely on interlacing fibers in set patterns. This process of building up layers gives designers and researchers more freedom than ever before, letting them make fabrics with complex shapes, better performance, and built-in features that are just right for their needs [89, 90].

The coming together of 3D printing and textiles is more than just a change in how things are made; it's a creative revolution. It lets textiles change from being passive materials to interactive, adaptive and smart systems that can react to changes in the environment or to what the user does. 3D printing is changing the way we use textiles by letting us make avant-garde fashion pieces, medical supports, and high performance, eco-friendly fabrics. It's where art meets engineering and imagination meets innovation. The main reasons for using 3D printing in textiles are:

- Freedom of making any type of lattice, chain mail, and biomimetic shapes [89, 92].
- Customization and quick prototyping for personalized items [89, 93]
- Combining features like sensing, conductivity, and shape adaptation [23, 26].
- Creation of textile hybrid composites that combine strength with flexibility. [89, 94].

7.5.2 *Methods of 3D Printing*

There are seven categories of AM (3D printing), according to ISO/ASTM 52900:2021 are: material extrusion, binder jetting, directed energy deposition, material jetting, powder bed fusion, sheet lamination and vat polymerization (ISO/ASTM 52900:2021) [91].

7.5.2.1 Material Extrusion

Material extrusion is the process of pushing something through a nozzle in a definite 3D pattern. Material extrusion process is mainly categorized into two types. Fused filament fabrication is the mostly used method in between the two.

Fused Filament Fabrication (FFF)/Fused Deposition Modeling (FDM): FFF is the most popular method because it is cheap and works with flexible polymers like PLA and TPU. In this method, a heated filament is extruded layer by layer, either as a separate piece or directly onto a textile substrate. FFF makes it possible to make hybrid composites that combine the strength of polymers with the softness of fabrics. The quality of the print and the strength of the bond depend mostly on the thickness of the layers, the speed of the print and the temperature of the extrusion [91].

7.5.2.2 Powder Bed Fusion (PBF)

Powder bed fusion (PBF) is also known as selective laser sintering (SLS). It uses a high energy laser to sinter powdered materials, usually nylon, to make complicated 3D shapes. This powder bed fusion method makes strong, flexible fabrics with

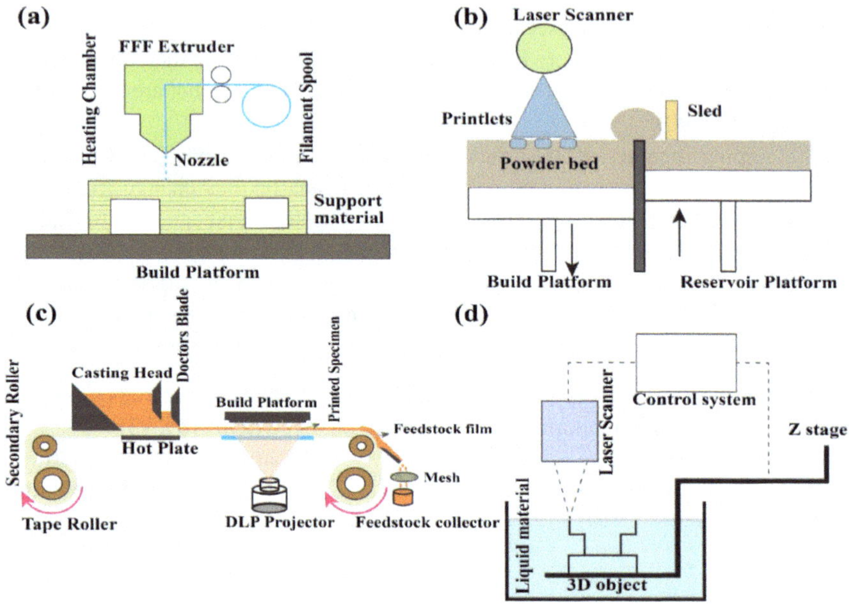

Fig. 7.5 **a** Fused filament fabrication process [95], **b** selective laser sintering (SLS) process [96]. **c** Digital light processing [97], **d** stereolithography (SLA) process [98]

detailed shapes that can be used in both fashion and technical settings. To get the best quality, the process needs to be finished with steps like removing the powder and finishing the surface [91].

7.5.2.3 Vat Photopolymerization (VPP)

7.5.2.3.1 Stereolithography (SLA)

SLA uses ultraviolet light to cure liquid photopolymer resin layer by layer to make high-resolution textile elements. Resin materials usually make things less elastic, but SLA is great at making smooth, detailed surfaces that are perfect for decorative pieces or sensor embedded parts that need to be precise (Fig. 7.5).

7.5.2.3.2 Digital Light Processing (DLP)

Digital Light Processing (DLP) is a high-resolution photopolymerization technique that employs a digital projector to flash entire layers of photosensitive resin simultaneously, enabling the rapid and precise fabrication of complex 3D structures. While traditional vat-based DLP systems often struggle with highly viscous composite

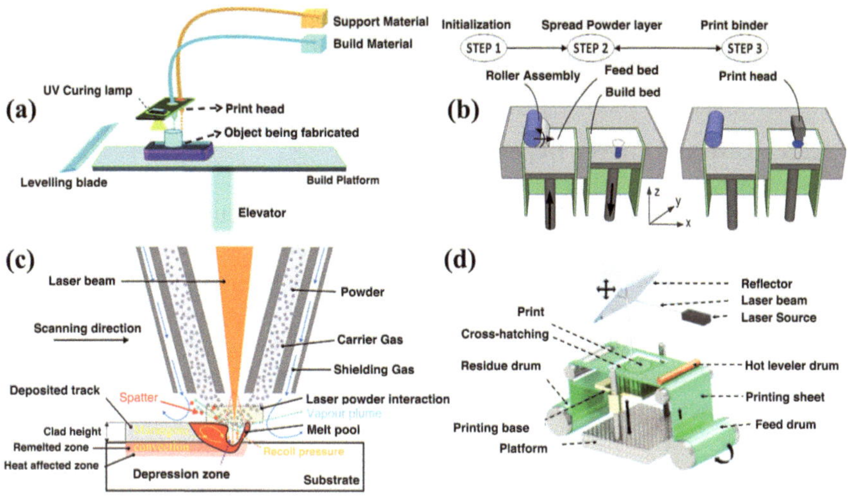

Fig. 7.6 Schematic of **a** material jetting [99]. **b** Binder jetting [100]. **c** Direct energy deposition [101]. **d** Sheet lamination [102]

resins and generate non-recyclable material waste, recent advances focus on optimizing interfacial dynamics at the liquid–solid boundary to improve printing speed, precision and material diversity. A notable interfacial strategy is the integration of tape-casting mechanisms to create a continuous film supply (CFS) [97].

7.5.2.4 Material Jetting

Material jetting is an advanced additive manufacturing technique in which tiny droplets of photopolymer resin are selectively deposited through multiple nozzles and immediately cured with UV light. Because droplets can be placed with micrometer-level precision. The process enables exceptionally smooth surfaces, fine details, and accurate geometries. It also allows simultaneous printing of different materials and full color parts, making it ideal for prototypes requiring realistic textures, transparency, or soft-hard material combinations. Recent developments focus on enhanced droplet control, improved printhead design, and expanding multi-material capabilities [99] (Fig. 7.6).

7.5.2.5 Binder Jetting

Binder jetting (BJ) is a powder bed 3D printing technique where a liquid binder is jetted onto a bed of powder (e.g., metal or ceramic) created layer by layer to bind particles into a green part. Post processing like sintering or infiltration; consolidates the part, reducing porosity and increasing strength. BJ is fast which allows large build volumes and doesn't need support structures but parts may shrink and change dimensions during heat treatment [100].

7.5.2.6 Directed Energy Deposition

Directed energy deposition (DED) is a metal additive manufacturing process in which a focused energy source such as a laser, electron beam, or plasma/arc melts feedstock (wire or powder) as it is deposited creating a molten pool that solidifies into a 3D structure. Over recent years, DED has been especially valued for its ability to repair components, build large and complex parts and deposit dissimilar materials. However, challenges include defect formation (e.g., porosity, lack of fusion) and the need for advanced monitoring and control to ensure part quality [101].

7.5.2.7 Sheet Lamination

Sheet lamination (SL) or laminated object manufacturing (LOM) builds 3D parts by bonding sheets layer by layer. In this case, low carbon steel is used. The study shows that welding motion during lamination affects density and tensile strength with optimized movement improving mechanical properties. SL operates at relatively low temperatures reducing energy use and minimizing porosity. Challenges remain in ensuring consistent layer bonding and avoiding thermal distortions. This method is suitable for structural components with moderate strength requirements [103].

New technologies like material jetting and direct ink writing are opening up even more possibilities by allowing multi-material deposition and the use of functional inks. These are important steps forward for smart textiles and wearable electronics [91].

7.5.3 Adhesion Mechanisms of 3D Printing

The quality of the bond between the polymer and the fabric is the most important factor in how well 3D printed textile composites work. There are a few different ways that this bonding works:

- **Mechanical Interlocking**: When the polymer melts cools, it gets stuck in the fabric pores and yarn interstices. The structure of knitted fabrics with open loops makes this effect even stronger. This gives it a lot of adhesion strength because of the friction and energy loss that happens when it breaks [104, 105].
- **Chemical Bonding**: Using coupling agents like silanes makes the surfaces of polymers and fibers more chemically compatible with each other. These treatments make it less likely for moisture to get in and make the interface properties better, which makes adhesion better [105–107].

- **Thermal Bonding**: When thermoplastic fibers melt in one place at the interface, they can fuse with the polymer layers that have already been deposited, making a strong bond. It is very important to control the temperature of the printing and the flow of the polymer for this fusion process [106, 108].
- **Interfacial Diffusion**: Interfacial diffusion serves as a specialized bonding mechanism, where polymer chains migrate and entangle across the contact point between 3D-printed materials and textile fibers. Unlike macroscopic thermal fusion, (which essentially joins materials at the surface) this process operates at a molecular level. It is driven by the mobility and intermingling of the chains themselves. Recent overviews of adhesion in woven fabrics emphasize that the resulting bond strength is fundamentally tied to the depth of this molecular integration [106]. As a result, finding the right balance of processing parameters like nozzle temperature and polymer flow are crucial for encouraging chain intermixing while shielding the fabric's structure from thermal damage [108].

7.5.3.1 Optimization of Adhesion of 3D Printing

Textile properties, polymer chemistry, printing parameters, and surface treatments all work together to control these adhesion mechanisms. To achieve durable and reliable polymer textile adhesion, several factors must be optimized.

- **Surface Pretreatment**: To prepare the fabric, researchers often turn to chemical functionalization or alkaline surface modifications [109] to strip away impurities and boost the substrate's reactivity. Other practical steps, such as plasma activation or even a thorough washing, help clear off surface contaminants so the fabric is easier to wet. Plasma treatment is particularly favoured because it raises surface energy which helps to make mechanical anchoring much more effective & without compromising the fiber's internal strength [110]. Beyond these physical treatments, applying specialized polymer coatings or adhesion promoters remains a reliable strategy for reinforcing the bond at the polymer-textile interface [110].
- **Material Compatibility**: Synthetic fibers like polyester and nylon bond better with thermoplastics because they have a stronger chemical affinity. Natural fibers like cotton, on the other hand, may need to be modified on the surface (for example, with plasma or chemical treatment) to improve adhesion because they are hydrophilic and porous [111–113].
- **Printing Parameter Optimization**: It's very important to be able to control the temperature of the nozzle, the speed of the print, the rate of extrusion, and the thickness of the layer. For example, the distance between the nozzle and the textile (z distance) needs to be set just right so that the polymer can get through without blocking the nozzle. Higher printing and bed temperatures, within the limits of the material, improve bonding by making polymers flow and fusing interfaces [110, 114].
- **Post Processing**: Methods like UV curing, thermal annealing, or hot pressing can help the polymer diffuse more deeply and stabilize the interface [115].

7.5.4 Process Parameters Shaping Mechanical Strength, Flexibility, Aesthetics, and Interactive Features

The overall performance of 3D printed textiles depends on more than just the choice of materials and the design of the structure. It also depends on the printing settings used during production. The final product's mechanical durability, flexibility, appearance, and interactive functions are all affected by things like temperature, layer thickness orientation, print speed, infill pattern, and post processing conditions.

7.5.4.1 Mechanical Strength

In applications like shoes, protective clothing, and orthopedic supports, mechanical strength is essential for maintaining dimensional stability, wash durability, and load-bearing capacity. To get the best mechanical strength out of 3D printed textiles, it is a must to carefully balance the processing parameters. The temperature of the nozzle, the thickness of the layers, the density of the infill, the direction of the build, and the rate of cooling all work together to determine the strength of the structure, the finish of the surface, and the bonding between the layers. Fine-tuning these variables guarantees strong adhesion between the polymer and textile, longer-lasting materials, and better mechanical performance for specific uses.

- **Nozzle Temperature**: The temperature of the printing nozzle significantly affects polymer flow, inter-layer adhesion and the bonding strength between the polymer and the textile substrate. While general printing temperatures for these filaments can range from 180 °C to 260 °C, achieving optimal mechanical properties requires material-specific tuning. For instance, to maximize the structural integrity and tensile strength of PLA, the recommended nozzle temperature interval is between 220 °C and 240 °C. On the other hand, flexible filaments like TPU typically require a controlled temperature range between 220 °C and 250 °C to ensure proper melting without excessive stringing. While higher temperatures generally improve layer adhesion and flexibility, they must be carefully controlled to prevent thermal damage or scorching to the underlying fabric [116, 117].
- **Layer Thickness**: The layer height directly influences the structural cohesion, vertical resolution and overall mechanical performance of Fused Deposition Modeling (FDM) printed components. While finer layer thicknesses (e.g., 0.1 mm) yield smoother surface finishes, superior vertical resolution and improved aesthetic quality, they simultaneously extend the overall print time. Furthermore, the relationship between layer thickness and mechanical strength depends heavily on the specific type of stress applied. Recent studies on FDM-printed PLA parts demonstrate that while finer layers enhance tensile strength due to improved inter-layer adhesion [121], thicker layers (e.g., 0.3 mm) can actually optimize impact resistance [122]. Therefore, layer thickness must be carefully calibrated based on the specific mechanical requirements of the final application.

- **Infill Density and Pattern**: The stiffness and weight of the internal structure are determined by the infill configuration. Dense infill patterns make things stronger and stiffer, while lower densities make things lighter and more flexible [120, 121].
- **Build Orientation**: The way the printing layers are oriented affects how strong they are in different directions. Vertical orientations often have lower tensile strength because the layers don't stick together as well, while horizontal orientations have higher strength. The print speed controls how fast the extrusion and movement happen during fabrication. Faster speeds speed up production but lower quality, while slower speeds usually make surface details better and layers stick together better [122, 123].

7.5.4.2 Flexibility

Flexibility is one of the most important things about textiles because it makes them comfortable to wear and strong enough to handle repeated bending, stretching, and washing. The choice of materials, the printing parameters and the structure of the substrate all work together to make 3D printed textiles flexible. Designers can make textiles that are soft and strong while still being comfortable and functional by choosing elastic polymers, optimizing layer configuration, aligning print direction with fabric stretch and using multi-material strategies.

- **Optimizing Polymer-Textile Composites**: The mechanical integrity of 3D-printed flexible polymers like TPU is highly dependent on fabrication parameters; specifically, lower print speeds and thinner layer heights significantly improve inter-layer adhesion and overall part hardness [124]. When integrating these polymers with fabrics to create advanced composites, the structural nature of the substrate is equally critical. Recent reviews highlight the immense potential of 3D woven and knitted spacer fabrics in technical textiles. Their unique three-dimensional, porous architecture provides an ideal foundation for extruded polymers to penetrate and anchor, resulting in highly resilient, flexible composites with enhanced structural performance [125].

7.5.4.3 Interactive Features

3D printing can make textiles that are both interactive and functional, which is an important part of making smart wearables. It has turned textiles into interactive platforms by adding sensors, materials that respond to stimuli and hybrid electronic integration. Smart textiles can now sense, respond and adapt in real time by precisely controlling materials, print parameters and design patterns. This combines functionality, comfort, and wearability to create the next generation of responsive, multifunctional fabrics.

- **Embedded Sensors**: 3D printing of conductive filaments with carbon black, graphene or silver nanoparticles directly onto textile substrates enables the fabrication of circuits, strain gauges or touch sensors. These built-in sensors make it possible to keep an eye on biomechanical activities and breathing in real time. Direct ink writing (DIW) 3D printing is often used to combine multiple layers of sensors while keeping the fabric flexible [11, 126, 127].
- **Improved Conductivity and Mechanical Optimization**: Achieving optimal mechanical performance in 3D-printed parts requires a comprehensive optimization of slicer settings. For demanding engineering plastics characterized by high melting points and large thermal expansion coefficients, research has demonstrated that structural integrity relies on the precise balance of six key variables: layer thickness, extrusion temperature, extrusion speed, filling speed, perimeter speed, and the speed of the first layer [128]. Beyond general mechanical stability, these fabrication parameters also critically influence specialized functional properties. For instance, the electrical conductivity of 3D-printed composites, such as PLA-Carbon Black, is highly sensitive to temperature and geometric settings. Conductivity is significantly enhanced by utilizing thinner printed layers and optimized nozzle temperatures, which minimizes internal voids and establishes robust continuous pathways for signal transmission [129].
- **Stimuli Responsive Materials**: Smart materials like hydrogels and shape memory polymers (SMPs) let textiles change shape in response to environmental triggers like heat, light, or moisture. The amount and direction of movement are determined by controlled print patterns and orientations. With 3D and 4D printing, complex functional textile structures can be made that can change shape in a precise way [130, 131].
- **Hybrid Integration**: By combining 3D printed structural parts with digitally printed electronic circuits, fabrics can be made that can sense, respond and protect at the same time. Textile hybrid electronics (THE) uses 3D printing to combine flexible coatings with rigid parts. This makes them work well while still being comfortable and breathable for wearable systems. Research has shown that it is possible to make printed fabric-based hybrid electric circuits that don't have the problems of being too big or hard to use. These circuits work perfectly with textiles, making wearable devices work better [127, 130, 132].

7.5.5 Applications of 3D Printed Functional Textiles

7.5.5.1 Fashion and Aesthetic Innovations

The fashion industry around the world is always looking for new technologies that can combine creativity, usefulness, and environmental friendliness.

7.5.5.1.1 Direct Integration of Functional Materials into Fashion

Functional printing materials also help fashion express itself. Thermochromic and photochromic inks let fabrics change color when the temperature or light changes. Conductive and metallic inks add glowing or reactive effects to the fabric. These materials make fashion more interactive, and eco-friendly inks made from natural or biodegradable polymers help the environment [133, 134]. 3D printed trims and accessories, like zippers, lace patterns, buttons, and jewelry, as part of the garment itself, can also be made. These printed features make things look better and last longer than regular trims that need separate stitching or adhesives [93, 135].

7.5.5.1.2 Virtual Fashion and Metaverse-Ready Digital Clothing

In the future, the combination of digital design, additive manufacturing, and eco-friendly materials is likely to change the way fashion is made. The growth of virtual fashion and metaverse-ready digital clothing suggests that physical and digital fashion will coexist in the future, thanks to the accuracy and flexibility of 3D printing. In the end, using printed functional textiles in fashion will lead not only to new ways to express the person artistically, but also to more responsible, personalized, and waste-efficient ways to make things [136].

7.5.5.2 Smart and Functional Textiles: Sensors, Wearables, Healthcare, and Protective Clothing

Printing technologies have enabled the field of smart and functional fabrics by allowing materials to sense, respond to, and change in response to their surroundings. Researchers have opened new possibilities in healthcare, wearable electronics, and protective clothing by printing conductive materials, responsive polymers, and multi-material composites directly onto flexible textile substrates. A recent study shows that printed electronics techniques such as inkjet, screen, gravure, offset, flexography, and even 3D printing can be used to add functionality to textiles [137].

7.5.5.2.1 Wearable Electronics and Printed Sensors

Printing methods such as inkjet, direct ink writing (DIW), and extrusion-based 3D printing enable the deposition of conductive inks, such as carbon black, graphene, silver nanoparticles, and conductive polymers like PEDOT: PSS, onto fabrics. These are sensors that measure strain, pressure, and temperature that are built into fabrics. Flexible conductive grids printed on elastic surfaces can be used in sports and rehabilitation to detect body movement. Printed electronics on textiles make lightweight, breathable, and flexible wearables that can power LEDs, microcontrollers, or wireless transmitters. They are also comfortable and easy to wash [138, 139].

7.5.5.2.2 Healthcare Applications

Additive manufacturing is rapidly advancing personalized healthcare through integrated sensors and customized treatments. Flexible, printed biosensors can be seamlessly integrated into garments for the continuous tracking of vital health metrics [140]. Simultaneously, 3D printing is transforming pharmaceutical applications by enabling highly controlled, patient-specific drug delivery systems [141]. For example, FDM technology is now used to fabricate personalized buccal patches for precise, unidirectional drug release inside the mouth [142]. Beyond drug delivery, wearable smart textiles feature printed electrodes capable of measuring ECG, EMG, and EEG signals without irritating gels, offering a reusable and highly comfortable solution for long-term patient monitoring [143].

7.5.5.2.3 Protective and Safety Clothing

The use of 3D printing and functional coating technologies in the military and industrial sectors has greatly improved the design of protective textiles. These methods allow for the addition of nanomaterials like graphene, carbon nanotubes, and metal oxide nanoparticles to fabric structures to give them important protective properties like flame resistance, UV resistance, and antimicrobial activity. Recent research has examined polymer-based nonwoven materials featuring printed conductive layers that offer efficient electromagnetic interference (EMI) shielding, underscoring the increasing significance of additive manufacturing in functional protective textiles [144]. Conductive coatings also protect against electromagnetic radiation and static discharge, making industrial and electronic work environments safer [145]. For military use, lightweight 3D printed fiber polymer composites are being made that are both flexible and resistant to impact. New research on graphene/Fe_3O_4 nanofibers shows that they are very good at blocking EMI, which opens up new possibilities for advanced defense textiles [146]. Also, protective clothing with built-in sensors that can detect dangerous gases or biological agents in real time is being made to make workers and soldiers safer and more aware of their surroundings [147].

7.5.5.3 Industrial Applications: Composites, Filtration, Automotive, and Energy Storage

3D printing technologies are changing the industrial textile sector in a big way. In this field, high performance, durability, and functionality are very important.

7.5.5.3.1 Printed Textile Composites

3D printing opens up exciting new ways to make traditional fabrics stronger without adding extra weight. Hybrid composites are stronger, more impact-resistant, and better at performing mechanical tasks than regular fabrics. They do this by combining carbon fiber textiles with printed polymer layers. This method is especially helpful for things like aerospace parts and high-performance sports gear, where both light weight and strength are important. With modern printing methods like co-extrusion and additive layering, precise bonding between fiber and polymer is possible, making sure the material is strong but still flexible enough for everyday use. Recent studies show that these hybrid composites can be designed to absorb energy while remaining strong, opening the door to the next generation of functional textiles [148, 149].

7.5.5.3.2 Filtration Applications

Functional inks and nanoparticle coatings enable fabrics to act as selective filters. Layers printed in 2D with antimicrobial nanoparticles make air filtration better [5]. 3D printing, along with functional inks and nanoparticle coatings, has made it possible for textiles to work as very effective selective filters. Researchers made 3D printed honeycomb filters out of recycled nylon and coated them with titanium dioxide (TiO_2) nanoparticles. In early tests, these filters removed up to 85% of greywater pollutants. With additive manufacturing, the size and shape of pores can be precisely controlled, improving the efficiency and selectivity of the filtration process [150]. 3D printed textiles coated with MXene, a 2D nanomaterial, have been shown to be about 90% effective at capturing nanoparticles in the 15–30 nm range for air filtration. With 3D printing, layer customization is possible as needed, improving both airflow and particle capture [151]. These advances show how 3D printing and functional coatings can work together to produce scalable, high-performance textile filters, useful for both industrial and environmental applications.

7.5.5.3.3 Automotive Sector

3D printing is increasingly used in automotive interiors to enhance the functionality and performance of fabrics. Direct printing of flame-retardant coatings and wear-resistant textures onto seat fabric is now a reality, which makes them safer and longer-lasting. Smart sensors built into cars via 3D printing can monitor the driver's posture and fatigue in real time. This lets the car make changes to its ergonomics to keep the driver comfortable and avoid accidents. Also, 3D printed polymer fiber composites are used in lightweight structural parts to make vehicles lighter and use less fuel without sacrificing safety or mechanical strength. Recent studies have shown that these new ideas work well to make cars safer and better at what they do [152, 153].

7.5.5.3.4 Energy Storage

Printed textiles are becoming increasingly important for flexible energy storage. For example, conductive inks are printed onto fabrics to make bendable batteries, supercapacitors, and energy harvesting layers for wearable electronics. A big step forward is 3D printed interdigitated electrodes, finger-like patterns that increase energy density and make it easier for ions to move. This 3D shape also makes the electrodes more flexible, making them ideal for wearable devices. Recent research shows that 3D printing provides very precise control over electrode design, enabling the creation of integrated, miniaturized energy-storage systems. Printed textiles are becoming lightweight, flexible energy-storage solutions that will power the future of wearable electronics. They are made by combining advanced conductive inks with 3D printed structures [154, 155].

7.6 Current Challenges and Future Outlook

Even though 2D and 3D printing are becoming more common in textiles, there are still a number of technical, economic, and practical problems that make it hard for businesses to use them on a large scale. To get consistent quality, strong adhesion, cost effectiveness, and environmental viability, these limitations must be overcome.

7.6.1 Production Challenges in 2D and 3D Printing on Textile Substrates

Despite rapid progress, printed textile technologies still face issues of scalability, speed, and cost.

7.6.1.1 Scalability and Production Speed

In functional 2D printing, continuous analog processes such as rotary screen, gravure, and flexography remain highly scalable, whereas digital methods like inkjet usually have lower throughput because of their "scan and print" motion [156]. Scaling up applications such as printed electronics is further complicated by the need to maintain very high registration accuracy on moving, stretchable, and "flexible substrates" or "flexible webs" [157]. On the 3D side, most systems used with textiles are suited primarily to prototypes or small batches; industrial deployment will require faster build rates, multi-nozzle or array heads, and continuous feed concepts that interface reliably with flexible substrates [89, 158].

7.6.1.2 Cost

Costs also remain significant. Functional 2D printing often relies on conductive inks containing silver nanoparticles, carbon nanotubes, or graphene, which are far more expensive than conventional dyes and pigments. Additionally, the maintenance of inkjet nozzles to prevent clogging from these valuable inks adds a substantial amount of operational costs, compared to standard colorant pastes [159]. For 3D printing, capital expenditure on industrial printers, specialized filaments, and ongoing maintenance, combined with relatively slow build speeds and long production times, leads to high per-part costs compared with established textile processes. Many studies, therefore, identify high material cost and low productivity as the main barriers to wider adoption of printed textiles at an industrial scale [160].

7.6.2 *Technical Challenges in 2D and 3D Printing on Textile Substrates*

Integrating printing with textiles involves complex interactions among inks, polymers, fabrics, and process conditions. Achieving stable adhesion and reproducible mechanical performance requires addressing issues related to surface structure, fiber chemistry, and thermal sensitivity.

7.6.2.1 Surface Roughness and Porosity

Surface roughness and porosity strongly affect both ink spreading and polymer deposition. Low viscosity functional inks can wick along yarns and into pores, reducing pattern resolution, and harming the performance of printed circuits, while the coffee ring effect can create non-uniform deposits as particles accumulate at droplet edges during drying [161]. For 3D polymer deposition, the uneven and porous textile topography can lead to non-uniform flow, trapped air and weak interfacial bonding, so pretreatments such as plasma activation, resin priming, or smoothing layers are often used to enhance wetting and adhesion; experiments show that surface roughness and porosity have a strong influence on peel strength in printed textile composites [110, 111]. Fiber chemistry further complicates adhesion, since hydrophobic fibers like polyester and polypropylene bond poorly with polar thermoplastics (for example, PLA) and aqueous functional inks, making surface modification or compatibilizers necessary to improve polarity and interfacial compatibility [106, 110, 162].

7.6.2.2 Thermal Sensitivity

Thermal sensitivity remains a major constraint in textile integration. While metallic conductors fundamentally require thermal sintering to achieve high conductivity [163], conventional sintering temperatures often exceed 150 °C. This exceeds the thermal tolerance of many textile fibers, such as cotton–spandex blends, risking melting, discoloration, or loss of mechanical strength. Consequently, this limitation drives the development of low-temperature curing strategies—including photonic sintering and reactive silver inks—tailored specifically for heat-sensitive flexible substrates [164]. Similarly, extrusion-based 3D printing relies on elevated nozzle and bed temperatures that can deform or melt thermoplastic textiles such as polypropylene or some nylons in direct contact with the print zone; optimizing printing temperature, bed conditions, cooling and print speed is therefore critical to avoid distortion or property loss [90, 165].

7.6.3 Research Trends and Future Directions

Improvements in materials science, process automation, and digital integration are likely to have an effect on the future of 2D and 3D printing in textile manufacturing. There are a number of promising research areas being looked into right now.

7.6.3.1 Development of Advanced Materials

Researchers are developing new polymer blends, nanocomposites, and biopolymers to make fabrics more elastic, stronger, and stickier. Smart materials like shape memory polymers (SMPs), electroactive polymers, and conductive composites are making it possible for textiles to be responsive and do more than one thing at a time for wearables, robotics, and healthcare. A recent study, for instance, emphasizes the use of SMPs for 4D textile applications [166, 167]. In 2D printing, the focus is on particle-free reactive inks and metal organic decomposition (MOD) inks. These inks exhibit high conductivity at low curing temperatures, addressing the thermal-sensitivity problem in textiles [168].

7.6.3.2 4D Printing and Adaptive Textiles

4D printing goes beyond static 3D printed shapes and includes changes that happen over time. For example, fabrics that change shape, stiffness, or color when they are exposed to heat, light, or moisture. These smart fabrics make it possible for clothes to adjust themselves, insulation to change shape, and surfaces to change shape [6, 158]. Similarly, in the realm of 2D printing, research is expanding into chromic inks (thermochromic/photochromic) that let textile surfaces change color. This fulfills the functional (camouflage, etc.) or aesthetic purpose [169].

7.6.3.3 Integration with Artificial Intelligence (AI) and Machine Learning

AI-driven modeling and machine learning algorithms can use process parameters to predict print quality, adhesion strength, and mechanical performance. In the future, these predictive systems could enable printers to optimize themselves by automatically adjusting settings to maintain consistent quality across different fabrics and materials [89].

7.6.3.4 Hybrid Manufacturing Systems

By combining additive manufacturing with traditional textile methods like weaving, knitting, or embroidery, hybrid structures can be made that are both flexible and strong. The most promising direction is textile hybrid electronics (THE). This approach combines the precision of 2D-printed circuits with the durability of 3D printed encapsulation [158, 170]. As a result, it produces strong, smart garments that are fully washable and durable.

7.7 Conclusion

The transition from conventional, resource-intensive wet processing to two-dimensional (2D) and three-dimensional (3D) printing represents a significant shift in textile surface functionalization. As modern textiles are increasingly required to provide protection, sensing, connectivity, and sustainability, traditional processes face limitations in precision and environmental performance. In contrast, printing technologies enable the targeted deposition of functional materials while preserving the softness and comfort of textile substrates. This chapter has shown that 2D printing methods such as screen and inkjet printing provide scalable, high-resolution surface modification for functionalities including conductivity, antimicrobial activity, UV protection, and responsiveness, while reducing material waste and chemical consumption. Complementarily, 3D printing introduces structurally active elements—such as reinforcements, sensors, and protective layers—transforming textiles from passive surfaces into multifunctional systems for applications spanning apparel, composites, interiors, and energy technologies. Although challenges related to durability, washability, and cost remain, advances in bio-based materials and digital design are accelerating adoption. Together, 2D and 3D printing constitute enabling technologies for a new generation of high performance, sustainable, and customizable textile systems.

AI-Usage Declaration

During the preparation of this chapter, the authors used AI tools to improve the language of the manuscript. After using this tool/service, the authors reviewed and edited the content as needed and take full responsibility for the content of the publication.

References

1. J. Ghosh et al., Grafting modification for textile functionalization: innovations and applications. Discov. Appl. Sci. **7**(1), 49 (2025). https://doi.org/10.1007/s42452024063645
2. A. Patti, Green advances in wet finishing methods and nanoparticles for daily textiles. Macromol. Rapid Commun. **46**(2), 2400636 (2025). https://doi.org/10.1002/marc.202400636
3. C. Comanescu, Recent advances in surface functionalization of magnetic nanoparticles. Coatings **13**(10), 1772 (2023). https://doi.org/10.3390/coatings13101772
4. K. Hassan et al., Functional inks and extrusion-based 3D printing of 2D materials: a review of current research and applications. Nanoscale **12**(37), 19007–19042 (2020). https://doi.org/10.1039/D0NR04933F
5. J. Ghosh, N.S. Rupanty, T. Noor, T.R. Asif, T. Islam, V. Reukov, Functional coatings for textiles: advancements in flame resistance, antimicrobial defense and self cleaning performance. RSC Adv. **15**(14), 10984–11022 (2025). https://doi.org/10.1039/D5RA01429H
6. H.C. Koch, D. Schmelzeisen, T. Gries, 4D textiles made by additive manufacturing on pre stressed textiles—an overview. Actuators **10**(2), 31 (2021). https://doi.org/10.3390/act10020031
7. M.A. Sayam, M. Al-Amin, R. Zhou, A.A. Mamun, A brief review of 3D printing technologies and materials used in smart textiles. Chem. Mater. Sci. **10**, 1710 (2024)
8. M. El-Shamy, H. Mohamed, Z. Gaafar, Y. Roshdy, A. Hassabo, Advancements in bioactive textiles: a review of antimicrobial fabric finishes and commercial products. J. Text. Color. Pol. Sci. **16**, 1281 (2024)
9. A. Hassabo et al., Evaluation of various printing techniques for cotton fabrics. J. Text. Color. Pol. Sci. **21**, 1190 (2023)
10. Y.Z.N. Htwe, M. Mariatti, Printed graphene and hybrid conductive inks for flexible, stretchable and wearable electronics: progress, opportunities and challenges. J. Sci. Adv. Mater. Dev. **7**(2), 100435 (2022). https://doi.org/10.1016/j.jsamd.2022.100435
11. B.K. Dejene, Inkjet printing of conductive nanomaterials on textiles for wearable electronics: advancements, challenges and future prospects. Mater. Today Adv. **28**, 100629 (2025). https://doi.org/10.1016/j.mtadv.2025.100629
12. R. Taurino et al., Enhancing textile properties using nanotechnology and 3D printing process. J. Sol Gel Sci. Technol. **114**(3), 916–933 (2025). https://doi.org/10.1007/s1097102506749x
13. V.S. Romaguera, M.B. Madec, S.G. Yeates, Inkjet printing of conductive materials for smart textiles and flexible electronics. MRS Proc. **1192**, 1192PP1404 (2009)
14. H. Anik, M. Akter, Advancements in textile printing from conventional to modern practices, pp. 317–348 (2025)
15. S. Mishra, M. Khan, D. Banerjee, A. Madala, Journey from 2D to 5D printing: a brief review. Adv. Med. Ind. Eng. **12**, 95–101 (2022). https://doi.org/10.1201/978100321674215
16. M.B.U. Rabbi, T. Ferdous, R.B. Dina, M.R. Islam, Recent advances in textile printing techniques, in *Sustainable Coloration Techniques in Textiles*, eds. by S. Maiti, M. Shahid, R.V. Adivarekar (Springer, Singapore, 2025), pp. 325–367
17. Q. Huang, Y. Zhu, Printing conductive nanomaterials for flexible and stretchable electronics: a review of materials, processes and applications. Adv. Mater. Technol. **4**(5), 1800546 (2019). https://doi.org/10.1002/admt.201800546
18. J. Yu, S. Seipel, V.A. Nierstrasz, Digital inkjet functionalization of water repellent textile for smart textile application. J. Mater. Sci. **53**(18), 13216–13229 (2018). https://doi.org/10.1007/s108530182521z
19. M. Prabhakar, A.K. Saravanan, A. Lenin, I. Leno, K. Mayandi, S.R. Piramanayagam, A short review on 3D printing methods, process parameters and materials". Mater. Today Proceed. **45**, 225 (2020)
20. M.I. Khan et al., Conductive polymers and their nanocomposites as adsorbents in environmental applications. Polymers **13**(21), 3810 (2021). https://doi.org/10.3390/polym13213810

21. S. Wu et al., 3D printing technology for smart clothing: a topic review. Materials **15**(20), 7391 (2022). https://doi.org/10.3390/ma15207391
22. Y. Shi, L.W. Taylor, A. Kulessa, V. Cheung, A.S.M. Sayem, Re-engineer apparel manufacturing processes with 3D weaving technology for efficient single step garment production. iScience **27**(8), 110315 (2024). https://doi.org/10.1016/j.isci.2024.110315
23. Z. Zhao, W. Liu, H. Liu, Flexible and durable direct ink writing 3D printed conductive fabrics for smart wearables. ACS Omega **10**(14), 14138–14149 (2025). https://doi.org/10.1021/acsomega.4c11367
24. Y. Choi, J. Kim, J. Lee, X. Chen, B. Seo, W. Choi, Recent progress on 2D material based smart textiles: materials, methods and multifunctionality. Adv. Eng. Mater. **27**(12), 2500188 (2025). https://doi.org/10.1002/adem.202500188
25. Z.O. Basyigit, Application technologies for functional finishing of textile materials, in *Textiles for Functional Applications* (IntechOpen, 2021). https://doi.org/10.5772/intechopen.95956
26. M.R. Islam, S. Afroj, J. Yin, K.S. Novoselov, J. Chen, N. Karim, Advances in printed electronic textiles. Adv. Sci. **11**(6), 2304140 (2024). https://doi.org/10.1002/advs.202304140
27. V.O. Mercadillo et al., Electrically conductive 2D material coatings for flexible and stretchable electronics: a comparative review of graphenes and MXenes. Adv. Funct. Mater. **32**(38), 2204772 (2022). https://doi.org/10.1002/adfm.202204772
28. A. Kamyshny, S. Magdassi, Conductive nanomaterials for 2D and 3D printed flexible electronics. Chem. Soc. Rev. **48**(6), 1712–1740 (2019). https://doi.org/10.1039/C8CS00738A
29. D. Valdec et al., Influence of printing substrate on quality of line and text reproduction in flexography. Appl. Sci. **11**(17), 7827 (2021). https://doi.org/10.3390/app11177827
30. J. Pathak, M. Gunjal, The study on exploring the evolution of offset printing in world. Int. J. Adv. Res. Sci. Commun. Technol. **61**, 126–136 (2025)
31. L.K. Bommineedi, N. Upadhyay, R. Minnes, Screen printing: an ease thin film technique, in *Simple Chemical Methods for Thin Film Deposition: Synthesis and Applications*, eds. by B.R. Sankapal, A. Ennaoui, R.B. Gupta, C.D. Lokhande (Springer, Singapore, 2023), pp. 449–507
32. A. Willfahrt, *Screen Printing Technology for Energy Devices* (Linköping University Electronic Press, 2019)
33. E.D. Dolden, Fundamental investigations into screen printing, PhD (University of Leeds, 2005). https://etheses.whiterose.ac.uk/id/eprint/26104/
34. ŁD. Sarapuk, J. Szalapak, G. Wróblewski, I. Wyżkiewicz, M. Słoma, M. Jakubowska, Rheology of inks for various techniques of printed electronics. Adv. Intell. Syst. Comput. **393**, 447–451 (2016). https://doi.org/10.1007/9783319239231_65
35. C. Xu, N. Willenbacher, How rheological properties affect fine line screen printing of pastes: a combined rheological and high speed video imaging study. J. Coat. Technol. Res. **15**, 912 (2018)
36. K. Bao, A. Lavrov, H. Nilsen, Numerical modeling of non-Newtonian fluid flow in fractures and porous media. Comput. Geosci. **21**, 9639 (2017)
37. H. Nguyen, D.N. Ngoc, *Incompressible Non Newtonian Fluid Flows* (2012)
38. R. Horwood, Towards a better understanding of screen print thickness control. Electrocom. Sci. Technol. **1**, 129 (1974)
39. B.P. Carr, Introduction to thick film fine line printing screens. Hybrid Circ. **11**(1), 4–7 (1994). https://doi.org/10.1108/eb044515
40. R. Kitsomboonloha, Modeling and applications of highly scaled gravure printing, Ph.D., University of California, Berkeley, United States (2014). https://www.proquest.com/docview/2895599810/abstract/2BD7A2A9D4944998PQ/1
41. P.J. Deng, W. Fang, J.D. Lu, Study about influencing of printing process on gravure printing ink transfer. Appl. Mech. Mater. **469**, 301–304 (2014). https://doi.org/10.4028/www.scientific.net/AMM.469.301
42. O. Şimşeker, Investigation of different solvents in flexographic printing ink's effects to print quality on coated and uncoated paper (2011). https://www.semanticscholar.org/paper/InvestigationofDifferentSolventsinFlexographic%C5%9Eim%C5%9Feker/b8bac8daba5ae6199c925ba359a4ca4f6903e252. Accessed 29 Nov 2025

43. A.B.P. Pazmiño, G.I. Rosero, M. Pérez, B. Lerner, Photopolymer flexographic printing plate mold for PDMS microfluidic manufacture. Polymers **17**(13), 1723 (2025). https://doi.org/10.3390/polym17131723
44. J.E. Döring, L. De Schoulepnikoff, Review on the art of anilox manufacturing—from the coated cylinder up to the gravure, pp. 542–549 (2011). https://doi.org/10.31399/asm.cp.itsc2011p0542
45. A.M.R. Carbonell, J. Ferri, E.G. Breijo, I. Montava, E.B. Belda, Influence of structure and composition of woven fabrics on the conductivity of flexography printed electronics. Polymers **13**(18), 3165 (2021). https://doi.org/10.3390/polym13183165
46. A. Joshi, Optimization of flexo process parameters to reduce the overall manufacturing cost. Int. J. Optim. Control Theor. Appl. **12**, 66–78 (2022)
47. K. You et al., On-demand picoliter-level-droplet inkjet printing for micro fabrication and functional applications. Small **20**(46), 2402638 (2024). https://doi.org/10.1002/smll.202402638
48. T. Cao, Z. Yang, H. Zhang, Y. Wang, Inkjet printing quality improvement research progress: a review. Heliyon **10**(10), e30163 (2024). https://doi.org/10.1016/j.heliyon.2024.e30163
49. D. Lohse, Fundamental fluid dynamics challenges in inkjet printing. Annu. Rev. Fluid Mech. **54**, 349–382 (2022)
50. D. Zhao, H. Zhou, Y. Wang, J. Yin, Y. Huang, Drop on demand (DOD) inkjet dynamics of printing viscoelastic conductive ink. Addit. Manufact. **48**, 102451 (2021). https://doi.org/10.1016/j.addma.2021.102451
51. N. Godard, S. Glinšek, A. Matavž, V. Bobnar, E. Defay, Direct patterning of piezoelectric thin films by inkjet printing. Adv. Mater. Technol. **4**(2), 1800168 (2019). https://doi.org/10.1002/admt.201800168
52. R. Bernasconi, S. Brovelli, P. Viviani, M. Soldo, D. Giusti, L. Magagnin, Piezoelectric drop on demand inkjet printing of high viscosity inks. Adv. Eng. Mater. **24**(1), 2100733 (2022). https://doi.org/10.1002/adem.202100733
53. J. Minase, T.F. Lu, B. Cazzolato, S. Grainger, A review, supported by experimental results, of voltage, charge and capacitor insertion method for driving piezoelectric actuators. Precis. Eng. **34**(4), 692–700 (2010). https://doi.org/10.1016/j.precisioneng.2010.03.006
54. S.G. Mallinson et al., Bubble nucleation pulse width in thermal inkjet actuators. J. Heat Mass Transf. **147**, 081602 (2025). https://doi.org/10.1115/1.4068607
55. J. Jiu, K. Suganuma, Metallic nanowires and their application. IEEE Trans. Comp. Packag. Manufact. Technol. **6**(12), 1733–1751 (2016). https://doi.org/10.1109/TCPMT.2016.2581829
56. A.M.D. Pascual, Carbon based nanomaterials. Int. J. Molecul. Sci. **22**(14), 7726 (2021). https://doi.org/10.3390/ijms22147726
57. N. Karim, S. Afroj, K. Novoselov, S. Yeates, All inkjet printed graphene silver composite ink on textiles for highly conductive wearable electronics applications. Sci. Rep. **8035**, 1–10 (2019). https://doi.org/10.1038/s4159801944420y
58. G.T.P. Azar et al., Selective electroless copper plating of ink jet printed textiles using a copper silver nanoparticle catalyst. Polymers **14**(17), 3467 (2022). https://doi.org/10.3390/polym14173467
59. N. Keawploy, R. Venkatkarthick, P. Wangyao, J. Qin, Screen printed textile electrodes using graphene and carbon nanotubes with silver for flexible supercapacitor applications. J. Metals Mater. Min. **30**, 39–44 (2020)
60. T. Elmaaty, B. Mandour, ZnO and TiO_2 nanoparticles as textile protecting agents against UV radiation: a review. AJOCS **4**(1), 1–14 (2018). https://doi.org/10.9734/AJOCS/2018/40329
61. M. Ivanuša et al., Eco friendly approach to produce durable multifunctional cotton fibres using TiO_2, ZnO and Ag NPs. Nanomaterials **12**(18), 83140 (2022). https://doi.org/10.3390/nano12183140
62. S. Pathania, J.J.L. Hmar, B. Verma, T. Majumder, V. Kumar, P. Chinnamuthu, Titanium dioxide (TiO_2) sensitized zinc oxide (ZnO)/conducting polymer nanocomposites for improving performance of hybrid flexible solar cells. J. Electr. Mater. **51**(10), 5986–6001 (2022). https://doi.org/10.1007/s11664022098150

63. X. Fan, L. Yahia, E. Sacher, X. Fan, L. Yahia, E. Sacher, Antimicrobial properties of the Ag, Cu nanoparticle system. Biology **10**(2), 20137 (2021). https://doi.org/10.3390/biology10020137
64. J.R. Beryl, J.R. Xavier, Halloysite for clay–polymer nanocomposites: effects of nanofillers on the anti-corrosion, mechanical, microstructure and flame retardant properties—a review. J. Mater. Sci. **58**(27), 10943–10974 (2023). https://doi.org/10.1007/s10853023087101
65. G.B. Tseghai et al., Integration of conductive materials with textile structures, an overview. Sensors **20**(23), 6910 (2020). https://doi.org/10.3390/s20236910
66. M.L. Catarino et al., The shift to bio based auxiliaries in textile wet processing: recent advances and industrial potential. Molecules **30**(19), 4016 (2025). https://doi.org/10.3390/molecules30194016
67. A.M. Thakker, D. Sun, Sustainable plant based bioactive materials for functional printed textiles. J. Text. Instit. **112**(8), 1324–1358 (2021). https://doi.org/10.1080/00405000.2020.1810474
68. B.H. Taymaz, H. Kamiş, M. Dziendzikowski, K. Kowalczyk, K. Dragan, V. Eskizeybek, Enhancing structural health monitoring of fiber reinforced polymer composites using piezoresistive Ti3C2Tx MXene fibers. Sci. Rep. **15**, 78338 (2025)
69. I.J. Gómez et al., Recent advances on 2D materials towards 3D printing. Chemistry **3**(4), 1314–1343 (2021). https://doi.org/10.3390/chemistry3040095
70. G. Hu, J. Kang, L.W.T. Ng, X. Zhu, R.C.T. Howe, C.G. Jones, M.C. Hersam, T. Hasan, Functional inks and printing of two-dimensional materials. Chem. Soc. Rev. **47**(9), 3265–3300 (2018). https://doi.org/10.1039/c8cs00084k
71. Z. Tang, K. Fang, M. Bukhari, Y. Song, K. Zhang, Effects of viscosity and surface tension of a reactive dye ink on droplet formation. Langmuir **36**, 1392 (2020). https://doi.org/10.1021/acs.langmuir.0c01392
72. B. He, S. Yang, Z. Qin, B. Wen, C. Zhang, The roles of wettability and surface tension in droplet formation during inkjet printing. Sci. Rep. **7**, 11841 (2017). https://doi.org/10.1038/s41598-017-12189-7
73. G. Mattana, A. Loi, M. Woytasik, M. Barbaro, V. Noël, B. Piro, Inkjet-printing: a new fabrication technology for organic transistors. Adv. Mater. Technol. **2**, 1700063 (2017). https://doi.org/10.1002/admt.201700063
74. W. Fung, *Coated and Laminated Textiles* (Woodhead Publishing, 2002)
75. N. Singh, J. Sheikh, Microencapsulation and its application in production of functional textiles. Indian J. Fibre Text. Res. **45**, 971–1426 (2020)
76. M.R. Islam et al., Fully printed and multifunctional graphene based wearable e textiles for personalized healthcare applications. iScience **25**(3), 03945 (2022). https://doi.org/10.1016/j.isci.2022.103945
77. Z. Xu et al., Real time *in situ* sensing of multiple water quality related parameters using micro electrode array (MEA) fabricated by inkjet printing technology (IPT). Sens. Actuat. B Chem. **237**, 1108–1119 (2016). https://doi.org/10.1016/j.snb.2016.09.040
78. A.P.M. Tavares, N.S. Ferreira, L.A.A.N.A. Truta, M.G.F. Sales, Conductive paper with antibody like film for electrical readings of biomolecules. Sci. Rep. **6**(1), 26132 (2016)
79. B.W. Lee, J. Kang, J.Y. Yoo, S.M. Won, Textile electronic systems for therapeutic applications. NPJ Flex Electron. **9**(1), 118 (2025)
80. G. Isapour, M. Lattuada, Bioinspired stimuli responsive color changing systems. Adv. Mater. **30**(19), 1707069 (2018). https://doi.org/10.1002/adma.201707069
81. D.H. Park, B.J. Park, J.M. Kim, Hydrochromic approaches to mapping human sweat pores. Acc. Chem. Res. **49**(6), 1211–1222 (2016). https://doi.org/10.1021/acs.accounts.6b00128
82. D. Pani, A. Achilli, A. Spanu, A. Bonfiglio, M. Gazzoni, A. Botter, Validation of polymer based screen printed textile electrodes for surface EMG detection. IEEE Trans. Neural Syst. Rehabil. Eng. **27**(7), 1370–1377 (2019). https://doi.org/10.1109/TNSRE.2019.2916397
83. R. Kolakovic, T. Viitala, P. Ihalainen, N. Genina, J. Peltonen, N. Sandler, Printing technologies in fabrication of drug delivery systems. Exp. Opin. Drug Deliv. **10**, 1711–1723 (2013). https://doi.org/10.1517/17425247.2013.859134

84. M. Nosonovsky, E. Bormashenko, Lotus effect: superhydrophobicity and self cleaning, in *Functional Properties of Bio Inspired Surfaces* (WORLD SCIENTIFIC, 2009), pp. 43–78. https://doi.org/10.1142/9789812837028_0003
85. F. Xiao et al., Multi response mixture design applied for the formulation of a textile green camouflage ink with reproduction of a leaf green. Text. Res. J. **93**(17–18), 4115–4125 (2023). https://doi.org/10.1177/00405175231171717
86. W. Li, Y. Wang, M. Li, L. Garbarini, F. Omenetto, Inkjet printing of patterned, multispectral and biocompatible photonic crystals. Adv. Mater. **31**, 1901036 (2019). https://doi.org/10.1002/adma.201901036
87. R. Yahaya et al., Development and characterisation of near infrared camouflage fabrics for military combat uniform. J. Ind. Text. **55**, 19055 (2025). https://doi.org/10.1177/15280837251319055
88. S. Krishnamoorthi, L. Prabhu, G. Kuriakose, D. Jose-lewis, J. Harikrishnan, Air cooling system based on phase change materials for a vehicle cabin. Mater. Today Proceed. **45**, 5991–5996 (2021). https://doi.org/10.1016/j.matpr.2020.09.412
89. M.T. Lemi, H.G. Lemu, E.M. Gutema, Review of recent advancements in 3D printing technologies for textile applications. Text. Leath. Rev. **8**, 72–104 (2025)
90. Y.Q. Xiao, C.W. Kan, Review on development and application of 3D printing technology in textile and fashion design. Coatings **12**(2), 267 (2022). https://doi.org/10.3390/coatings12020267
91. J.P. Manaia, F. Cerejo, J. Duarte, Revolutionising textile manufacturing: a comprehensive review on 3D and 4D printing technologies. Fash. Text. **10**(1), 20 (2023). https://doi.org/10.1186/s40691023003397
92. Y. Wu, C. An, Y. Guo, 3D printed graphene and graphene/polymer composites for multifunctional applications. Materials **16**(16), 5681 (2023). https://doi.org/10.3390/ma16165681
93. A. Ansari, N. Rane, 3D printed textile and fashion. Int. J. Home Sci. **11**(2), 30–32 (2025)
94. S. Park, W. Shou, L. Makatura, W. Matusik, K. Fu, 3D printing of polymer composites: Materials, processes and applications. Matter **5**(1), 43–76 (2022). https://doi.org/10.1016/j.matt.2021.10.018
95. A. Dey, I.N.R. Eagle, N. Yodo, A review on filament materials for fused filament fabrication. JMMP **5**(3), 69 (2021)
96. Y.A. Gueche, N.M.S. Ballester, S. Cailleaux, B. Bataille, I. Soulairol, Selective laser sintering (SLS), a new chapter in the production of solid oral forms (SOFs) by 3D printing. Pharmaceutics **13**(8), 1212 (2021)
97. W.A. Sarwar, J.H.Kang, H.I. Yoon, Optimized zirconia 3D printing using digital light processing with continuous film supply and recyclable slurry system. Materials **14**(13), 3446 (2021). https://doi.org/10.3390/ma14133446
98. J. Huang, Q. Qin, J. Wang, A review of stereolithography: processes and systems. Processes **8**(9), 1138 (2020). https://doi.org/10.3390/pr8091138
99. S. Chokshi et al., Medical 3D printing using material jetting: technology overview, medical applications and challenges. Bioengineering **12**(3), 249 (2025). https://doi.org/10.3390/bioengineering12030249
100. X. Fang, Y. Zu, Q. Ma, J. Hu, State of the art of metal powder bonded binder jetting printing technology. Discov. Mater. **3**(1), 15 (2023). https://doi.org/10.1007/s4393902300050w
101. M.M. Imran, A.C. Idris, L.C. De Silva, Y.B. Kim, P.E. Abas, Advancements in 3D printing: directed energy deposition techniques, defect analysis and quality monitoring. Technologies **12**(6), 86 (2024). https://doi.org/10.3390/technologies12060086
102. S.D. Nath, S. Nilufar, An overview of additive manufacturing of polymers and associated composites. Polymers **12**(11), 2719 (2020). https://doi.org/10.3390/polym12112719
103. T.L.H. Ngoc, H.T.X. Chi, P.S. Minh, V.T. Nguyen, T.M.T. Uyen, Movement strategy influences on the characteristics of low carbon steel generated by the lamination object manufacturing method. Metals **14**(3), 356 (2024)

104. Y. Ling et al., High mechanical interlocking and hydrogen bonding based carboxymethyl cellulose/SiO_2 composite structures for interfacial bonding of reinforced polyimide fiber/EPDM composites. Compos. B Eng. **304**, 112652 (2025). https://doi.org/10.1016/j.compositesb.2025.112652
105. C. Zhang, J. Zhou, M. Nie, Q. Wang, F. Yang, R. Han, Interfacial reinforcement strategy of basalt fiber/polymer composite constructed by interlocking interface composed of ZnO nanowires. Colloids Surf. A Physicochem. Eng. Aspects **681**, 132778 (2024). https://doi.org/10.1016/j.colsurfa.2023.132778
106. T.N. Kočevar, 3D printing on textiles: overview of research on adhesion to woven fabrics. TEKS **63**, 164–177 (2023)
107. F. Rauf, M. Umair, K. Shaker, Y. Nawab, T. Ullah, S. Ahmad, Investigation of chemical treatments to enhance the mechanical properties of natural fiber composites. Int. J. Polym. Sci. **2023**, 1–13 (2023). https://doi.org/10.1155/2023/4719481
108. T. Kozior, T. Blachowicz, A. Ehrmann, Adhesion of three dimensional printing on textile fabrics: inspiration from and for other research areas. J. Eng. Fibers Fabrics **15**, 1558925020910875 (2020). https://doi.org/10.1177/1558925020910875
109. Y. Han, C. Yun, Effect of the alkaline surface modification of the substrate fabric on the peel strength of 3D printed composite fabrics. Fash Text. **12**(1), 25 (2025). https://doi.org/10.1186/s40691025004369
110. D. Popescu, C.G. Amza, 3D printing onto textiles: a systematic analysis of the adhesion studies. 3D Print. Addit. Manufact. **11**(2), e586–e606 (2024). https://doi.org/10.1089/3dp.2022.0100
111. Y. Han, C. Yun, Effect of substrate fabric characteristics on the peel strength of 3D printed composite fabrics. Fash Text. **11**(1), 40 (2024)
112. K.J. Jomboh et al., Properties and applications of natural, synthetic and hybrid fiber reinforced polymer composite: a review. AIMSMATES **11**(4), 774–801 (2024)
113. S. Navaratnam, K. Selvaranjan, D. Jayasooriya, P. Rajeev, J. Sanjayan, Applications of natural and synthetic fiber reinforced polymer in infrastructure: a suitability assessment. J. Build. Eng. **66**, 105835 (2023). https://doi.org/10.1016/j.jobe.2023.105835
114. G.G. Berk, B. Karacan, I. Balkis, Embedding 3D printed filaments with knitted textiles: investigation of bonding parameters. Cloth. Text. Res. J. **40**(3), 171–186 (2022). https://doi.org/10.1177/0887302X20982927
115. I. Ahmed, M. Ali, M. Elsherif, H. Butt, UV polymerization fabrication method for polymer composite based optical fiber sensors. Sci. Rep. **13**(1), 10823 (2023). https://doi.org/10.1038/s41598023339916
116. F.R. López, M.M.L. Pavón, E.C. Correa, M.H. Molina, Effects of nozzle temperature on mechanical properties of polylactic acid specimens fabricated by fused deposition modeling. Polymers **16**(13), 1867 (2024). https://doi.org/10.3390/polym16131867
117. S. Li, K. Wang, J.P.M. Correia, Y. Liu, S. Ahzi, Investigating gradient temperature control for enhanced interfacial bonding behavior in material extrusion 3D printing continuous fiber reinforced polymer composites. Eur. J. Mech. A. Solids **107**, 105349 (2024). https://doi.org/10.1016/j.euromechsol.2024.105349
118. Y. Liu et al., Effects of printing layer thickness on mechanical properties of 3D printed custom trays. J. Prosth. Dent. **126**(5), 671.e1-671.e7 (2021). https://doi.org/10.1016/j.prosdent.2020.08.025
119. A. Milovanović et al., Layer thickness influence on impact properties of FDM printed PLA material. Proced. Struct. Integ. **56**, 190–197 (2024). https://doi.org/10.1016/j.prostr.2024.02.055
120. Ö. Akçay, A. Arı, Effect of infill density and infill pattern on mechanical properties of 3D printed PLA produced by FFF. Fırat Üniversitesi Mühendislik Bilimleri Dergisi **37**(1), 223–232 (2025)
121. S.S. Ambati, R. Ambatipudi, Effect of infill density and infill pattern on the mechanical properties of 3D printed PLA parts. Mater. Today Proceed. **64**, 804–807 (2022). https://doi.org/10.1016/j.matpr.2022.05.312

122. R. Bartosiak, F. Kaźmierczyk, P. Czapski, The influence of filament orientation on tensile stiffness in 3D printed structures—numerical and experimental studies. Materials **16**(15), 5391 (2023). https://doi.org/10.3390/ma16155391
123. E. Linul, New insights into tailoring anisotropy driven shell design for enhanced compression performance in additively manufactured structures. Prog. Addit. Manuf. **12**, 1394 (2025)
124. R. Lesmana, D. Mardiyana, D.I. Sumarno, Analysis of the effect of print speed and layer height on the hardness of TPU 95A filament 3D printed products. J. Konv. Ener. Manuf. **64**, 2 (2025)
125. B.K. Dejene, A.D. Gudayu, Exploring the potential of 3D woven and knitted spacer fabrics in technical textiles: a critical review. J. Ind. Text. **54**, 15280837241253614 (2024). https://doi.org/10.1177/15280837241253614
126. M.A. Abtew, A.S. Ahmmed, M.G. Tadesse, S.U. Zaman, Sensor embedded and electronics textiles: a review on advanced materials, manufacturing and applications. Text. Res. J. **121**, 00405175251338152 (2025). https://doi.org/10.1177/00405175251338152
127. K. Pak et al., Fabrication of multifunctional wearable interconnect E textile platform using direct ink writing (DIW) 3D printing. NPJ Flex Electron. **9**(1), 48 (2025)
128. J. Wang, H. Hu, Z. Liu, Y. Shi, Y. Huang, Optimization of 3D printing nozzle parameters and the optimal combination of 3D printer process parameters for engineering plastics with high melting points and large thermal expansion coefficients. Materials **18**(3), 500 (2025). https://doi.org/10.3390/ma18030500
129. N. Naboulsi, F. Majid, F.A. Hmazi, E. Baghaz, H.E.A. Belghiti, J.D. Kechagias, Electrical conductivity and microstructural features of 3D printed PLA CB composites fabricated at different temperatures. Next Mater. **9**, 101122 (2025). https://doi.org/10.1016/j.nxmate.2025.101122
130. W. Wu, J. Wang, G. Li, 3D/4D printing of stimuli responsive polymers in biomedical engineering: materials, stimulations and applications. Mater. Sci. Eng. R. Rep. **166**, 101071 (2025). https://doi.org/10.1016/j.mser.2025.101071
131. Y. Zhang, Z. Lu, C. Wu, Y. Xu, Anisotropic stimuli responsive polymeric materials: chemistry and applications. Front. Mater. **12**, 1533330 (2025). https://doi.org/10.3389/fmats.2025.1533330
132. H. Tu, Z. Li, Z. Chen, Y. Gao, F. Xuan, A step forward for smart clothes: printed fabric based hybrid electronics for wearable health monitoring. Sensors **24**(21), 6991 (2024). https://doi.org/10.3390/s24216991
133. K.F. Alshammari et al., Development of thermochromic ink using the anthocyanidin based red cabbage extract for anticounterfeiting applications. ACS Omega **7**(51), 48215–48223 (2022). https://doi.org/10.1021/acsomega.2c06314
134. N. Idris, N.D.N. Affandi, I.Z. Borhan, M.I.A. Kadir, R. Yahaya, L. Indrie, Thermochromic behaviour and comfort properties of printed woven fabric. Coatings **15**(6), 692 (2025). https://doi.org/10.3390/coatings15060692
135. S. Lu, J. Zhang, M. Xi, N. Li, Z. Wang, 3D printable plasmonic titanium nitride nanoparticles enhanced thermoplastic polyurethane composite for improved photothermal de icing and infrared labeling (2024). arXiv. https://doi.org/10.48550/ARXIV.2410.22934
136. M. Glogar, S. Petrak, M.M. Naglić, Digital technologies in the sustainable design and development of textiles and clothing—a literature review. Sustainability **17**(4), 1371 (2025). https://doi.org/10.3390/su17041371
137. G. Baysal, Flexible and stretchable printable conductive inks for wearable textile applications. Tekstil ve Mühendis **31**(133), 49–62 (2024). https://doi.org/10.7216/teksmuh.1342520
138. Q. Wu et al., 3D printed sensors for wearable electronics and smart gesture recognition. Adv. Mater. Technol. **9**(6), 2302048 (2024). https://doi.org/10.1002/admt.202302048
139. C.M. Vidhya, Y. Maithani, J.P. Singh, Recent advances and challenges in textile electrodes for wearable biopotential signal monitoring: a comprehensive review. Biosensors **13**(7), 679 (2023). https://doi.org/10.3390/bios13070679
140. I. Jeerapan, S. Khumngern, Printed devices for wearable biosensors: laboratory to emerging markets. IEEE Flex. Electron. **2**(5), 358–365 (2023). https://doi.org/10.1109/JFLEX.2023.3272624

141. J. Bernatoniene, J. Stabrauskiene, J.A. Kazlauskaite, U. Bernatonyte, D.M. Kopustinskiene, The future of medicine: how 3D printing is transforming pharmaceuticals. Pharmaceutics **17**(3), 390 (2025). https://doi.org/10.3390/pharmaceutics17030390
142. H. Chen, X. Li, Y. Gong, T. Bu, X. Wang, H. Pan, Unidirectional drug release from 3D printed personalized buccal patches using FDM technology. Int. J. Pharm. **645**, 123382 (2023). https://doi.org/10.1016/j.ijpharm.2023.123382
143. A.B. Nigusse, D.A. Mengistie, B. Malengier, G.B. Tseghai, L.V. Langenhove, Wearable smart textiles for long term electrocardiography monitoring—a review. Sensors **21**(12), 4174 (2021). https://doi.org/10.3390/s21124174
144. J.T. Orasugh, O.J. Botlhoko, L.T. Temane, S.S. Ray, Progress in polymer nonwoven textile materials in electromagnetic interference shielding applications. Funct. Comp. Mater. **5**(1), 5 (2024). https://doi.org/10.1186/s42252024000546
145. T. Blachowicz, D. Wójcik, M. Surma, M. Magnuski, G. Ehrmann, A. Ehrmann, Textile fabrics as electromagnetic shielding materials—a review of preparation and performance. Fibers **11**(3), 29 (2023). https://doi.org/10.3390/fib11030029
146. U. Anwar et al., Graphene's role in enhancing Fe_3O_4 nanofibers: a comparative exploration of room temperature impedance characteristics and EMI shielding performance. RSC Adv. **15**(20), 16098–16109 (2025). https://doi.org/10.1039/D5RA00006H
147. A. Osman, J. Lu, 3D printing of polymer composites to fabricate wearable sensors: a comprehensive review. Mater. Sci. Eng. R. Rep. **154**, 100734 (2023). https://doi.org/10.1016/j.mser.2023.100734
148. A.A. Zia et al., Impact resistance of 3D printed continuous hybrid fiber reinforced composites. Polymers **15**(21), 4209 (2023)
149. J. Szary, M. Barburski, J. Świniarski, Multi layered composite for custom production: integrating 3D printed core with fiber reinforced composites. J. Mater. Sci. **60**(37), 16934–16951 (2025)
150. S.K. Saha, M.A. Jalil, N. Sakib, T. Islam, Fused filament fabrication of recycled nylon-TiO_2 honeycomb filters for grey wastewater filtration. Micro Nano Lett. **20**(1), e70009 (2025). https://doi.org/10.1049/mna2.70009
151. P. Upadhyay, S. Ippolito, B.S. Mohammadlou, M.S. Waring, Y. Gogotsi, Nanoparticle air filtration using MXene coated textiles. J. C **11**(1), 13 (2025). https://doi.org/10.3390/c11010013
152. B. Choudhuri, M.F. Uddin, R. Sen, Review on 3D printing technology in automotive industry and electric vehicle. Proceed. Instit. Mech. Eng. D J. Autom. Eng. **64**, 7273 (2025)
153. P. Cheng et al., 3D printed continuous fiber reinforced composite lightweight structures: a review and outlook. Comp. B Eng. **250**, 110450 (2023). https://doi.org/10.1016/j.compositesb.2022.110450
154. A. Gi, N. Gupta, P. Saha, P. Bhattacharya, 3D printing of MXenes based electrodes for energy storage applications. Recent. Prog. Mater. **05**(02), 1–23 (2023)
155. R. Chen et al., 3D printed interdigital electrodes for electrochemical energy storage devices. J. Mater. Res. **36**(22), 4489–4507 (2021)
156. S. Simske et al., New findings in security printing and imaging. NIP Dig. Fabricat. Confer. **25**, 158–160 (2009)
157. K. Jain, M. Klosner, M. Zemel, S. Raghunandan, Flexible electronics and displays: high resolution, roll to roll, projection lithography and photoablation processing technologies for high throughput production. Proceed. IEEE **93**(8), 1500–1510 (2005). https://doi.org/10.1109/JPROC.2005.851505
158. J.P. Manaia, F. Cerejo, J. Duarte, Revolutionising textile manufacturing: a comprehensive review on 3D and 4D printing technologies. Fash Text. **10**(1), 20 (2023)
159. D. Wang, S. Wu, X. Cheng, Z. Yan, H. Tan, Challenges and countermeasures of inkjet printing technology in textile printing application. Front. Hum. Soc. Sci. **4**(8), 214–220 (2024)
160. D.B. Sitotaw, D. Ahrendt, Y. Kyosev, A.K. Kabish, Additive manufacturing and textiles—state of the art. Appl. Sci. **10**(15), 5033 (2020). https://doi.org/10.3390/app10155033

161. J. Niu, L. Qi, H. Lian, J. Luo, R. Zhang, X. Chao, Revisiting the inhomogeneity in drop on demand printing of graphene: an effective route for overcoming the coffee ring effect. Surf. Interf. **46**, 104036 (2024). https://doi.org/10.1016/j.surfin.2024.104036
162. M. Gorlachova, B. Mahltig, 3D printing on textiles: an investigation on adhesion properties of the produced composite materials. J. Polym. Res. **28**(6), 207 (2021)
163. H.O. Onyishi, C. Oluah, An overview of the state of the art and applications of sintered metals (2020)
164. Y. Mou, Y. Zhang, H. Cheng, Y. Peng, M. Chen, Fabrication of highly conductive and flexible printed electronics by low temperature sintering reactive silver ink. Appl. Surf. Sci. **459**, 249–256 (2018). https://doi.org/10.1016/j.apsusc.2018.07.187
165. O.S. Nesheim, S.W. Eikevåg, M. Steinert, C.W. Elverum, Identifying thermal effects during 3D printing by comparing in layer infrared pre and postheating of carbon fiber reinforced polyamide 6. Addit. Manufact. **101**, 104705 (2025). https://doi.org/10.1016/j.addma.2025.104705
166. S.Y. Jang, C. Chung, J. Ha, A comparative study on the fashion design process utilizing shape memory textiles and conventional textiles: implications for the industry and education. Fash Text. **12**(1), 12 (2025)
167. S.T. Gobena, A.D. Woldeyohannes, Comparative review on the application of smart material in additive manufacturing: 3D and 4D printing. Discov. Appl. Sci. **6**(7), 353 (2024)
168. A.J. Kell, K. Wagner, X. Liu, B.H. Lessard, C. Paquet, Advanced applications of metal–organic decomposition inks in printed electronics. ACS Appl. Electron. Mater. **6**(1), 1–23 (2024). https://doi.org/10.1021/acsaelm.3c00910
169. M. Hossain, Adaptive camouflage textiles with thermochromic colorant and liquid crystal for multidimensional combat background, a technical approach for advancement in defence protection. Am. J. Mater. Eng. Technol. **9**, 31–47 (2021)
170. E.A.F. Urquiza, Advances in additive manufacturing of polymer fused deposition modeling on textiles: from 3D printing to innovative 4D printing—a review. Polymers **16**(5), 700 (2024). https://doi.org/10.3390/polym16050700

Chapter 8
Emerging Nanomaterials in Surface Modification and Functionalization of Textiles

Shah Md. Maruf Hasan, Md. Kamrul Hassan Chowdhury, Md. Hasibul Hasan, Ibrahim Hossain, Md. Nahid-Ull-Islam, Mahmuda Akter, and Siam Sarker

Abstract Recent developments in nanomaterials are transforming the textile business by enabling sophisticated surface modification and functionalization of textiles, endowing them with novel and enhanced qualities. Through material manipulation at the nanoscale, all of these technologies have the potential to provide textiles with a wide range of functionalities, from enhanced water resistance and durability to antibacterial activity, UV protection, and even electrical conductivity. This chapter describes the most important kinds of nanomaterials for textile modification, their varied integration techniques into textiles, and their numerous industrial uses in fields including smart textiles, sportswear, and health care. Additionally covered in this chapter are the environmental problems, cost-effectiveness, and scalability challenges associated with the usage of nanomaterials. Lastly, it discusses the potential applications of nano-enabled textiles in cutting-edge industries like wearable electronics and responsive materials, as well as future approaches for creating sustainable, multipurpose textiles. The function and surface modification of textiles are covered in detail in this chapter, underscoring the potential of nanomaterials to completely transform the textile sector.

Keywords Nanomaterials · Functional textiles · Surface modifications

8.1 Introduction

Nanoparticles are a viable option for industrial mass-production systems because of their distinctive surface characteristics, low cost, and ability to add new functions to products. Functionalizing fibers and textiles with materials like zinc oxide, titania,

S. Md. Maruf Hasan · Md. Kamrul Hassan Chowdhury · Md. Hasibul Hasan · I. Hossain · Md. Nahid-Ull-Islam · M. Akter · S. Sarker (✉)
Department of Apparel Engineering, Bangladesh University of Textiles, Dhaka, Bangladesh
e-mail: siamsarker100@gmail.com

M. N. Morshed and M. Shahid (eds.), *Advances in Surface Modification and Functionalization in Textiles*, Textile Science and Clothing Technology,
https://doi.org/10.1007/978-3-032-23377-6_8

and silver has improved their macroscopic qualities. In the textile business, adding nanoparticles to synthetic fibers can help with problems including UV radiation resistance, flammability, and the growth of microorganisms [1]. Wearable electronics, often known as "next generation electronics," are a major outcome of efforts in nanotechnology and nanoscience. Conducting polymers, metal organic frameworks, nanostructured carbon derivatives, and nanostructured metal and oxide materials are being used to create multifunctional textiles, which are accomplished by surface modification using nanomaterials. Textile surfaces can be altered using methods like Chemical Vapor Deposition (CVD) and Physical Vapor Deposition (PVD). Energy harvesting, electromagnetic interference filters, and wearable sensors for environmental quality and medical monitoring are all potential uses for these wearable fabrics [2]. Nanotechnology has transformed the textile industry by enhancing antimicrobial properties, UV protection, self-cleaning abilities, and overall mechanical strength, as well as facilitating the development of smart textiles. Advanced fabrication techniques like electrospinning, coating, and printing have improved fabric durability and functionality. The combination of electronics and nanomaterials allows for the creation of flexible, responsive wearable technology. Unlike traditional methods, which often lead to transient effects, nanotechnology enables the achievement of permanent features such as hydrophobicity, antibacterial properties, and conductivity without compromising fabric texture or breathability. This advancement addresses the ongoing challenge of functionalizing textiles while maintaining comfort. Textiles can be made water-resistant by creating nano whiskers out of hydrocarbons that are three orders of magnitude smaller than an ordinary cotton fiber. The fabric incorporates nano whiskers to produce a peach fuzz look [3]. Similar to the Lotus effect, the high surface tension created by the gaps between individual whiskers, which are bigger than water molecules but less than a drop of water, allows the water to stay on the surface [4, 5]. As the whiskers penetrate gases, they remain breathable. Adding gel-forming chemicals or applying a nanoparticulate film to the cloth can also create three-dimensional surface structures that repel water [6]. For instance, cotton fibers can be coated with nanoparticulates using an audio frequency plasma of fluorocarbon derivatives [7]. Superhydrophobicity can be produced by roughening the fabric's surface without compromising the material's softness or ability to withstand abrasion. Textiles can also be made hydrophobic by using silica (SiO_2) nanoparticles (NPs) in conjunction with water-repellent substances [8]. A further driving force behind the research into water-repellent materials is bioinspired design. Duck feathers, for instance, are made up of several scale structures with preening oil to keep water out. By applying chitosan to cotton and polyester fabrics using a surface solution precipitation method and then modifying with a silicon compound to produce a low surface energy, the microstructure of duck feathers was replicated [9].

Hydrogen bonds in cellulose molecules contribute to the structure of cotton fibers. Nanocoatings can enhance textile comfort and wrinkle resistance. Traditional resin treatments increase hydrophobicity but may decrease fiber strength and dyeability. Treatments with nanoparticles (NPs), particularly on cotton and silk, improve wrinkle resistance through cross-linking with cellulose, especially using carboxylic acid and TiO_2 NPs. The best crease recovery angles are obtained with BTCA and succinic acid, and treated fabrics exhibit a softer texture compared to untreated ones [10, 11]. The fashion industry values the integration of optical technologies into textiles and apparel. Photonic materials, such as films, nanoadditives, and optical fibers, have been used in textiles and clothing to improve both aesthetics and functionality. Optical technologies are mostly used to customize the appearance of fabrics or garments by adjusting light intensity, color, and pattern. Optical films consisting of periodic dielectric multilayers can be directly coated on fabrics, providing a highly reflecting and multicolored look with varying color perceptions based on the angle of observation. Holographic films can serve similar tasks and offer a more complicated 3D visual impression [12, 13].

Nanoparticles on textiles face challenges in their dispersion before coating due to van der Waals and electrostatic double-layer forces. Stable dispersions can be achieved through dispersing agents, effective grinding, and covalent surface functionalization by organic compounds and monomers. The durability of nanoparticles on textiles is a major issue, as they lack surface functional groups. Researchers propose surface functionalization of fibers or nanoparticles through physical and chemical methods to enhance binding ability and ensure long-term durability [14].

8.2 Nanomaterials Used in Textile Functionalization

8.2.1 Nanoparticles

Nanoparticles are a group of atoms or molecules of an element with sizes ranging from 1 to 100 nm, and they can have varied morphologies such as amorphous, crystalline, spherical, or needle-like morphologies. The nanometer-scale dimensions result in new physical and chemical characteristics that are significantly different from their bulk form, with higher surface area, reactivity, and optical properties [15]. In textile applications, metal (e.g., Ag, Cu), metal oxide (ZnO, TiO_2, CuO), and polymeric nanoparticles are most commonly used (Fig. 8.1) [16, 17]. These particles interact with textile fibers at the molecular level, leading to enhanced functionalities (figure) such as antimicrobial properties (silver nanoparticles), UV-blocking (zinc oxide and titanium dioxide), fire resistance (MgO), and self-cleaning behavior (silica or TiO_2), etc. [18]. The theory of nanofunctionality is usually related to surface energy interaction, photocatalysis, and electrostatic interaction with fiber surfaces. Due to their lower cost, simplicity in integration, and process compatibility, nanoparticles offer a scalable and promising solution to the creation of future-generation functional and smart textiles.

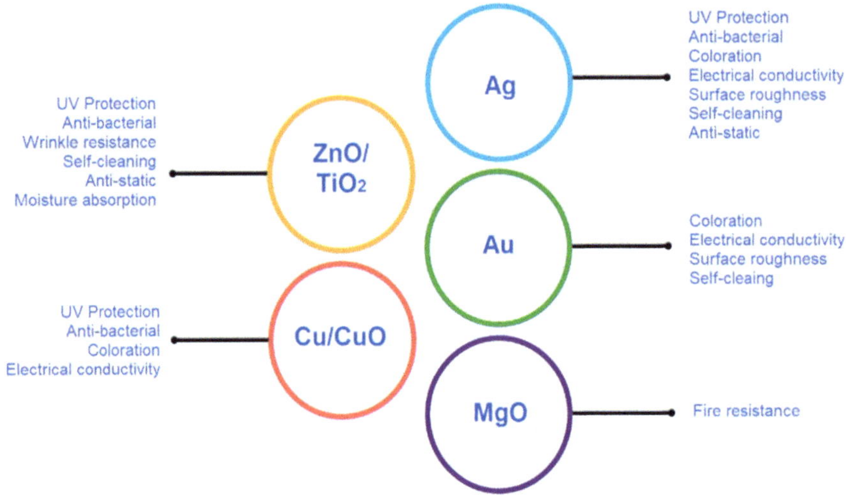

Fig. 8.1 Different nanoparticles are used to impart functional properties on textile substrates [19]. Copyright © 2023 M. Syduzzaman et al., published by Wiley-VCH GmbH (under the Creative Commons Attribution International License)

8.2.2 Nanotubes

Nanotubes are hollow tubes composed of atoms on the nanoscale, typically a few nanometers in diameter and several micrometers in length [20]. Among them, carbon nanotubes (CNTs) have been studied extensively for their new arrangement of carbon atoms in a cylindrical lattice form. CNTs, particularly of single-walled (SWCNTs) and multi-walled (MWCNTs) forms (Fig. 8.2), have superior mechanical strength, thermal stability, and electrical conductivity [21]. Because of the superior aspect ratio and inherent mechanical strength, CNTs can significantly toughen textile fabrics and also impart electrical conductivity [22]. These features make CNTs highly suitable for future use in smart clothing, wearable sensors, and electromagnetic interference (EMI) shielding [23]. The functionality and integration of CNTs in textile matrices are often accounted for using theoretical models such as percolation theory and interface interaction theory. These models explain the formation of conductive networks in the fibers or coatings, which decide the functional performance of CNT-enhanced textile materials.

8.2.3 Nanofibers

Nanofibers are fibrous materials (diameter at least < 100 nm), typically produced following various techniques, such as electrospinning, force spinning, self-assembly, melt blowing, and solution blow spinning [25–29]. Nanofibers are becoming popular

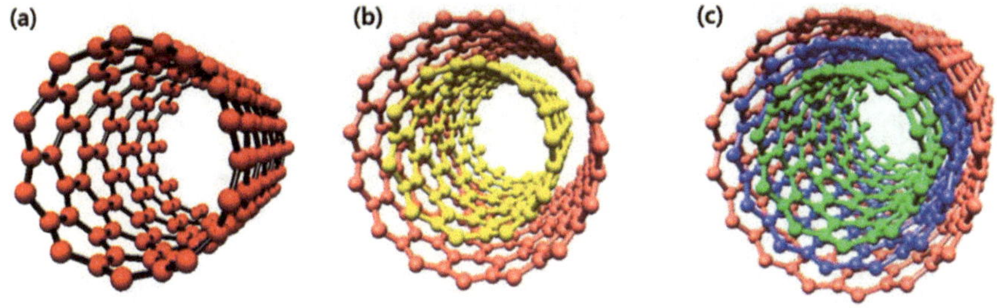

Fig. 8.2 Schematic representation of CNT structure: **a** SWCNT, **b** DWCNT, **c** MWCNT [24] 2021 by the authors. Licensee MDPI, Basel, Switzerland (Creative Commons Attribution (CC BY) license)

in energy, medical, and textile sectors due to their unique active chemistry, high permeability, low basis weight, and excellent filtration capabilities [30]. For example, Wang et al. [31] developed coaxial nanofiber yarn-based triboelectric nanogenerators (TENGs) that enhance the sensitivity and flexibility of wearable sensors. Their work enabled the creation of smart, breathable, and washable devices for real-time health monitoring.

In drug delivery applications, electrospun nanofibers (Fig. 8.3) offer greater flexibility in incorporating different polymers and drugs, higher loading capacity, and improved encapsulation efficiency, making them an excellent choice for loading antibacterial agents. Beyond drug delivery, nanofibers can be used directly as fabric materials or as functional layers in composite structures for filtration, biomedical textiles, and energy storage [32]. However, due to the small diameter of the fibers, the resulting nanofiber web is often very thin, typically under one micron, with limited mechanical strength. This restricts the use of conventional web handling techniques. To address this limitation, nanofibers are often integrated into layered composite structures by combining them with suitable substrate materials. These substrates are selected to provide the necessary mechanical support and complementary functionalities [33]. Researchers have reported numerous studies considering versatile application sectors and their future directions of nanofibers.

8.2.4 *Nanocomposites*

A nanocomposite is a solid material composed of multiple phases, where at least one of the phases has dimensions in the nanometer scale (typically 1–100 nm). These materials are engineered to achieve a synergistic combination of properties by incorporating nanoscale components into a bulk matrix. This nanoscale structuring enables nanocomposites to retain the key characteristics of the individual

Fig. 8.3 SEM picture of a typical electrospun layer of polymer nanofibers [34].

constituents while introducing new or enhanced functionalities not found in any single component alone [35]. Typically, a nanocomposite consists of two main parts: a continuous matrix phase and a dispersed reinforcing phase. The matrix is responsible for maintaining the material's structural integrity and facilitating stress transfer, which can be metallic, inorganic, non-metallic, or polymeric. The reinforcing phase is usually fibrous in nature, such as glass fibers or organic fibers, and is distributed throughout the matrix. This multiphase architecture allows nanocomposites to overcome conventional material limitations and exhibit improved mechanical, thermal, electrical, or barrier properties [36]. Recent developments in surface nanostructured textiles have introduced promising biomedical applications by utilizing nanocomposites made from nanoparticles and biopolymers. These advanced systems offer a sustainable alternative for surface modification, where biopolymer films embedded with nanoparticles are coated onto various textile substrates. The synergy between environmentally friendly biopolymers and functional nanoparticles enhances the material's barrier performance and imparts antimicrobial and antioxidant properties, making them ideal for medical and healthcare-related uses. A schematic representation of such nanoparticles–biopolymer composite development is illustrated in Fig. 8.4 [37].

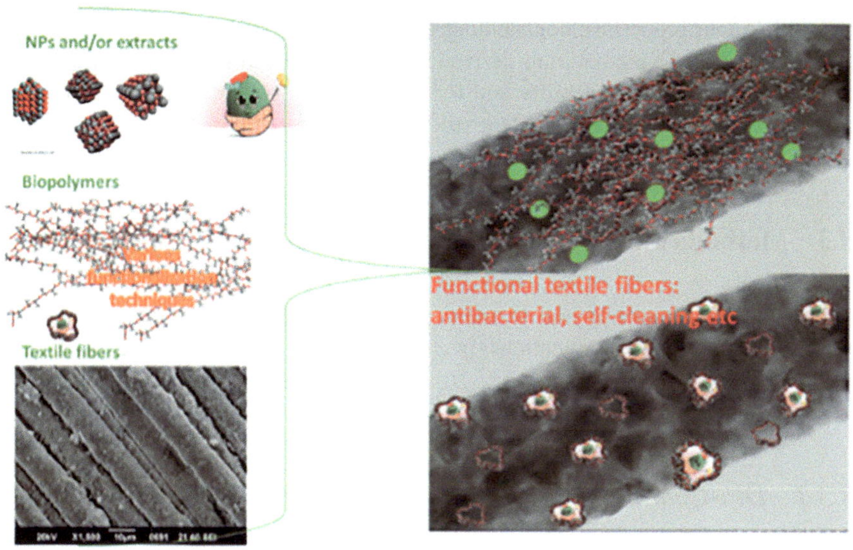

Fig. 8.4 Nanoparticles–biopolymers composite materials schematic development [37]. Reprint with permission from Vrinceanu et al. [37], copyright @ 2022 SAGE

8.2.5 *Nanocellulose and Bio-based Nanomaterials*

Cellulose is an affordable, renewable, abundant, and environmentally friendly biomaterial derived from plants [38]. Nanocellulose is the nanoscale fibrillar form of natural cellulose and is emerging as a promising alternative to petroleum-based polymers, which are often non-degradable. Due to its nontoxicity, biocompatibility, and biodegradability, nanocellulose is gaining attention for use in paper and textile coatings [39–42]. It is extracted from plant biomass through mechanical and/or chemical pretreatment processes that separate cellulose from lignin and hemicellulose. The resulting cellulose is processed into cellulose nanofibers (CNF) or cellulose nanocrystals (CNC), which possess excellent mechanical and functional properties. Nanocellulose finds applications across various fields, including biomedical, energy, composites, water treatment, hydrogels, and sustainable fibers and packaging [43]. By exploiting its surface chemistry and nanoscale, nanocellulose can contribute to a range of benefits such as reinforcement onto fabrics or fibers, development of barrier coating for protection, and functionalization of textiles. nanocellulose can enhance these materials in terms of their mechanical properties, resistance, wettability, and thermal stability [44]. Another material, Chitosan, analogous to cellulose, has garnered significant attention for the fabrication of modified nanostructures exhibiting amplified application potential. This biopolymer, derived through the deacetylation of chitin, presents itself as a biodegradable, biocompatible, nontoxic, and renewable polysaccharide bearing amine functional groups [45]. Lignin nanoparticles, a byproduct of the paper and pulp industry, exhibit excellent UV-blocking and antioxidant properties, contributing to protective clothing and outdoor textiles [46]. Starch-based nanomaterials derived from natural starch sources are particularly valued for their biodegradability and for their use in moisture-responsive or smart coatings, aligning with sustainable textile goals.

8.3 Methods of Integrating Nanomaterials into Textiles

Integrating nanomaterials into textiles can significantly enhance their properties, leading to innovative applications in various fields. Here are some prominent methods.

8.3.1 *Dip-Coating*

Dip-coating is one of the most popular and an effective procedure for integrating nanomaterials into textiles, as it has rarely had a negative impact on textile qualities. This procedure involves dipping and passing the fabric through a nanomaterial solution. The fabric is then squeezed using various rollers to remove any surplus solution, followed by curing or drying. Dardari et al. developed a catalyst-based dip-coating on manmade PES fabric, depositing graphene oxide (GO) and decorating the fabric using Cu^0 and Ag^0 nanoparticles. They applied dip-coating and performed in situ reduction of Cu^{2+} and Ag^+ to Cu^0 and Ag^0 on the surface. They explored the fabric's particular conditions with GO and other nanomaterials. The functional fabric was utilized to reduce Rhodamine B (20 ppm), 4-nitrophenol (0.03×10^{-3} m), and Methyl Orange (20 ppm). Furthermore, the catalyst exhibited good catalytic activity. In kinetics, RhB, 4-NP, and MO had reaction constants of 0.569, 0.564, and 0.42 min^{-1}, respectively [47, 48]. In this approach, the absorption capacity of the textiles and the solid content of the liquid both assist to the effective integration of nanomaterials [49]. For example, a nanowool-containing cotton fabric was created by immersing the cotton fabric in a nanopowder wool solution to provide wrinkle healing capabilities. The cotton cloth was dipped in the solution five times, yielding a pick-up of 65 wt%. The fabric was then dried and treated to see if its characteristics had improved. The pores of the cotton fabric were discovered to be closed by the nanowool following the successful dip-coating procedure, which improved the fabric's wrinkle resistance [47].

8.3.2 *Electrospraying*

Certain substances' nanomaterials can also be integrated using the electrospraying procedure, which involves a syringe and a collecting plate. The syringe feed contains a polymer solution that is exposed to a high electric voltage, resulting in charged particles in the capillary apex of the syringe. The particles at the tip can withstand the liquid's surface tension and begin to spread, forming a conical shape. The jet of solution breaks down and liquids begin to drop down as a spray when a low-viscosity polymer solution is employed in this process known as electrospraying [50]. Güne Soglu et al. electrosprayed fluorocarbon resins anddimethyloldihydroxylethyleneurea (DMDHEU) reagents on cotton fabrics to provide oil and water-repellent properties [51]. It was observed that the polymer solution of fluorocarbon resins and DMDHEU avoided any superfluous agglomeration on the surface of the fabrics using this procedure, which was confirmed by scanning electronic microscopic testing. The fluorocarbon resins and DMDHEU increased the cotton fabric's oil and water repellency via electrospraying. Ibili and Daşdemir demonstrated superhydrophobic and antibacterial characteristics for medicinal applications utilizing a one-step electrospraying technique [52]. They also conducted a comparison to the pad-dry-cure procedure. Coatings using silver chloride-titanium dioxide (AgCl-TiO_2) and dendritic polymers with methyl end groups were used. Electrospraying has significantly increased water repellency, resulting in a contact angle shift of 125.4° to 168.5°. The report also indicated that the functionalized fabrics exhibited more than 99% antibacterial activity against Escherichia coli and Staphylococcus aureus.

8.3.3 Sol–Gel Process

Nanomaterials can be integrated into textiles using the sol–gel technique, which operates on hydrolysis and condensation. Metal oxide nanomaterials such as SiO_2 and TiO_2 are synthesized by the sol–gel technique from precursor molecules. Hydrolysis and condensation of water and alcohol are the primary mechanisms involved in the synthesis and production of homogeneous colloidal metal oxide nanoparticles [53]. To integrate the sol–gel solution into the textile, a basic pad-dry-cure or dip-coating process can be used [54]. After deposition, the nanomaterial solution transforms into gel, aggregates, and develops a three-dimensional network during the drying phase. After drying and curing, the gel left a thin film on the substrate surface [55, 56]. Many investigations have been conducted on TiO_2 coating on textile fabric using the sol–gel technique. Allaka and Yepuri established an affordable approach for creating superhydrophobic cotton and silk fabrics through the combination of sol–gel and dip-coating techniques [57]. They primarily focused on the need for hydrophobic and highly luminescent clothing in applications such as COVID-19 frontline clothes and military apparel. According to their XRD results, the anatase phase of titania in the coated fabrics has been clearly identified. Wetting property experiments revealed contact angles of 146° for cotton and 136° for silk, which are quite acceptable for hydrophobic surfaces. In terms of diffuse reflectance spectroscopy, it showed 84% reflectivity in UV and 92% in IR wavelengths. The coated materials demonstrated outstanding hydrophobicity and reflectivity, making them appropriate for applications such as low-maintenance garments. Daoud et al. investigated the use of TiO2 nanoparticles to cotton fabric to improve its UV protection and antibacterial capabilities. TiO_2 nanosols were manufactured from titanium tetraisopropoxide via a sol–gel technique, including acetic acid, nitric acid (HNO_3), and ethanol. Finally, the coating procedure was completed using the dip-paddry-cure method [58–60].

8.3.4 Chemical Vapor Deposition (CVD)

The vapor deposition process involves condensing the vapor of nanomaterials through condensation, chemical reaction, or conversion before depositing it on the substrate. This method includes chemical vapor deposition (CVD) as one of its categories [54]. CVD typically creates the coating on the substrate. In this method, the coating material's gaseous state is delivered into a reacting chamber at a specific temperature, where it reacts with the substrate and deposits on its surface [47]. Egami et al. created a conductive cotton fabric with a thin coating of polypyrrole nanoparticles (< 50 nm) using the CVD process [61]. Simsek and Karaman used initiated chemical vapor deposition (iCVD) to create thin PHFBA films on textiles [62]. PHFBA, with its short perfluoroalkyl group, can primarily function as a low-surface-energy treatment for hydrophobic textile integration. The iCVD process reached accumulation rates of up to 83 nm min^{-1}, substantially double those of PECVD. Polyester and cotton fabrics became superhydrophobic after being coated with PHFBA and decorated with SiO_2 nanoparticles, which can induce dual-scale roughness. The finished treated fabrics showed a significant angle of contact for water ($>165°$), demonstrating hydrophobicity in principle.

8.3.5 Electroless Deposition

Electroless plating, commonly known as chemical plating, is an autocatalytic process that deposits metal ions on a substrate. To ensure stability, this method uses a reducing agent as well as additional additives. Plating is a way of applying nanostructures to surfaces rather than merely covering them. Metals such as Ag, Fe, Cu, Ni, and Al are commonly employed to build a layer on the surface by chemically reducing these metal ions and deposition. Electroless plating is the deposition of metals using chemical reactions rather than electrical energy. Metal-plated fabrics have a variety of applications, including EMI shielding, self-cleaning, antibacterial characteristics, and electrical conductivity [63–65]. Frunza et al. employed electroless deposition to impart crystalline ZnO particles to polyester fabrics, with a comparison of the materials, i.e., poly(lactic acid) (PLA), polyamide (PA), and hemp [66]. XRD and SEM analysis indicated wurtzite-type ZnO particles (20–500 nm in diameter and up to 1 μm in length) deposited into textile fibers. The polyester fabrics exhibited superhydrophobic characteristics, with water contact angles exceeding 150° as determined using the sessile drop method. They used the Cassie-Baxter model to describe hydrophobic behavior. Furthermore, to activate the EMI shielding characteristic in the polyethylene terephthalate (PET) fabric, Lu et al. produced an Ag/PET fabric via electroless deposition assisted by ultrasonic agitation [56]. The PET fabric was first treated with various chemicals to prepare it for electroless coating. After that, the pretreated cloth was placed in the silver-plating mixture without any additional catalyst. After retaining the ultrasonic frequency and pH in the plating bath, the samples were taken up and rinsed with toluene, ethanol, and water, then dried in an oven at 50 °C for 1 h. The Raman spectra and thermogravimetric measurement tests indicated that the PET fabric had a robust Ag coating of 6.7%, which ultimately boosted the EMI shielding property [56].

8.3.6 Plasma Treatment

The plasma coating approach is preferable to the standard approach because it is environmentally conscious, requiring less energy and material input [54]. Applying an electric field to a gas can produce plasma, the fourth state of a material. This ionized plasma can react with the surface of the fabric, improving its functionality. Gas flow, frequency, the proper combination and preference for gases are the important parameters for the efficient plasma coating of the fabric [67]. In brief, plasma polymerization produces a nanocoating on the substrate after gas-phase stimulation and plasma surface contact [61]. There are numerous articles available on plasma polymerization-assisted textile coatings. Horrocks et al., for example, coated clay with hexamethylene disiloxane silicon to create an atmospheric plasma treated cotton fabric with flame-retardant properties [68]. Again, Alongi and Malucelli used cold oxygen plasma treatment to coat the cotton fabric with silica and talcite nanoparticles to improve the fire resistance [69]. Another study investigated plasma printing and direct one-step printing of silver nanoparticles on flexible substrates [70]. The researcher created Ag nanoparticles by spraying a silver nitrate solution into a nonthermal plasma jet and depositing them on polyester fibers. FE-SEM, XRD, and EDX analysis revealed spherical nanoparticles with a face-centered cubic crystalline framework. The best SERS performance was obtained at a plasma voltage of 6 kV, with the possibility for swab-based analyte screening.

8.4 Functional Properties of Nanomaterial-Modified Textiles

8.4.1 Water and Oil Repellency

The development of textiles modified with nanomaterials to repel water and oil has attracted significant interest due to their potential applications in enhanced-performance textiles, protective apparel, and self-cleaning fabrics. Conventional liquid-repelling fabrics use fluorinated chemicals with low surface energy. The pursuit of superhydrophobicity and oleophobicity using alternative nanomaterials, including silica nanoparticles, graphene-based coatings, and polymer nanocomposites, has been prompted by environmental concerns about perfluorinated compounds (PFCs) [71]. Superhydrophobicity is modeled after the lotus leaf effect, which occurs when the textile surface has a contact angle with water of more than 150°, thereby inhibiting wetting and encouraging self-cleaning. According to recent research, low-energy nanocoatings in conjunction with hierarchical surface structures can provide long-lasting and multipurpose repellency [72]. To improve the stability and performance of repellent textiles, cutting-edge methods such as layer-by-layer assembly, electrospinning, and plasma-assisted deposition have been used. Furthermore, bioinspired modifications using fluorine-free polymers and polydimethylsiloxane (PDMS) have become viable, environmentally friendly substitutes.

Applications in the industrial, medical, and outdoor clothing sectors are expanded by the incorporation of nanotechnology into textile engineering, which permits improved performance in harsh conditions [73]. Zhang et al. [74] introduced a fluorine-free, scalable spray-coating strategy to fabricate superhydrophobic and super oleophobic polymerized organo silanes/Al_2O_3 nanoparticles (POS/Al_2O_3NPs) coatings. This work offers an economical solution utilizing the hydrolytic condensation of TEOS and HDTMS in the presence of AlO_3NPs to address issues, including air cushion failure under hot water and intricate preparation procedures. Zhang et al. [75] report a multifunctional Fe_3O_4 magnetic nanoparticle (MNP)-coated fluorinated carbon fabric (CF) membrane for dye removal, ionic adsorption, and oil–water separation. The dipping process, commonly referred to as the immersion method, is an effective surface modification method, shown in Fig. 8.5 [76]. The type of substrate being treated impacts the final properties. For instance, the dipping approach has been employed in paper coating to improve paper surfaces by making them more waterproof or hydrophobic (water-repellent).

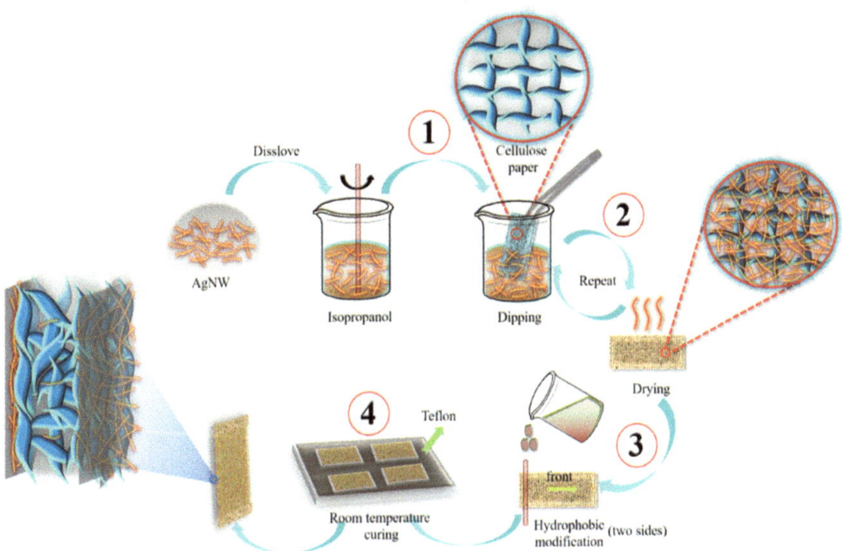

Fig. 8.5 Fabrication process of H-AgNW/cellulose paper via a dip-coating process and hydrophobic modification [76]. © 2019 by the authors. Licensee MDPI, Basel, Switzerland (Creative Commons Attribution (CC BY) license)

8.4.2 Antimicrobial and Antiviral Properties

Nanoparticles have attracted attention due to their unique properties in antimicrobial and antiviral applications. As global concern about infectious diseases grows, innovative materials that prevent microbial and pathogenic viruses are being developed. Nanoparticles present intriguing solutions for the treatment of infections because of their small size, large surface area, and reactivity. It accomplishes this by interacting with viral proteins to prevent replication, rupturing cell membranes, and producing reactive oxygen species. Metal-based nanoparticles, such as zinc oxide (ZnO), copper (Cu), titanium dioxide (TiO_2), and silver (Ag), are popular for their intrinsic antimicrobial and antiviral properties [77, 78]. Ions generated by nanoparticles when they penetrate into textiles damage microbial cell structures, impede the function of enzymes, and induce oxidative stress, which kills or inhibits the growth of microorganisms.

These properties make nanoparticles particularly useful in sectors such as food processing, healthcare, and personal protective equipment (PPE) where cleanliness and infection prevention are important. For instance, silver nanoparticles interact with bacterial proteins to interfere with cellular functions, causing membrane deterioration and cell death. Nanoparticles are incorporated into textiles to improve their antibacterial properties and lower the risk of illnesses and unpleasant smells [79, 80]. Figure 8.6 shows the overall mechanism of action in bacteria, including the stages of

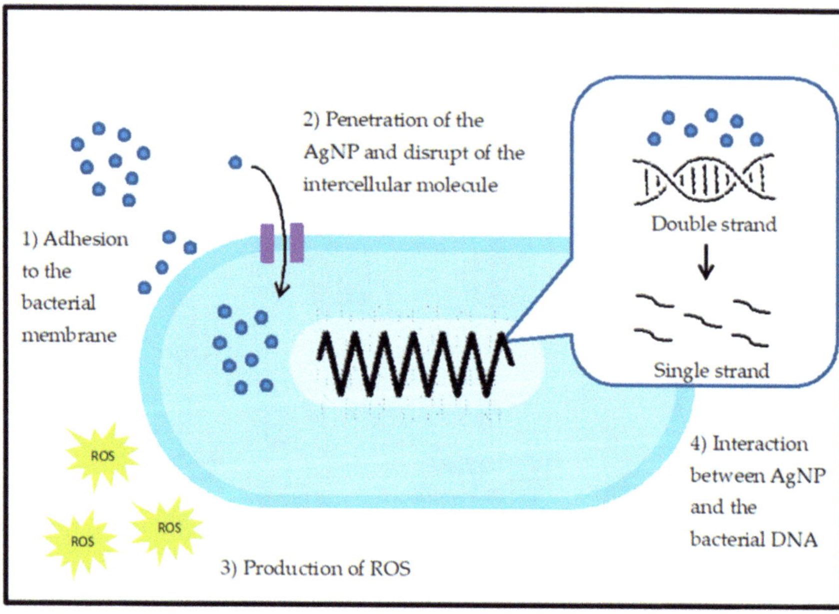

Fig. 8.6 Antibacterial activities of silver nanoparticles [81]. © 2020 by the authors. Licensee MDPI, Basel, Switzerland (Creative Commons Attribution (CC BY) license)

nanoparticle adhesion, membrane difficulties, internal damage, and ROS production that contribute to cellular toxicity.

Salleh et al. [81] worked on silver nanoparticles (AgNPs) due to their novel properties, such as their antimicrobial, antiviral, and antifungal activities. These features make AgNPs important in a variety of industries, including home goods, textiles, and healthcare. Physical, chemical, or biological approaches can be employed for producing AgNPs, however, biological synthesis has advantages since it uses natural agents and materials that are not toxic and are more ecologically friendly. When AgNPs come into interaction with microorganisms, they attach to their cell membrane by electrostatic interactions, breaking down the membrane's structure and enhancing its permeability. AgNPs release silver ions, which connect to negatively charged proteins and modify their usual structure, which is a further critical phase in this process. The manufacture of reactive oxygen species (ROS) is a further significant mechanism that contributes to oxidative stress in microbial cells and ultimately to cell death.

8.4.3 UV Protection

UV-protective clothes with nanoparticles can provide strong protection against harmful UV radiation. The integration of nanoparticles into textiles has emerged as an effective approach to enhance UV protection. Moreover, recent research has noted the utilization of nanoparticles, especially ZnO, SiO_2, and TiO_2, due to their superior UV-blocking capabilities [82, 83]. TiO_2 nanoparticles provide excellent UV protection when deposited on the surface of polyester fabrics, effectively absorbing and scattering UV rays upon exposure to sunlight. Additionally, nanoparticles such as ZnO contribute to UV resistance through their ability to absorb and reflect UV radiation. There are various treatment routes for textile fabrics to be implemented for coating the surface in order to provide high protection against harmful UV rays [82, 84]. Different nanomaterial-based coatings, including graphene, spherical nanoparticles, and inorganic nanotubes, have been applied to textile fabrics to enhance UV protection. These coatings are achieved through techniques like immersion, electrophoretic deposition, vacuum filtration, dip-coating, and layer-by-layer methods, ensuring durability and effectiveness. UV protection in textiles relies on either UV absorption by semiconducting materials, which can be enhanced by band gap engineering through doping with metal nanoparticles, or UV reflection, which is influenced by material crystallinity and nanocomposite-based coatings. Attia et al. developed a one-dimensional coating based on halloysite nanotubes (HNTs) functionalized with natural extracts, which significantly enhanced UV protection in textiles, increasing UPF by 57% when applied to linen fabrics via supramolecular interactions [85]. A transparent nanocomposite layer composed of cellulose nanocrystals derived from natural cotton fibers and wrapped around halloysite nanotubes (HNTs) was developed, demonstrating enhanced UV protection by increasing the UPF from 39 to 58 [86]. A green, scalable method was developed to synthesize graphene sheets from sugar beet leaves, providing effective UV protection. The resulting transparent, free-standing graphene thin film blocked approximately 70% of UV rays while allowing visible light transmission, providing a promising, renewable solution for UV filtering applications [87].

8.4.4 Flame Retardancy and Thermal Stability

Flame-retardant textiles with nanoparticles offer enhanced fire resistance and safety for the wearer. Over the fabric, different types of nanoparticles, such as alumina, silica, and titanium dioxide act as protective layers when applying those particles [88]. The most noticeable is the creation of char, in which nanoparticles form a shielding carbonaceous layer that protects the cloth and prevents more combustion [89]. Another way that nanocoating creates a thermally stable barrier with sluggish heat transfer and restricted oxygen diffusion is heat shielding. An insulating layer produced by intumescent coatings stops flames from spreading [69]. Flame-retardant nanoparticles mitigate combustion by suppressing the release of flammable vapors and generating inert gases through a gas-phase mechanism. Furthermore, specific nanoparticle systems enhance the thermal conductivity of textiles, promoting efficient heat dissipation and thereby delaying the onset of ignition [90]. It has been demonstrated that some nanoparticles possess flame-retardant qualities. Metal oxides, including TiO_2, Al_2O_3, and MgO, enhance thermal stability by catalyzing char formation and creating a protective coating [91]. Phosphorus-based nanomaterials, such as nano-DOPO derivatives, act as radical scavengers, interrupting combustion and increasing char yield [92]. Flame-retardant and thermally stable textiles are used in various industries. Protective clothing textiles are essential for safeguarding firefighters, military personnel, and workers in hazardous environments in different industries like automotive, aerospace, and construction by lowering the risk of burn injuries. Most defensive clothing uses flame-retardant fabrics to protect against fire conditions [93]. Makhlouf et al. [94] developed a cotton fabric that is flame-retardant and has antibacterial coatings, PVA-based formulations with nano chitosan (nCH), melamine phosphate (MP), and melamine salt of chitosan phosphate (MCHP). The coating fabrics constituted noticeable self-extinguishing behavior, achieving a high LOI of 58.2%. Carosio et al. [95] one-step deposition of green polyelectrolytes onto acrylic fabric surfaces to impart flame-retardant properties is the first and only dedicated work to date on surface treatment of acrylic fabrics for improving their fire performance. They combine chitosan and montmorillonite in a single pot to create a multilayered coating. After mixing the oppositely charged polyelectrolytes to produce polyelectrolyte complexes [PECs], centrifugal forces compact the prepared solution. The gel is then applied to the fabric surfaces using a simple doctor-blade coating process, as shown in Fig. 8.7.

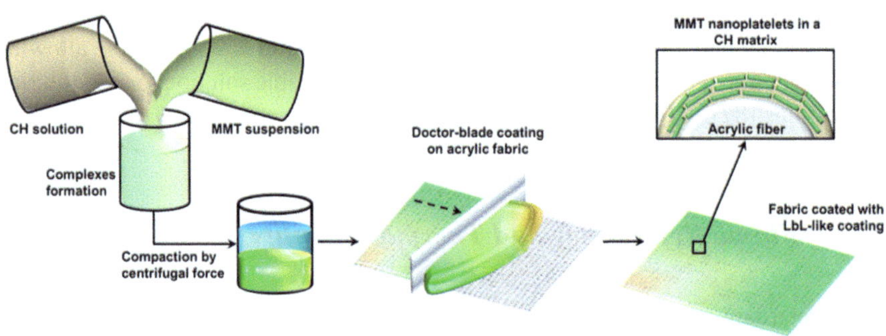

Fig. 8.7 Schematic description of the process adopted to coat acrylic fabric: chitosan (CH) solution and montmorillonite (MMT) suspension are mixed in controlled conditions and then centrifuged; the resulting gel-like precipitate is then doctor-bladed on the fabrics [95]. © 2018 by the authors. Licensee MDPI, Basel, Switzerland (Creative Commons Attribution (CC BY) license)

8.4.5 Conductivity and Sensing

Nanoparticles provide conductivity and sensing when incorporated into textiles, allowing for the creation of novel fabrics with several uses. Applications for these fabrics, including e-textiles, have expanded to include environmental sensing, wearable electronics, and health monitoring. Incorporating conductive nanoparticles, such as graphene, carbon nanotubes, and conductive polymers, has been shown to improve the electrical characteristics of textiles without sacrificing their breathability and flexibility. There are several methods for giving textiles conductivity and sensing capabilities. To create highly conductive wearable e-textiles, Karim et al. [96] use graphene-silver composite inks that are inkjet printed. Graphene-coated cotton textiles were also studied by Badawi et al. [97] for improved durability and conductivity. Polyaniline and poly (3,4-ethylenedioxythiophene)/poly(4-styrenesulfonate) were combined by Pattanarat et al. [98] to create a textile-based sensor that demonstrated great sensitivity to ammonia. Attia et al. [99] developed the use of nano MnO_2/polyaniline composites for textile applications like electronics.

8.4.6 Smart and Multifunctional Properties

Nanomaterial-modified textiles have been incorporated to advance the development of innovative, multifunctional fabrics, offering enhanced properties such as anti-static and wrinkle-resistant performance. Textile fiber contains nanoparticles, including metal oxides, carbon black, and TiO_2, to generate static electricity [100]. The conductive nature of these nanoparticles guarantees the dissipation of static charges. They alter the surface characteristics of textiles, lowering the amount of static generated by friction. In contrast, wrinkle resistance uses cross-linking agents and shape-memory polymers in nanoparticles to create a protective layer on the textile's surface, thereby reducing friction between threads and minimizing the likelihood of wrinkles [101]. In addition to keeping fabrics smooth and clean after numerous washings and uses, these adjustments also lessen the need for ironing, extending the fabric's life and functionality. To guard against electronic gadgets, these textiles are utilized in a variety of everyday apparel applications [102].

Triboelectric nanogenerators are smart, energy-efficient devices that have also been developed on wearable textiles. The complete flexible and foldable WTNG's fabrication method is shown in Figs. 8.8a and b an Ag-coated textile substrate, ZnO NR array-templated PDMS nanopatterns functioning as a triboelectric active layer at the bottom side were produced by hydrothermally growing ZnO and dip-coating a PDMS solution. A field-emission scanning electron microscopy (FE-SEM) image of nanopatterned PDMS on an Ag-coated plain weave textile is shown in Fig. 8.8c. A photographic image (Fig. 8.8d) exhibits the excellent flexibility and foldability of the WTNG with the nanopatterned PDMS on the Ag-coated textile substrate [103].

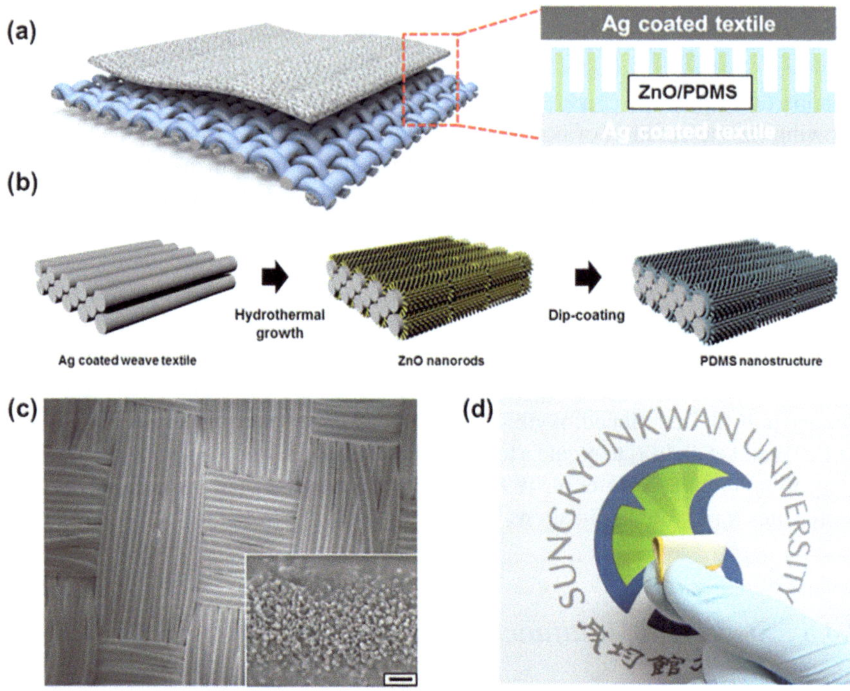

Fig. 8.8 **a** WTNG schematic illustration. **b** The nanopatterned PDMS structure's manufacturing method. **c** FE-SEM pictures of the PDMS-nanopatterned bottom textile. A high-resolution picture of the ZnO NR-templated PDMS nanopatterns is shown in the inset. **d** A photograph of the foldable, flexible WTNG [103]. Reprint with permission from (Seung, W. et al.), copyright @ 2015 ACS Publications

8.5 Applications of Functional Nanomaterial-Modified Textiles

In the field of advanced materials, nanoparticle deposition on textile substrates has been a state-of-the-art approach [104]. Indeed, it conveys the ability to give textile substrates a wide range of functions, including flame retardancy [105], UV protection [106], stain resistance [107], ultrahigh hydrophobic [108], and many more to accommodate a variety of applications in many industries. Nanoparticles can therefore effectively increase the textile materials' characteristics, usefulness, and functional properties [47].

8.5.1 Sportswear and Activewear

Nanoparticles have remodeled the sportswear business by improving moisture management and other functional qualities [109]. Additionally, antibacterial nanoparticles can help treated materials resist undesirable smells by inhibiting bacterial development and activity [110]. In sports like cycling, jogging, and swimming, nano finishes can help create lightweight, high-performance textiles with sufficient agility. Furthermore, materials treated with UV-protective coatings can act as a barrier, shielding athletes from harmful radiation and extending the life of their gear [111].

The sportswear industry is heavily influenced by nanotechnology, which improves product quality, usefulness, and longevity. Sports textiles have been treated with nanoparticles such as carbon nanotubes (CNTs), graphene, MXene, Ag, and others to improve their mechanical, chemical, and physical properties. Nanocomposites, such as nylon and polyolefin nanocomposites, can help to reduce the brittleness of products. It can thus be used to make engine covers as well [111].

In general, nanoclays improve flexural strength and flame retardancy in materials. Breathability has always been an important consideration in sportswear. Sportswear can be self-cleaning due to nanotechnology [47]. Nanoparticles such as Al_2O_3, SiO_2, and TiO_2 can limit UV radiation exposure by reducing light absorption, which is critical for outdoor activities. Sports textiles can benefit from nanofibers, such as functional zinc oxide polyurethane, which provide both UV protection and antibacterial properties [111]. Sportswear can also be self-cleaning thanks to nanotechnology. Photocatalytic nanoparticles, such as TiO_2 and ZnO, can provide hydrophobic surfaces for stain removal [112]. Schoeller Textile AG's self-cleaning, high-performance nanosphere-based fabrics are one example of how nanoparticles are used in sports [113]. Furthermore, using conductive nanoparticles opens up new avenues for intelligent sensing and applications, such as motion tracking and real-time data collection for vital sign monitoring [114, 115] Since sportswear is designed to survive harsh conditions, phase change materials (PCMs) had to be included through coaxial and composite electrospinning techniques [109, 111]. PCMs, which are classified as paraffin or nonparaffin, can control and maintain body temperature during physical exercise. Sportswear containing multifunctional nanoparticles can defend against a variety of conditions [116]. Mountain climbing, swimming, skiing, archery, football, hockey, cricket, and canoeing may all be adapted to a range of situations [110].

8.5.2 Healthcare Industry

The incorporation of nanoparticles onto textile substrates used in the healthcare business confers amazing properties and capabilities that improve patient health and sanitary safety [117]. It not only protects patients by repelling stains and bacteria, but also provides countless simple, user-friendly solutions to common problems in the human body [47]. For example, several medical gadgets can become infection-resistant merely by removing an antibacterial coating [118]. Accurate medication distribution to the right places can be achieved with a low risk of side effects. Advanced nanocoatings on medical devices can improve biocompatibility, enabling highly sensitive biosensing and precise disease diagnosis via biomarkers. Furthermore, it enables point-of-care testing, facilitating early diagnosis in patients [117]. Researchers created gold-liquid–metal-based textiles that are mechanically sensitive and use conductive gold nanoparticles to overcome their limited biocompatibility, exhibiting prominent electrochemical performance across a range of biological applications, along with refined sensitivity for detecting and quantifying blood glucose levels and sweat content [47]. In contrast to viral vectors, non-viral nanostructures (NPs) are less prone to immunological reactions, less carcinogenic, and safer. They are simpler to prepare and have no restrictions on gene size or danger of viral recombination. Non-viral nanostructures are ideal for gene delivery because they can penetrate deep into membranes, possess a high surface-to-volume ratio, are small, and are positively charged [119, 120]. Non-viral nanostructure gene therapy is thought to be safer than viral vector treatment, as it poses fewer oncogenic risks and rarely triggers immune reactions. With no risk of virus recombination and no restrictions on the amount of the gene to be given, non-viral methods are easier to prepare. Because of their high surface-to-volume ratio, small size, and positive charge, nanoparticles can pass through membranes and deliver genes. Figure 8.9 shows the main nanosystems employed in gene therapy [121].

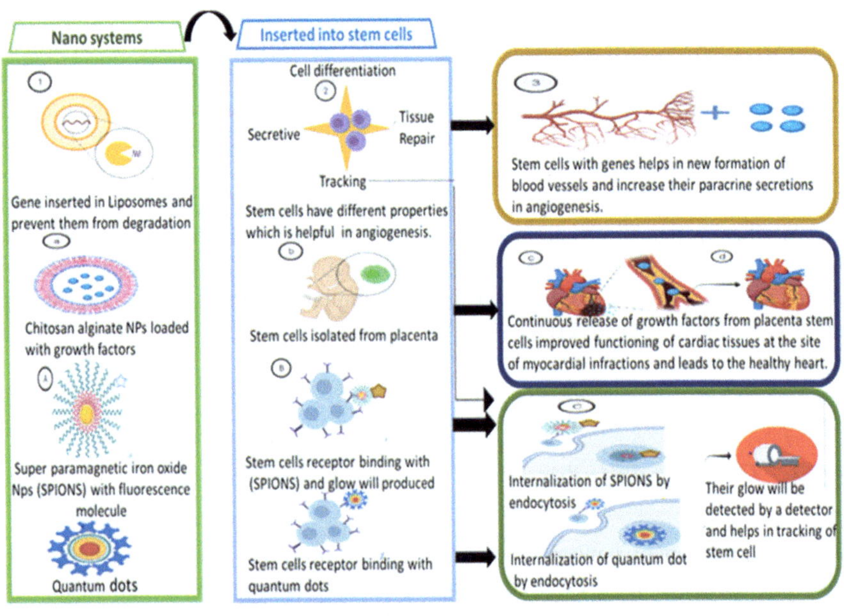

Fig. 8.9 Various nanosystems used in gene therapy [121]. © 2021 by the authors. Licensee MDPI, Basel, Switzerland (Creative Commons Attribution (CC BY) licensed)

In health care and the medical industry, medical equipment must be kept germ-free and sterile by law. To solve this issue, Novi et al. [122] developed a highly effective and long-lasting antimicrobial nanocomposite textile which showed remarkable antibacterial efficiency against common bacterial and fungal illnesses, scaling down to 99.9999% within 24 h, as well as its non-irritating covering held up well after 100 washing cycles. This precoating technique was introduced using an antibacterial Zn nanocomposite-coated cloth, which aids in the in situ growth of Zn nanoparticles. To make it compatible, they utilized both synthetic and natural fibers. Su et al. [123] proposed an environmentally friendly method for in situ manufacture of Ag-coated textiles for efficient healthcare applications. The researchers successfully applied Ag-based coatings to cotton fabric using an environmentally friendly atmospheric-pressure plasma-assisted in situ Ag-reduction approach. The study discovered that AgO nanoparticle-coated textiles had a factor of approximately 105 against both gram-negative (E. coli) and gram-positive (S. aureus) bacteria. Moreover, in the trial they effectively removed the already existing biofilms. With much reduced Ag concentrations and remarkable biocompatibility, the treated cloth proved non-cytotoxic to human tissue cells. A thorough investigation of the influencing factors, including core–shell nanoparticles, citric acid concentration, and curing temperature, revealed that under optimal conditions (1% core–shell nanoparticles, 5% citric acid, drying at 80 °C, curing at 160 °C for 2 min), the treated fabrics exhibited exceptional antibacterial activity against Gram-negative E. coli and Gram-positive S. aureus, even after 20 washing cycles.

Butola and Kumar [124] created an in situ synthesis of silver nanoparticles using green chemistry for the production of chitosan and polysaccharide-mediated cellulosic functional textiles. The material demonstrated antioxidant and strong antibacterial properties against E. coli and S. aureus. Fouda et al. [125] demonstrated the biosynthesis of ZnO nanoparticles for use in diverse medical textile applications. Medicinal textiles have made tremendous advances in the field of nanoparticles, taking on a more important role. This study reported the use of bioactive macromolecules derived from the fungus Aspergillus terreus to produce ZnO nanoparticles. These nanoparticles were used in multifunctional fabrics, giving antibacterial characteristics and UV protection while reducing cytotoxicity. Proteins produced by the utilized fungus have an affinity for encapsulating ZnO nanoparticles. Furthermore, the cytotoxicity assessment revealed that using ZnO nanoparticles on cotton fabric is safe at the prescribed dosages, with negligible negative effects. The treated fabrics demonstrated strong antibacterial efficiency against both Gram-positive and Gram-negative bacteria, as well as better UV protection with higher UVA and UVB blocking values. Moreover, nano finishes can enhance pharmaceutical solubility and stability, enabling more effective drug delivery across biological barriers [126].

8.5.3 Military and Defense

Professionals in the military and defense industries must deal with a variety of weather conditions at work [127]. Because of the materials used, military apparel is typically hefty and bulky [128]. Nano finishing technologies are a vital asset in the military and defense industries, providing nearly all of the essential properties for the tools and materials utilized in these fields. In the military and defense industries, improving the resilience and protective qualities of textiles is one of the most significant uses of nano finishing [128]. Fabrics with strong resistance to tearing, punctures, and abrasion can be made possible via nanotechnology [129]. These materials are suitable for use in tactical equipment and military uniforms, and their resilience to weather conditions and longevity guarantee the wearers' performance and safety [130].

Today's military operations must operate in a variety of harsh conditions, such as water depths, fire, heat, hypoxia, and extreme cold [127]. In addition to causing physical and mental strain, these extremely harsh conditions can significantly affect troop health [131]. Battles in the past have demonstrated the importance of weather and terrain-specific difficulties. Operations in mountainous areas showed significant hazards of hypoxia and hypothermia because of clothing [130]. Due to high levels of sunlight, desert operations are susceptible to heat-related and direct UV exposure, necessitating the use of specialized thermally stable apparel. Prolonged exposure to sunlight, which contains UV radiation, poses severe and chronic health risks to human skin, eyes, and the immune system. In response to these concerns, UV-protective fabrics that block UV rays while allowing visible light transmission have garnered significant interest. The ultraviolet protection factor (UPF) of a modified silk fabric is detailed, including its UV shielding mechanism, transmittance, and overall performance, as illustrated in Fig. 8.10.

The performance of firearms and ammunition can be improved in the weapon room by using nano finishing. Nanostructured materials can also be used to create bulletproof body armor and vehicle protection materials that are lighter and stronger, improving overall safety and mobility in combat scenarios. Nano finishing offers a wide range of potential uses in defense and military applications. In supporting the operational capabilities and safety of military troops and equipment, their capacity to improve durability, protection, weapon performance, communication systems, stealth capabilities, and vehicle efficiency outweighs many crucial functions [128].

8.5.4 Fashion and Wearable Textiles

Nanoparticles have demonstrated a wide range of possible applications in the field of wearable textiles, including numerous capabilities and improvements [133, 134]. Using a variety of methods, the nanoparticles are incorporated into textiles to improve

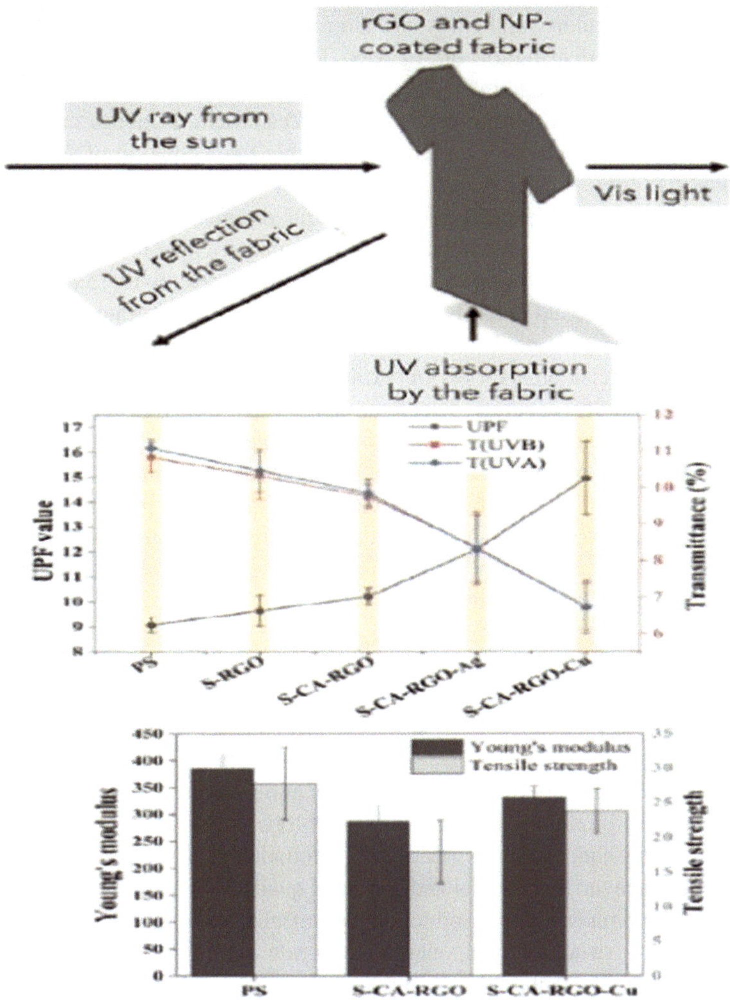

Fig. 8.10 Functional textiles with incorporation of rGO, Cu, and Ag NP for military applications [132]. Reprint with permission from (Bhattacharjee, S. et al.), copyright @ 2020 John Wiley & Sons

their characteristics [135]. Nanoparticles guarantee thermal and electrical conductivity, the most crucial wearable feature, with increased flexibility and durability. Recognizing the progress made in the fields of smart wearables and electronic textiles (e-textiles), with an emphasis on next-generation wearable technology, has already resulted from mastering the fusion of mechanical and electrical characteristics with pliable yet conductive fibers. A simple technique for creating and adding ultra-stretchable Ag NPs to polyurethane (PU) hybrid conductive fibers was presented by Kwon et al. [136]. They included in situ chemical reduction, adjusting solvent diffusion, and adding a durable self-healing polymer as an encapsulating layer. By using higher-molecular-weight solvents and regulated diffusivity, they were able to control the spatial distribution of Ag NPs. This resulted in a percolated Ag NPs network that demonstrated high conductivity (30,485 S cm^{-1}) even at 300% tensile strain. In addition to exhibiting mechanical and electrical resilience during biaxial stretching, the self-bondable properties can enable the formation of self-weavable interconnects [47].

Textile-based wearable electronics demonstrated remarkable deformability, flexibility, and portability. A stretchable fiber strain sensor with a low initial resistance (0.9 Ω cm^{-1}), a broader strain range of $\approx$ 220%, higher sensitivity (5.8 $\times$ 104), less hysteresis, and improved stability (>5000 cycles) was demonstrated by Kim et al. [137]. The sensors permitted for real-time gesture monitoring and posture maintenance during exercise on a variety of textiles, and they didn't retain any hazardous materials. Sensitivity, accurate detection, and power supply in e-textiles are still areas that require work, though. Nevertheless, nanoparticles can consistently aid in the development of conductive textiles, opening up possibilities for smart fabric applications such as illness detection, health monitoring, GPS tracking, and responsive apparel. One exciting area where science has given fashion a very simple living by combining innovation with the functionality of everyday clothing is the use of nanoparticles in wearable textiles.

8.5.5 Automotive and Aerospace Industry

In terms of performance, durability, and application variety, nano finishes provide a quite major and unique contribution [138]. Modern automobile technology thus depends on many nano finishing processes as a foundation. The finishing or coating is meant to enhance the mechanical and chemical qualities of automotive textile products [139]. Nanomaterials are used to increase mechanical qualities, lessen material weight, reinforce structural components, and create a strong structure. Furthermore, lowering the weight of automobile components has a bigger long-term impact on fuel economy and carbon footprint reduction [140]. Figure 8.11 shows the manufactured dragonfly-inspired wing, which is similar to its natural version in terms of length (around 44 mm), weight (26.22 mg), and average thickness (187 μm). However, more weight reduction is required to match the weight of the natural wing, which is around 15.8 mg. This can be accomplished by eliminating superfluous glue. By improving the mask design and the mold production process, dimensional parameters such as length, chord, area, and thickness may be improved [141].

Advanced composite materials can be produced for usage in various automobile device components [19]. Additionally, nano finishes involve the use of nanoscale polymers and additive materials for composites and materials with structural qualities and thermal resistance [142]. Furthermore, employing materials with nano finishes enables the vehicle body panels and chassis components to have high-performance qualities that are needed for durability. On top of that, surface coating offers essential properties to the automotive sector. Actually, the nanocoatings could provide

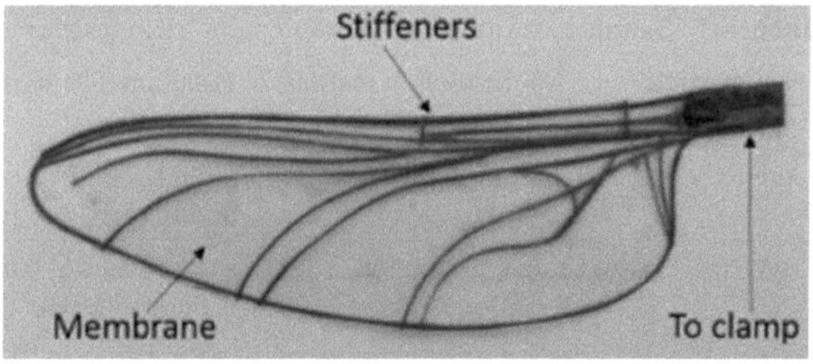

Fig. 8.11 Dragonfly-inspired nanocomposite flapping wing for micro air vehicles [141]. Reprint with permission from (Kumar, D. et al.), copyright @ 2019 Springer Nature

improved wear resistance and protection even in quite hot environments [16, 17]. Moreover, these elements can efficiently increase the lifetime of the materials and save money and maintenance in several ways. By shielding engine parts from corrosion and friction, nanocoating raises engine performance. It can also lower fuel consumption and raise engine efficiency [143]. In the automotive industry, nano finishing helps to create tiny and strong components [144, 145]. It also provides space, though, for smaller, lighter, and stronger electronic components. The ability to transport heat can also be improved by nano finishes since they can have a favorable impact on thermal characteristics [146]. It may thus also control a better thermal management system with an efficient cooling mechanism for engines and batteries, which might also improve the comfort level for the clients. The automotive sector finds many smart applications these days [145]. Various nanosensing applications in cutting-edge environmental monitoring, traffic safety, and navigation involve nano finishing and coatings [146].

The ballistic characteristics of propellants can be greatly enhanced by metallic nanoparticles, particularly coated Al nanoparticles. Solid lubricant coatings made of polymers and containing various nanofillers, such as graphene and MoS_2, can significantly enhance mechanical and tribological properties, making them suitable for use in aircraft systems, including gears, satellite components, air bearings, and more (Table 8.1).

8.6 Testing and Evaluation of Nanomaterials in Surface Modification and Functionalization of Textiles

8.6.1 Health Risks Assessment

Long-term exposure to accidental nanomaterials from welding fumes or combustion engines has been associated with significant health consequences, such as an increased risk of cardiovascular disease, fibroids, lung cancer, and chronic obstructive pulmonary disease [156]. Nanoparticles can enter the circulatory system after deposition in the lungs, higher doses can cause systemic translocation and lung damage,

Table 8.1 Summary of mostly used nano particles in textiles

Nanoparticle type	Application in textiles	Functional benefit	References
AgNPs	Antimicrobial finishing	Hold back the growth of bacteria and fungi	[147]
TiO_2	• UV protection • Self-cleaning	• Work as UV barrier • Photocatalytic degradation of stains	[54]
ZnO	• UV protection • Antimicrobial finishing	• Ingests UV rays • Shows antimicrobial action	[148]
SiO_2	• Crease resilience • Water deterrence	• Lessens creasing • Refines hydrophobicity	[149]
CuNPs	Antimicrobial and curative textiles	Through ion release, eliminates infectious agents and microbes	[150]
CNTs	Conductive materials, sensors, and strengthening	Mechanical power and conductivity of electricity	[151]
AuNPs	• Biosensors • Smart textiles	Optical and conducting attributes	[152]
Fe_3O_4	Textiles with electromagnetic and receptive properties	Facilitates fashionable clothing to have magnetic responses	[153]
Chitosan	• Antimicrobic finishing • Wound-curing textiles	Biodegradable and antibacterial	[154]
Nanoclay	Fire resistance and barrier features	Enhances thermal endurance and gas barrier	[155]

while lower doses lead to minor effects. Smaller particles and slower breathing facilitate more deposition. Dermal exposure, particularly from damaged skin, is another route of absorption, and gastrointestinal exposure can occur via ingested particles that bypass the mucociliary escalator. Research on nanostructured food, drink, and medications is limited, though most are promptly eliminated. Testing standards such as EN ISO 105-E04 evaluate textile skin contact for 4 h at 37 °C, while a variant test mimics immersion in artificial sweat solutions at 40 °C with mechanical agitation, utilizing both acidic and alkaline solutions with acrylic balls for 30 min [157]. Another anticipated test in the standard procedure is the mechanical release from textiles via a "linting" mechanism, based on EN ISO 9073-10:2004. A pilot study revealed silver presence in skin layers after cutaneous exposure to a cloth with silver nanoparticles, tested on human participants wearing sleeves for eight hours daily over five days. This method, under development at CEN/TC248, may assist textile testing labs in assessing the transfer of nanoparticles to the skin and evaluating health risks posed by nanotextiles [158].

8.6.2 Safety and Toxicity Assessment

Understanding diverse synthesis methods of nanomaterials (NMs) is crucial due to their varying physical properties and biological impacts. Researchers must avoid assuming NM consistency across trials or solely relying on industrial data, as discrepancies in NM characteristics can lead to inaccuracies. To improve NM characterization and lab comparisons, standard reference materials (SRMs) and comprehensive methods are essential. Although NMs are often linked to higher toxicity, further studies are necessary to evaluate their physical properties during cell delivery. Standardized measurement techniques and accessible SRMs are vital for consistent data and assessments [159]. Currently, no specific guidelines exist for choosing clothing to mitigate dermal exposure to nanoaerosols, though some standards include testing with nanometer-sized particles. Respirators may be required when other controls fail to prevent exposure, as existing limits for non-nanoscale particles might not adequately protect against nanoparticle exposure. Properly selected and fit-tested NIOSH-certified respirators can be beneficial [160]. Monitoring nanoparticles is challenging, but Advanced Nuclear Analytical Techniques (NATs) provide effective tools for studying chemistry, nano compounds, and toxicology, with a key focus on understanding the toxicological characteristics of nano compounds in nanotoxicology [161].

8.6.3 Regulations and Standards

The document "Nanomaterials in REACH" underscores the necessity of ensuring the safety of nanomaterials throughout their lifecycle for sustainable use and acceptance in the market. Despite REACH not specifically covering nanomaterials, it is imperative for producers, importers, and consumers to ensure that their nanomaterials do not pose risks to human health or the environment. Additionally, it references ISO standards relevant to nanotechnologies, including ISO/TR 27628 and ISO/TS 27687, which are issued by ISO/TC 229 [162, 163]. The British Standards Institution (BSI), the American Society of Mechanical Engineering (ASME), the American Society for Testing and Materials (ASTM), the Institute of Electrical and Electronics Engineers (IEEE), and the National Institute of Standards and Technology (NIST) are among the other organizations that have standards either in the process of being developed or published. A huge number of concerned researchers and experts from a wide variety of areas have suggested the establishment of institutes for the testing and standardization of nanomaterial impacts [164]. To investigate the nanotoxicological characteristics of goods, specialized organizations are forming committees, such as the Nano-Associated Safety Assessment (RT-NASA). Peer review, toxicity evaluation, toxicokinetics, need assessment, physicochemical properties, and risk communication are the six steps in the process. Researchers conduct toxicity assessments on target organs and toxicokinetics, while investigators oversee and collaborate on nano safety studies. This method ensures clear responsibilities, experiments, and expectations, thereby improving the management framework for researchers [164, 165].

8.7 Sustainable Practices of Nanomaterial Integration in Textiles

As more people become aware of the importance of sustainability, the world is moving toward it. The incorporation of nanoparticles into textiles also demonstrates this tendency [166, 167]. Plasma finishing may be used to lessen the overall impact of chemical and thermal finishes. Plasma finishing could be used to lessen the overall environmental impact of chemical and thermal finishes [168]. Palaskar et al. developed a multifunctional cotton fabric using a sustainable plasma treatment method [169]. Such a treatment imparts flame-retardant, UV-protective, and antibacterial properties to the fabric. Among the numerous metal nanoparticles, silver nanoparticles are the most widely studied for sustainable textile finishing. For example, Raza et al. developed a sustainable method for producing silver nanoparticles, then introduced bio-based reducing agents [170]. The group experimented with reducing $AgNO_3$ using black tea extract, chitosan, and starch to synthesize Ag nanoparticles on the surface of cotton fabric. The silver nano finishing resulted in the maximum conductivity of the cloth when combined with black tea essence. This procedure paves the way for long-term garment finishing. Vankar and Shukla also used a biosynthetic method to manufacture Ag nanoparticles on cotton fabric [171]. The researchers used lemon leaf extract to decrease Ag nanoparticles in situ and create Ag nanoparticle-coated cotton fabric. Habib et al. developed a citric acid cross-linking agent to aid in the finishing of commercially dyed denim fabric and confer antibacterial activity [47]. The nano finish exhibited remarkable resistance to multiple wash cycles, making it suitable for use as a surgical gown. Recent improvements in textile manufacturing have resulted in a diverse range of goods with improved interactive capabilities. Microencapsulation is a process in which a membrane or film shields the functional material, or core, from the external environment [172]. With the correct stimuli, the functional nanomaterial is released from the core, producing the intended effect. Singh et al. created scent textiles using the microencapsulation technique, and the cotton textile substrates demonstrated antibacterial capabilities as well [173].

8.7.1 Green Synthesis and Eco-Friendly Nanomaterials

Eco-friendly nanomaterials are thought to have a significant influence on redefining the current environmental engineering and science fields. Many nanomaterials are now being researched and developed for use in treating water and contaminated areas due to their specific activity against stubborn pollutants. Nowadays, there is a growing global concern about the development of environmental engineering applications. Metal oxide nanoparticles have attracted significant attention due to their unique properties, including tiny size, a high surface-area-to-volume ratio, surface modifiability, outstanding magnetic properties, and high biocompatibility. Consequently, the manufacturing of nanomaterials should be economically viable, environmentally sustainable, and widely accepted by society [174]. According to recent studies, the synthesis, characterization, and application of nanomaterials have become increasingly important over the last few decades. "Greener," ecologically friendly procedures in chemistry and chemical technology are becoming increasingly popular and necessary as a result of global environmental contamination problems. Eco-friendly solutions are gaining favor in the modern world. The ultimate product's environmental friendliness and the use of a sustainable technique are critical. To address the downsides, a green method for nanoparticle production has emerged, using naturally biodegradable materials to limit their environmental impact [175–177].

The decomposition of any organic compounds by the green approach (plant extracts) is mostly related to the presence of polyphenols in the biodegradable treatment material. A variety of plant extracts and their chemical ingredients are the subject of extensive research. Metal oxide nanoparticles have been established to act as a next-generation, economical disinfecting agent, which finds enormous potential in medical concerns, contaminant treatment, and other industrial uses due to the achievement of unique properties such as antibacterial, antifungal, UV-filtering properties, high catalytic and photochemical activity [178]. Green synthesis of iron nanoparticles is an approach in which plant extracts, microorganisms (bacteria, fungi, algae, and yeast), and enzymes play an important part in the synthesis of naturally biodegradable materials. Plant extract-based synthesis is favorable because it decreases the risk of future contamination by lowering the reaction time and retaining the cell structure [179].

Nanoparticle production is gaining popularity in a variety of sectors, including microorganisms, algae, fungi, and plants. Microbes have a defensive mechanism against reactive ions in their surroundings that converts them into stable atoms. However, large amounts of nanoparticles can cause cell death [179]. Algae are quickly creating nanoparticles with diverse properties for a variety of uses [180]. Fungi are attractive due to their ease of managing biomass, low cost, and effective production of extracellular enzymes. However, they may cause contamination and genetic modification of organisms [181]. Greener synthesis is easier, less expensive, and more reproducible, often resulting in more stable materials. This approach is gaining popularity in greener surroundings because of its simplicity of scaling up for bigger production volumes and a lower risk of contamination. A significant amount of study has been carried out using plants and their parts for nanoparticle synthesis, including extracts from green tea leaf, Terminalia chebula fruit, oolong and black tea leaf, banana peel, Colocasia esculenta leaves, sorghum bran, eucalyptus leaf, and Tridax procumbens [175, 182–189]. Metallic iron nanoparticles are manufactured as reducing and capping agents for a wide range of applications. Phytochemical screening tests identify polyphenols with a variety of reducing properties. These compounds alter precursors and thereby facilitate nanoparticle production. Phytochemical compounds like phenol derivatives and flavonoids are necessary for biological functions, including antioxidants and free radical scavenging. Nanoparticles are created according to their final shape and stability [190].

8.7.2 Biodegradable and Recyclable Nanomaterials

Biodegradable nanoparticles break down into nontoxic substances under typical surroundings; therefore, lowering their long-term ecological consequence. Typically, the sources of these materials involve organic polymers, plant extracts, or biologically derived metals [191]. Chitin-derived substance, Chitosan, is a comprehensive antimicrobial agent in textiles. The nanoparticles of it are completely biodegradable, biocompatible, and hold inherent antibacterial characteristics, which make them the best possible option for the healthcare industry and hygiene products [167, 192]. Wood, cotton, or agricultural wastes are the main providers of the CNCs. These nature-derived, fully biodegradable components improve mechanical and barrier properties in textile substrates [193, 194]. These polysaccharides are environmentally friendly coverings that provide enclosure to active agents, control odor, and come up with flame resistance [195, 196]. Nanoparticles such as AgNPs, ZnO, and TiO_2 are produced by green synthesis, which uses microorganisms or plant extracts. These nanoparticles elude precarious symptoms and are persistently impervious to the environment [197–199].

Recyclable nanomaterials can be recovered and reused. In a textile-based circular economy, they play a vital role [200]. Recoverable by heat or solvents, these particles have hydrophobic and anti-static qualities [201]. Graphene can be removed and reused while maintaining its functionality, despite its non-biodegradability [202, 203]. TiO_2 nanoparticles are durable and maintain their antibacterial and UV-blocking properties even after several recycling cycles, making them photocatalytic nanomaterials [204].

8.7.3 Closed-Loop Systems for Textile Recycling

Closed-loop systems set to reuse materials from the expiration of textile substrates to fabricate new fibers or products, keeping down waste. Injecting suitable nanomaterials ensures safe and operative recycling [205]. Textile-substrate integrity is largely enhanced by nanomaterials through expanding strength, resistance to wear, and microbial shielding. Materials like Nanoclay and Graphene help to improve tensile and abrasion attributes, stretching textile lifetime [206–208]. Smart nanomaterials, for instance, RFID-driven nanoparticles and quantum dots, assist in tracing the origin, composition, and background of processing of textile materials. These tracers enhance grouping and regaining, activating high-quality outcomes [209, 210]. Biodegradable nanocomposites combine with nature-derived polymers with nanomaterials to fabricate disposable textiles with preserved functionality, convenient for a closed-loop lifecycle [211, 212].

8.8 Challenges and Future Trends and Innovations of Nanomaterials in Textile Surface Modification and Functionalization

- **Economic and Cost Factors**: Producing nanomaterials, such as carbon nanotubes, graphene, and metallic nanoparticles, is challenging due to scaling challenges, high production costs, and sluggish commercial adoption. Future trends include low cost, eco-friendly nanomaterials; efficient manufacturing techniques such as roll-to-roll coating, electrospinning, and digital textile printing; and financial incentives, such as government grants, R&D funding, and partnerships with the tech and fashion industries, to boost commercial viability. These advancements aim to reduce costs and enhance the efficiency of textile manufacturing.
- **Reproducibility, Consistency, and Longevity**: Textile nanomaterials face durability issues, standards gaps, and batch-to-batch variability. These problems can hinder repeatability, detach or deteriorate over time, and impair the usefulness of textiles. Surface engineering strategies, encapsulation techniques, and the development of testing methodologies to enhance stability and adhesion are among future developments. These developments are intended to increase lifespan and set trustworthy standards for durability testing in many sectors.
- **Wearable Electronics and IoT Integration**: Wearable technology faces challenges related to power supply and energy storage, signal stability and interference, and user comfort. IoT ecosystems, flexible energy harvesters, transistors, and nanosensors are examples of upcoming innovations. These technologies aim to provide safe, adaptable, and lightweight energy sources; ensure dependable signal transmission and processing; and improve wearability and biocompatibility. Future innovations include solar cells for smart textiles, nanogenerators, piezoelectric fibers, and IoT ecosystems for intelligent interaction and real-time data analytics.
- **Development of Multifunctional, Sustainable Smart Fabrics and Advanced Fabrication Techniques**: Nanomaterials have the potential to impact the environment and create issues with material compatibility. They can be applied to

eco-friendly coatings that offer several benefits, including antibacterial and UV protection. However, complex production can be time-consuming and ineffective. Future developments in environmentally friendly nanocoatings include chitosan, cellulose nanocrystals, green-synthesized metal nanoparticles, and other nontoxic, biodegradable substances. Modular stacking and hybrid nanocomposites can also be used for simultaneous performance with advanced techniques, including electrospinning, 3D printing, and creative weaving.

8.9 Conclusion

The integration of advanced nanomaterials into textiles has marked the beginning of a new era in the development of high-performance, multifunctional fabrics. These innovations have enabled a wide range of applications across diverse sectors. This chapter has explored various classes of nanomaterials, including nanoparticles, nanotubes, nanofibers, nanocomposites, and bio-based nanocellulose, that are significantly enhancing textile functionality. When combined with advanced incorporation techniques such as dip-coating, electrospraying, sol–gel processing, and plasma treatment, these nanomaterials impart textiles with superior properties. These include improved water- and oil-repellency, antimicrobial activity, UV protection, thermal resistance, and responsiveness to electronic stimuli. As a result, nanomaterial-enhanced textiles are increasingly used in sectors such as sportswear, health care, military, fashion, and aerospace, highlighting their versatility and broad applicability. Importantly, textiles are evolving from passive materials into active systems capable of sensing, energy harvesting, and communication. This shift represents a significant advancement toward smart, responsive, and interactive textile solutions. However, the widespread adoption of nanomaterials in textiles presents several challenges. Concerns related to long-term durability, human and environmental health, regulatory compliance, and economic feasibility remain critical. Addressing these issues requires comprehensive testing, standardized evaluation protocols, and robust life cycle assessments to ensure safe, effective, and responsible implementation. Sustainability has become an integral focus in this domain. Current research emphasizes the development of green synthesis methods, biodegradable nanomaterials, and closed-loop recycling systems to reduce environmental impact and promote circularity. Looking ahead, the convergence of nanotechnology with wearable electronics and the Internet of Things (IoT) is expected to further transform the textile industry. Continued research into scalable, cost-effective, and reproducible fabrication techniques will be key to translating laboratory-scale innovations into commercially viable products. In summary, nanomaterials have transformed textile surface modification from a passive enhancement strategy into a platform for multifunctional and sustainable innovation. As science, engineering, and design continue to converge, the future of textiles promises to be not only smart and functional but also environmentally conscious and human-centered.

AI-Usage Declaration

During the preparation of this chapter, the authors used AI tools to improve the language of the manuscript. After using this tool/service, the authors reviewed and edited the content as needed and take full responsibility for the content of the publication.

References

1. P.J. Rivero, A. Urrutia, J. Goicoechea, F.J. Arregui, Nanomaterials for functional textiles and fibers. Nanoscale Res. Lett. **10**, 1–22 (2015)
2. D.K. Subbiah, S. Balasubramanian, A.J. Kulandaisamy, K. Jayanth Babu, A. Das, J.B.B. Rayappan, Surface modification of textiles with nanomaterials for flexible electronics applications, in *Advances in Functional Finishing of Textiles*. ed. by M. Shahid, R. Adivarekar (Springer Singapore, Singapore, 2020), pp.1–42
3. E. Russell, Nanotechnologies and the shrinking world of textiles. Text. Horiz. **9**(10), 7–9 (2002)
4. L. Gao, T.J. McCarthy, The "lotus effect" explained: two reasons why two length scales of topography are important. Langmuir **22**(7), 2966–2967 (2006)
5. A. Marmur, The lotus effect: superhydrophobicity and metastability. Langmuir **20**(9), 3517–3519 (2004)
6. E.M. El-Khatib, Antimicrobial and self-cleaning textiles using nanotechnology. Res. J. Text. Appar. **16**(3), 156–174 (2012)
7. J. Zhang, P. France, A. Radomyselskiy, S. Datta, J. Zhao, W. van Ooij, Hydrophobic cotton fabric coated by a thin nanoparticulate plasma film. J. Appl. Polym. Sci. **88**(6), 1473–1481 (2003)
8. G.Y. Bae, B.G. Min, Y.G. Jeong, S.C. Lee, J.H. Jang, G.H. Koo, Superhydrophobicity of cotton fabrics treated with silica nanoparticles and water-repellent agent. J. Colloid Interface Sci. **337**(1), 170–175 (2009)
9. Y. Liu, X. Chen, J.H. Xin, Hydrophobic duck feathers and their simulation on textile substrates for water repellent treatment. Bioinspir. Biomim. **3**(4), 046007 (2008)
10. H. Chien, H. Chen, C. Wang, The study of non-formaldehyde crease-resist finishing fabrics treated with the compound catalyst of nanometer grade TiO_2 under UV light and different polycarboxylic acid. J. Hwa Gang Text. **10**(2), 104–114 (2003)
11. C.C. Wang, C.C. Chen, Physical properties of crosslinked cellulose catalyzed with nano titanium dioxide. J. Appl. Polym. Sci. **97**(6), 2450–2456 (2005)
12. M. O'Mahony, S. Braddock-Clarke, *Techno Textiles 2: Revolutionary Fabrics for Fashion and Design* (Thames & Hudson, London, 2005)
13. Fd.C. Vasconcellos, A.K. Yetisen, Y. Montelongo, H. Butt, A. Grigore, C.A. Davidson, J. Blyth, M.J. Monteiro, T.D. Wilkinson, C.R. Lowe, Printable surface holograms via laser ablation. ACS Photonics **1**(6), 489–495 (2014)
14. M.P. Gashti, E. Pakdel, F. Alimohammadi, Nanotechnology-based coating techniques for smart textiles. In: *Active Coatings for Smart Textiles* (Elsevier, 2016), pp. 243–268
15. T.-D. Ngo, Chapter 8—nanotechnological intervention in the wool composites, in *Wool Fiber Reinforced Polymer Composites*. ed. by S. Thomas, S. Jose (Woodhead Publishing, 2022), pp.141–154
16. M.I.H. Mondal, *Protective Textiles From Natural Resources* (Woodhead Publishing, 2022)
17. M.I.H. Mondal, F. Ahmed, M.M. Islam, M.N. Pervez, J. Saha, 11—metal and metal oxides nanoparticles in healthcare and medical textiles, in *Medical Textiles from Natural Resources*. ed. by M.I.H. Mondal (Woodhead Publishing, 2022), pp.341–371

18. M. Syduzzaman, A. Hassan, H.R. Anik, I.S. Tania, T. Ferdous, F.F. Fahmi, Unveiling new frontiers: Bast fiber-reinforced polymer composites and their mechanical properties. Polym. Compos. **44**(11), 7317–7349 (2023)
19. M. Syduzzaman, A. Hassan, H.R. Anik, M. Akter, M.R. Islam, Nanotechnology for high-performance textiles: a promising frontier for innovation. ChemNanoMat **9**(9), e202300205 (2023). https://doi.org/10.1002/cnma.202300205
20. N. Giribabu, P.V. Rao, N. Salleh, 25 - nanoparticle and targeted systems for colon cancer therapy, in *Nanoarchitectonics for Smart Delivery and Drug Targeting*. ed. by A.M. Holban, A.M. Grumezescu (William Andrew Publishing, 2016), pp.695–713
21. W. Zhou, X. Bai, E. Wang, S. Xie, Synthesis, structure, and properties of single-walled carbon nanotubes. Adv. Mater. **21**(45), 4565–4583 (2009). https://doi.org/10.1002/adma.200901071
22. J. Di, X. Zhang, Z. Yong, Y. Zhang, D. Li, R. Li, Q. Li, Carbon-nanotube fibers for wearable devices and smart textiles. Adv. Mater. **28**(47), 10529–10538 (2016)
23. S. Akram, M. Ashraf, A. Javid, H.A. Abid, S. Ahmad, Y. Nawab, A. Rasheed, Z. Xue, A. Nosheen, Recent advances in electromagnetic interference (EMI) shielding textiles: a comprehensive review. Synth. Met. **294**, 117305 (2023)
24. T.V. Patil, D.K. Patel, S.D. Dutta, K. Ganguly, A. Randhawa, K.-T. Lim, Carbon nanotubes-based hydrogels for bacterial eradiation and wound-healing applications. Appl. Sci. **11**(20), 9550 (2021). https://www.mdpi.com/2076-3417/11/20/9550
25. C.J. Ellison, A. Phatak, D.W. Giles, C.W. Macosko, F.S. Bates, Melt blown nanofibers: fiber diameter distributions and onset of fiber breakup. Polymer **48**(11), 3306–3316 (2007)
26. E.S. Medeiros, G.M. Glenn, A.P. Klamczynski, W.J. Orts, L.H. Mattoso, Solution blow spinning: a new method to produce micro-and nanofibers from polymer solutions. J. Appl. Polym. Sci. **113**(4), 2322–2330 (2009)
27. S. Padron, A. Fuentes, D. Caruntu, K. Lozano, Experimental study of nanofiber production through forcespinning. J. Appl. Phys. **113**(2), 024318 (2013)
28. J.C. Stendahl, M.S. Rao, M.O. Guler, S.I. Stupp, Intermolecular forces in the self-assembly of peptide amphiphile nanofibers. Adv. Func. Mater. **16**(4), 499–508 (2006). https://doi.org/10.1002/adfm.200500161
29. J. Xue, T. Wu, Y. Dai, Y. Xia, Electrospinning and electrospun nanofibers: methods, materials, and applications. Chem. Rev. **119**(8), 5298–5415 (2019)
30. Kenry, C.T. Lim, Nanofiber technology: current status and emerging developments. Prog. Polym. Sci. **70**, 1–17 (2017). https://doi.org/10.1016/j.progpolymsci.2017.03.002
31. Y. Wang, L. Chu, S. Meng, M. Yang, Y. Yu, X. Deng, C. Qi, T. Kong, Z. Liu, Scalable and ultra-sensitive nanofibers coaxial yarn-woven triboelectric nanogenerator textile sensors for real-time gait analysis. Adv. Sci. **11**(28), 2401436 (2024). https://doi.org/10.1002/advs.202401436
32. R. Goyal, L.K. Macri, H.M. Kaplan, J. Kohn, Nanoparticles and nanofibers for topical drug delivery. J. Control. Release **240**, 77–92 (2016)
33. K. Graham, M. Gogins, H. Schreuder-Gibson, Incorporation of electrospun nanofibers into functional structures. Int. Nonwovens J. **13**(2), 1558925004os–1551300209 (2004). https://doi.org/10.1177/1558925004os-1300209
34. C. Wolf, M. Tscherner, S. Köstler, V. Ribitsch, Optochemical sensors based on polymer nanofibers with ultra-fast response characteristics. Proc. IEEE Sens. **2014**, 950–953 (2014). https://doi.org/10.1109/ICSENS.2014.6985159
35. E. Omanović-Mikličanin, A. Badnjević, A. Kazlagić, M. Hajlovac, Nanocomposites: a brief review. Heal. Technol. **10**(1), 51–59 (2020). https://doi.org/10.1007/s12553-019-00380-x
36. R.-M. Wang, S.-R. Zheng, Y.G. Zheng, *Polymer Matrix Composites and Technology* (Elsevier, 2011)
37. N. Vrinceanu, B. Ştefan, C.M. Rîmbu, A.S. Neculai-Valeanu, S. Ferrándiz, M. Suchea, Nanoparticle/biopolymer-based coatings for functionalization of textiles: recent developments (a minireview). Text. Res. J. **92**, 004051752110706 (2022). https://doi.org/10.1177/00405175211070613

38. A. Payen, Mémoire sur la composition du tissu propre des plantes et du ligneux. Comptes rendus **7**, 1052–1056 (1838)
39. L. Colombo, L. Zoia, M.B. Violatto, S. Previdi, L. Talamini, L. Sitia, F. Nicotra, M. Orlandi, M. Salmona, C. Recordati, Organ distribution and bone tropism of cellulose nanocrystals in living mice. Biomacromol **16**(9), 2862–2871 (2015)
40. H. Kargarzadeh, I. Ahmad, S. Thomas, A. Dufresne, *Handbook of Nanocellulose and Cellulose Nanocomposites* (Wiley, 2017)
41. R.J. Moon, A. Martini, J. Nairn, J. Simonsen, J. Youngblood, Cellulose nanomaterials review: structure, properties and nanocomposites. Chem. Soc. Rev. **40**(7), 3941–3994 (2011)
42. M. Roman, Toxicity of cellulose nanocrystals: a review. Ind. Biotechnol. **11**(1), 25–33 (2015)
43. C. Zinge, B. Kandasubramanian, Nanocellulose based biodegradable polymers. Eur. Polymer J. **133**, 109758 (2020)
44. R. Hendrickx, G. Desmarais, M. Weder, E.S. Ferreira, D. Derome, Moisture uptake and permeability of canvas paintings and their components. J. Cult. Herit. **19**, 445–453 (2016)
45. M.N.R. Kumar, A review of chitin and chitosan applications. React. Funct. Polym. **46**(1), 1–27 (2000)
46. H. Sadeghifar, A. Ragauskas, Lignin as a UV light blocker—a review. Polymers **12**(5), 1134 (2020)
47. H.R. Anik, S.I. Tushar, S. Mahmud, A.H. Khadem, P. Sen, M. Akter, Into the revolution of nanofusion: merging high performance and aesthetics by nanomaterials in textile finishes. Adv. Mater. Interfaces **12**(1), 2400368 (2025)
48. M. El Ouardi, M. Arab, M. Saadi, A. Baqais, H. Ait Ahsaine, Vanadate-based photocatalytic materials: a perspective on synthesis approaches and pollutants photocatalytic degradation. Nano Mater. Sci. (2024). https://doi.org/10.1016/j.nanoms.2024.11.002
49. M.L. Gulrajani, *Advances in the Dyeing and Finishing of Technical Textiles* (Elsevier, 2013)
50. J.M. Deitzel, J. Kleinmeyer, D. Harris, N.B. Tan, The effect of processing variables on the morphology of electrospun nanofibers and textiles. Polymer **42**(1), 261–272 (2001)
51. C. Güneşoglu, D. Kut, M. Orhan, Performing the electrospraying process for the application of textile nano finishing particles. Text. Res. J. **80**(2), 106–115 (2010)
52. H. Ibili, M. Daşdemir, AgCl-TiO_2/dendrimer-based nanoparticles for superhydrophobic and antibacterial multifunctional textiles. J. Text. Inst. **114**(5), 861–873 (2023)
53. J. Hu, *Active Coatings for Smart Textiles* (Woodhead Publishing, 2016)
54. M. Montazer, T. Harifi, *Nanofinishing of Textile Materials* (Woodhead Publishing, 2018)
55. M. Hossain, A.K. Biswas, S. Parvez, S. Bain, Advanced functionalization of textiles by nanofinishing: a review, in *Proceedings of the International Conference on Mechanical, Industrial and Materials Engineering, Rajshahi, Bangladesh* (2013)
56. Y. Lu, S. Jiang, Y. Huang, Ultrasonic-assisted electroless deposition of Ag on pet fabric with low silver content for EMI shielding. Surf. Coat. Technol. **204**(16–17), 2829–2833 (2010)
57. G. Allaka, V. Yepuri, Inexpensive hydrophobic and infrared reflective coatings on the cotton and silk fabrics using sol-gel dip coating technique. Mater. Today Proc. (2023)
58. W.A. Daoud, J.H. Xin, Nucleation and growth of anatase crystallites on cotton fabrics at low temperatures. J. Am. Ceram. Soc. **87**(5), 953–955 (2004)
59. W.A. Daoud, J.H. Xin, Synthesis of single-phase anatase nanocrystallites at near room temperatures. Chem. Commun. **16**, 2110–2112 (2005)
60. W.A. Daoud, J.H. Xin, Y.-H. Zhang, Surface functionalization of cellulose fibers with titanium dioxide nanoparticles and their combined bactericidal activities. Surf. Sci. **599**(1–3), 69–75 (2005)
61. Y. Egami, K. Suzuki, T. Tanaka, T. Yasuhara, E. Higuchi, H. Inoue, Preparation and characterization of conductive fabrics coated uniformly with polypyrrole nanoparticles. Synth. Met. **161**(3–4), 219–224 (2011)
62. B. Şimşek, M. Karaman, Initiated chemical vapor deposition of poly (hexafluorobutyl acrylate) thin films for superhydrophobic surface modification of nanostructured textile surfaces. J. Coat. Technol. Res. **17**(2), 381–391 (2020)

63. A.P. Periyasamy, M. Venkataraman, D. Kremenakova, J. Militky, Y. Zhou, Progress in sol-gel technology for the coatings of fabrics. Materials **13**(8), 1838 (2020)
64. A.P. Periyasamy, K. Yang, X. Xiong, M. Venkataraman, J. Militky, R. Mishra, D. Kremenakova, Effect of silanization on copper coated milife fabric with improved EMI shielding effectiveness. Mater. Chem. Phys. **239**, 122008 (2020)
65. K. Yang, A.P. Periyasamy, M. Venkataraman, J. Militky, D. Kremenakova, J. Vecernik, R. Pulíček, Resistance against penetration of electromagnetic radiation for ultra-light cu/ni-coated polyester fibrous materials. Polymers **12**(9), 2029 (2020)
66. L. Frunza, N. Preda, E. Matei, S. Frunza, C.P. Ganea, A.M. Vlaicu, L. Diamandescu, A. Dorogan, Synthetic fabrics coated with zinc oxide nanoparticles by electroless deposition: structural characterization and wetting properties. J. Polym. Sci. Part B Polym. Phys. **51**(19), 1427–1437 (2013)
67. M. Gulrajani, G. Deepti, Emerging techniques for functional finishing of textiles. Indian J. Fibre Text. Res. **36**, 388–397 (2011)
68. A. Horrocks, S. Nazaré, R. Masood, B. Kandola, D. Price, Surface modification of fabrics for improved flash-fire resistance using atmospheric pressure plasma in the presence of a functionalized clay and polysiloxane. Polym. Adv. Technol. **22**(1), 22–29 (2011)
69. J. Alongi, G. Malucelli, Cotton flame retardancy: state of the art and future perspectives. RSC Adv. **5**(31), 24239–24263 (2015)
70. M. Shariat, Plasma jet printing of silver nanoparticles on polyester fabric as a surface-enhanced Raman scattering substrate. Optik **277**, 170698 (2023)
71. T. Yeerken, G. Wang, H. Li, H. Liu, W. Yu, Chemical stable, superhydrophobic and self-cleaning fabrics prepared by two-step coating of a polytetrafluoroethylene membrane and silica nanoparticles. Text. Res. J. **89**(23–24), 4827–4841 (2019)
72. J. Yong, F. Chen, Q. Yang, J. Huo, X. Hou, Superoleophobic surfaces. Chem. Soc. Rev. **46**(14), 4168–4217 (2017)
73. K. Wang, Y. Shen, M. Duan, T. Wu, J. Zhao, J. Wang, Non-fluorinated durable water repellent and stain resistant coating, in *Proceedings of the 3rd International Conference on Materials Chemistry and Environmental Engineering (CONF-MCEE 2023), Stanford, CA, USA* (2023)
74. B. Zhang, W. Xu, Q. Zhu, B. Hou, Scalable, fluorine free and hot water repelling superhydrophobic and superoleophobic coating based on functionalized Al_2O_3 nanoparticles. J. Mater. Sci. Technol. **66**, 74–81 (2021)
75. J. Zhang, Y. Shao, C.-T. Hsieh, Y.-F. Chen, T.-C. Su, J.-P. Hsu, R.-S. Juang, Synthesis of magnetic iron oxide nanoparticles onto fluorinated carbon fabrics for contaminant removal and oil-water separation. Sep. Purif. Technol. **174**, 312–319 (2017)
76. F. Ren, H. Guo, Z.-Z. Guo, Y.-L. Jin, H.-J. Duan, P.-G. Ren, D.-X. Yan, Highly bendable and durable waterproof paper for ultra-high electromagnetic interference shielding. Polymers **11**(9), 1486 (2019)
77. Y. Feng, L. Liu, J. Zhang, H. Aslan, M. Dong, Photoactive antimicrobial nanomaterials. J. Mater. Chem. B **5**(44), 8631–8652 (2017)
78. P. Prasher, M. Singh, H. Mudila, Silver nanoparticles as antimicrobial therapeutics: current perspectives and future challenges. 3 Biotech **8**, 1–23 (2018)
79. M. Chakravarty, A. Vora, Nanotechnology-based antiviral therapeutics. Drug Deliv. Transl. Res. **11**, 748–787 (2021)
80. M. Lopez-Carrizales, K.I. Velasco, C. Castillo, A. Flores, M. Magaña, G.A. Martinez-Castanon, F. Martinez-Gutierrez, In vitro synergism of silver nanoparticles with antibiotics as an alternative treatment in multiresistant uropathogens. Antibiotics **7**(2), 50 (2018)
81. A. Salleh, R. Naomi, N.D. Utami, A.W. Mohammad, E. Mahmoudi, N. Mustafa, M.B. Fauzi, The potential of silver nanoparticles for antiviral and antibacterial applications: a mechanism of action. Nanomaterials **10**(8), 1566 (2020)
82. O. Cf, *Transforming Our World: The 2030 Agenda for Sustainable Development* (United Nations, New York, 2015)
83. M. Montazer, S. Seifollahzadeh, Enhanced self-cleaning, antibacterial and UV protection properties of nano TiO_2 treated textile through enzymatic pretreatment. Photochem. Photobiol. **87**(4), 877–883 (2011)

84. M. Li, G. Li, J. Jiang, Z. Zhang, X. Dai, K. Mai, Ultraviolet resistance and antimicrobial properties of ZnO in the polypropylene materials: a review. J. Mater. Sci. Technol. **31**(4), 331–339 (2015)
85. N.F. Attia, H.E. Ahmed, A.A. El Ebissy, S.E. El Ashery, Green and novel approach for enhancing flame retardancy, UV protection and mechanical properties of fabrics utilized in historical textile fabrics conservation. Prog. Org. Coat. **166**, 106822 (2022)
86. T.E. Elmetwaly, S.S. Darwish, N.F. Attia, R.R. Hassan, A.A. El Ebissy, A.S. Eltaweil, A.M. Omer, H.R. El-Seedi, S.E. Elashery, Cellulose nanocrystals and its hybrid composite with inorganic nanotubes as green tool for historical paper conservation. Prog. Org. Coat. **168**, 106890 (2022)
87. N.F. Attia, J. Park, H. Oh, Facile tool for green synthesis of graphene sheets and their smart free-standing UV protective film. Appl. Surf. Sci. **458**, 425–430 (2018)
88. N.F. Attia, A. Mohamed, A. Hussein, A.-G.M. El-Demerdash, S.H. Kandil, Greener bio-based spherical nanoparticles for efficient multilayer textile fabrics nanocoating with outstanding fire retardancy, toxic gases suppression, reinforcement and antibacterial properties. Surf. Interfaces **36**, 102595 (2023)
89. A.R. Horrocks, Developments in flame retardants for heat and fire resistant textiles—the role of char formation and intumescence. Polym. Degrad. Stab. **54**(2–3), 143–154 (1996)
90. A.B. Morgan, C.A. Wilkie, *Flame Retardant Polymer Nanocomposites* (Wiley, 2007)
91. M. Li, M. Prabhakar, J.-i Song, Effect of synthesized lignin-based flame retardant liquid on the flame retardancy and mechanical properties of cotton textiles. Ind. Crops Prod. **212**, 118283 (2024)
92. G. Malucelli, Biomacromolecules and bio-sourced products for the design of flame retarded fabrics: current state of the art and future perspectives. Molecules **24**(20), 3774 (2019)
93. A.K.R. Choudhury, *Flame Retardants for Textile Materials* (CRC Press, 2020)
94. G. Makhlouf, A. Abdelkhalik, H. Ameen, Preparation of highly efficient chitosan-based flame retardant coatings with good antibacterial properties for cotton fabrics. Prog. Org. Coat. **163**, 106627 (2022)
95. F. Carosio, J. Alongi, Flame retardant multilayered coatings on acrylic fabrics prepared by one-step deposition of chitosan/montmorillonite complexes. Fibers **6**(2), 36 (2018)
96. N. Karim, S. Afroj, S. Tan, K.S. Novoselov, S.G. Yeates, All inkjet-printed graphene-silver composite ink on textiles for highly conductive wearable electronics applications. Sci. Rep. **9**(1), 8035 (2019)
97. N.M. Badawi, K.M. Batoo, S. Hussain, N. Agrawal, M. Bhuyan, S. Bashir, R. Subramaniam, R. Kasi, Design and characterization of electroconductive graphene-coated cotton fabric for wearable electronics. Coatings **13**(9), 1601 (2023)
98. K. Pattanarat, N. Petchsang, T. Osotchan, I.-M. Tang, Y.-H. Kim, R. Jaisutti, High conductivity and durability textile gas sensor-based polyaniline-decorated-poly (3, 4-ethylenedioxythiophene)/poly (4-styrenesulfonate) for ammonia detection. ACS Appl. Polym. Mater. **4**(12), 9006–9014 (2022)
99. R. Attia, N. Yousif, R. Helal, N. Ali, Fabricating electronic textile using nano MnO_2/polyaniline composites for capacitor device. J. Ind. Text. **51**(7), 1161–1180 (2022)
100. A. Basuoni, H. El-Sayed, Developing durable anti-static hydrophilic acrylic fabrics with improved dyeability with cationic and anionic dyes. Emergent Mater. **6**(4), 1339–1350 (2023)
101. M.M. Rahman, M. Mashud, M.M. Rahman, *Advanced Technology in Textiles: Fibre to Apparel* (Springer Nature, 2023)
102. S. Bhattacharjee, C.R. Macintyre, X. Wen, P. Bahl, U. Kumar, A.A. Chughtai, R. Joshi, Nanoparticles incorporated graphene-based durable cotton fabrics. Carbon **166**, 148–163 (2020)
103. W. Seung, M.K. Gupta, K.Y. Lee, K.-S. Shin, J.-H. Lee, T.Y. Kim, S. Kim, J. Lin, J.H. Kim, S.-W. Kim, Nanopatterned textile-based wearable triboelectric nanogenerator. ACS Nano **9**(4), 3501–3509 (2015)
104. B. Deopura, R. Alagirusamy, M. Joshi, B. Gupta, *Polyesters and Polyamides* (Elsevier, 2008)

105. N.F. Attia, M.S. Morsy, Facile synthesis of novel nanocomposite as antibacterial and flame retardant material for textile fabrics. Mater. Chem. Phys. **180**, 364–372 (2016)
106. N.R. Dhineshbabu, S. Bose, UV resistant and fire retardant properties in fabrics coated with polymer based nanocomposites derived from sustainable and natural resources for protective clothing application. Compos. B Eng. **172**, 555–563 (2019)
107. M.M. Hassan, S.J. Leighs, Effect of surface treatments on physicomechanical, stain-resist, and UV protection properties of wool fabrics. Appl. Surf. Sci. **419**, 348–356 (2017)
108. M. Shahid, S. Maiti, R.V. Adivarekar, S. Liu, Biomaterial based fabrication of superhydrophobic textiles—a review. Mater. Today Chem. **24**, 100940 (2022)
109. A. Arputharaj, V. Nadanathangam, S.R. Shukla, Development of multi-functional cotton surface for sportswear using nano zinc oxide. J. Nat. Fibers **17**(3), 346–358 (2020)
110. B. Patel, T.N. Shaikh, S. Chaudhari, Sustainable augmentation in lyocell knitted sportswear functionality by in-situ synthesis of biogenetically reduced silver nanoparticles. Hybrid Adv. **3**, 100075 (2023)
111. A. Subic, *Materials in Sports Equipment* (Woodhead Publishing, 2019)
112. S. Hussain, F. Wahab, M. Usama, A. Arshad, H. Khan, S. Gul, Materials chemistry: properties and applications. Chem. Mod. Approach **29**, 34–216 (2024)
113. H. Mehling, L.F. Cabeza, Heat and cold storage with PCM. Heat Mass Transf. **15**, 11–55 (2008)
114. P.M. Conn, *Imaging and Spectroscopic Analysis of Living Cells: Optical and Spectroscopic Techniques*, vol. 504 (Academic Press, 2012)
115. B.S. Priyadarshini, R. Mitra, U. Manju, Titania nanoparticle-stimulated ultralow frequency detection and high-pass filter behavior of a flexible piezoelectric nanogenerator: a self-sustaining energy harvester for active motion tracking. ACS Appl. Mater. Interfaces **15**(39), 45812–45822 (2023)
116. R.K. Rajamony, M. Samykano, Applications of nano-enhanced phase change materials in textiles, in *Nano Enhanced Phase Change Materials: Preparation, Properties and Applications* (Springer, 2023), pp. 201–222
117. J. Barman, A. Tirkey, S. Batra, A.A. Paul, K. Panda, R. Deka, P.J. Babu, The role of nanotechnology based wearable electronic textiles in biomedical and healthcare applications. Mater. Today Commun. **32**, 104055 (2022)
118. T. Lim, H.J. Kim, S. Won, C.H. Kim, J. Yoo, J.H. Lee, K.S. Son, I.-W. Nam, K. Kim, S.Y. Yeo, Liquid metal-based electronic textiles coated with au nanoparticles as stretchable electrode materials for healthcare monitoring. ACS Appl. Nano Mater. **6**(10), 8482–8494 (2023)
119. M. Singh, M. Briones, G. Ott, D. O'Hagan, Cationic microparticles: a potent delivery system for DNA vaccines. Proc. Natl. Acad. Sci. **97**(2), 811–816 (2000)
120. W. Tan, K. Wang, X. He, X.J. Zhao, T. Drake, L. Wang, R.P. Bagwe, Bionanotechnology based on silica nanoparticles. Med. Res. Rev. **24**(5), 621–638 (2004). https://doi.org/10.1002/med.20003
121. S. Anjum, S. Ishaque, H. Fatima, W. Farooq, C. Hano, B.H. Abbasi, I. Anjum, Emerging applications of nanotechnology in healthcare systems: grand challenges and perspectives. Pharmaceuticals **14**(8), 707 (2021)
122. V.T. Novi, A. Gonzalez, J. Brockgreitens, A. Abbas, Highly efficient and durable antimicrobial nanocomposite textiles. Sci. Rep. **12**(1), 17332 (2022)
123. T.-L. Su, T.-P. Chen, J. Liang, Green in-situ synthesis of silver coated textiles for wide hygiene and healthcare applications. Colloids Surf. A **657**, 130506 (2023)
124. B. Butola, A. Kumar, Green chemistry based in-situ synthesis of silver nanoparticles for multifunctional finishing of chitosan polysaccharide modified cellulosic textile substrate. Int. J. Biol. Macromol. **152**, 1135–1145 (2020)
125. A. Fouda, E. Saad, S.S. Salem, T.I. Shaheen, In-vitro cytotoxicity, antibacterial, and UV protection properties of the biosynthesized zinc oxide nanoparticles for medical textile applications. Microb. Pathog. **125**, 252–261 (2018)
126. J.S. Meena, S.B. Choi, S.-B. Jung, J.-W. Kim, Electronic textiles: new age of wearable technology for healthcare and fitness solutions. Materials Today Bio **19**, 100565 (2023)

127. N. Kumar, A. Dixit, *Nanotechnology for Defence Applications* (Springer, 2019)
128. P. Kiekens, S. Jayaraman, *Intelligent Textiles and Clothing for Ballistic and NBC Protection: Technology at the Cutting Edge* (Springer Science & Business Media, Dordrecht, 2012)
129. R. Revaiah, T. Kotresh, B. Kandasubramanian, Technical textiles for military applications. J. Text. Inst. (2020)
130. G.T.P. Azar, S. Danilova, L. Krishnan, Y. Fedutik, A.J. Cobley, Selective electroless copper plating of ink-jet printed textiles using a copper-silver nanoparticle catalyst. Polymers **14**(17), 3467 (2022)
131. H.N. Siphesihle, M.P. Hlengiwe, S. Singh, A framework for using nanotechnology in military gear, in *Nanomaterials: Biomedical, Environmental, and Engineering Applications* (2018), pp. 239–255
132. S. Bhattacharjee, C.R. Macintyre, P. Bahl, U. Kumar, X. Wen, K.F. Aguey-Zinsou, A.A. Chughtai, R. Joshi, Reduced graphene oxide and nanoparticles incorporated durable electroconductive silk fabrics. Adv. Mater. Interfaces **7**(20), 2000814 (2020)
133. R. Revaiah, T. Kotresh, B. Kandasubramanian, Studies on properties of coated para-aramid fabric samples developed for military applications–post UVA-340 exposure. J. Text. Inst. **115**(4), 581–592 (2024)
134. R. Revaiah, T. Kotresh, B. Kandasubrmanian, Military customized fabric hyperbaric chamber-design and safety aspects. J. Text. Inst. **113**(5), 869–881 (2022)
135. S. Veeralingam, A. Gandrothula, S. Badhulika, Tungsten oxysulfide nanoparticles interspersed nylon based e-textile as a low cost, wearable multifunctional platform for ultra-sensitive tactile sensing and breath sensing applications. Mater. Res. Bull. **160**, 112133 (2023)
136. C. Kwon, D. Seong, J. Ha, D. Chun, J.H. Bae, K. Yoon, M. Lee, J. Woo, C. Won, S. Lee, Self-bondable and stretchable conductive composite fibers with spatially controlled percolated Ag nanoparticle networks: Novel integration strategy for wearable electronics. Adv. Func. Mater. **30**(49), 2005447 (2020)
137. H. Kim, A. Shaqeel, S. Han, J. Kang, J. Yun, M. Lee, S. Lee, J. Kim, S. Noh, M. Choi, In situ formation of ag nanoparticles for fiber strain sensors: toward textile-based wearable applications. ACS Appl. Mater. Interfaces **13**(33), 39868–39879 (2021)
138. K. Virmani, C. Deepak, S. Sharma, U. Chadha, S.K. Selvaraj, Nanomaterials for automotive outer panel components: a review. Eur. Phys. J. Plus **136**(9), 921 (2021)
139. A. Bashari, M. Shakeri, A.R. Shirvan, S.A.N. Najafabadi, Functional finishing of textiles via nanomaterials, in *Nanomaterials in the Wet Processing of Textiles* (2018), pp. 1–70
140. B. Banerjee, *Rubber Nanocomposites and Nanotextiles: Perspectives in Automobile Technologies* (Walter de Gruyter GmbH & Co KG, 2019)
141. D. Kumar, P.M. Mohite, S. Kamle, Dragonfly inspired nanocomposite flapping wing for micro air vehicles. J. Bionic Eng. **16**, 894–903 (2019)
142. S. Chakrabarty, K. Jasuja, Applications of nanomaterials in the textile industry, in *Nanoscale Engineering of Biomaterials: Properties and Applications* (Springer, 2022), pp. 567–587)
143. F.M. Mwema, T.-C. Jen, L. Zhu, *Thin Film Coatings: Properties, Deposition, and Applications* (CRC Press, 2022)
144. W. Fung, J.M. Hardcastle, *Textiles in Automotive Engineering* (Elsevier, 2000)
145. I.G. Schwarz, S. Kovačević, I. Kos, Physical–mechanical properties of automotive textile materials. J. Ind. Text. **45**(3), 323–337 (2015)
146. J.P. Arenas, Applications of acoustic textiles in automotive/transportation, in *Acoustic Textiles* (2016), pp. 143–163
147. S. Shahidi, J. Wiener, Antibacterial agents in textile industry. Antimicrob. Agents **2012**, 387–406 (2012)
148. C.R. Mendes, G. Dilarri, C.F. Forsan, Vd.M.R. Sapata, P.R.M. Lopes, P.B. de Moraes, R.N. Montagnolli, H. Ferreira, E.D. Bidoia, Antibacterial action and target mechanisms of zinc oxide nanoparticles against bacterial pathogens. Sci. Rep. **12**(1), 2658 (2022). https://doi.org/10.1038/s41598-022-06657-y
149. R. Dastjerdi, M. Montazer, A review on the application of inorganic nano-structured materials in the modification of textiles: focus on anti-microbial properties. Colloids Surf. B **79**(1), 5–18 (2010)

150. A.P. Ingle, N. Duran, M. Rai, Bioactivity, mechanism of action, and cytotoxicity of copper-based nanoparticles: a review. Appl. Microbiol. Biotechnol. **98**, 1001–1009 (2014)
151. M.D. Lima, N. Li, M. Jung de Andrade, S. Fang, J. Oh, G.M. Spinks, M.E. Kozlov, C.S. Haines, D. Suh, J. Foroughi, Electrically, chemically, and photonically powered torsional and tensile actuation of hybrid carbon nanotube yarn muscles. Science **338**(6109), 928–932 (2012)
152. P.K. Jain, X. Huang, I.H. El-Sayed, M.A. El-Sayed, Noble metals on the nanoscale: optical and photothermal properties and some applications in imaging, sensing, biology, and medicine. Acc. Chem. Res. **41**(12), 1578–1586 (2008)
153. L. Pan, Y. Xie, H. Yang, M. Li, X. Bao, J. Shang, R.-W. Li, Flexible magnetic sensors. Sensors **23**(8), 4083 (2023)
154. E.I. Rabea, M.E.-T. Badawy, C.V. Stevens, G. Smagghe, W. Steurbaut, Chitosan as antimicrobial agent: applications and mode of action. Biomacromol **4**(6), 1457–1465 (2003)
155. F. Hussain, M. Hojjati, M. Okamoto, R.E. Gorga, Polymer-matrix nanocomposites, processing, manufacturing, and application: an overview. J. Compos. Mater. **40**(17), 1511–1575 (2006)
156. T.S. Peixe, E. Souza Nascimento, K.L. Schofield, A.S.A. Arcuri, R.P. Bulcão, Nanotoxicology and exposure in the occupational setting. Occup. Dis. Environ. Med. **03**(03), 35–48 (2015). https://doi.org/10.4236/odem.2015.33005
157. L. Almeida, D. Ramos, Health and safety concerns of textiles with nanomaterials. IOP Conf. Ser. Mater. Sci. Eng. **254**, 102002 (2017). https://doi.org/10.1088/1757-899x/254/10/102002
158. C. Bianco, S. Kezic, M.J. Visser, O. Pluut, G. Adami, P. Krystek, Pilot study on the identification of silver in skin layers and urine after dermal exposure to a functionalized textile. Talanta **136**, 23–28 (2015). https://doi.org/10.1016/j.talanta.2014.12.043
159. S.M. Hussain, L.K. Braydich-Stolle, A.M. Schrand, R.C. Murdock, K.O. Yu, D.M. Mattie, J.J. Schlager, M. Terrones, Toxicity evaluation for safe use of nanomaterials: recent achievements and technical challenges. Adv. Mater. **21**(16), 1549–1559 (2009). https://doi.org/10.1002/adma.200801395
160. B. Mubeen, A.N. Ansar, R. Rasool, I. Ullah, S.S. Imam, S. Alshehri, M.M. Ghoneim, S.I. Alzarea, M.S. Nadeem, I. Kazmi, Nanotechnology as a novel approach in combating microbes providing an alternative to antibiotics. Antibiotics **10**(12), 1473 (2021)
161. C. Chen, Y.-F. Li, Y. Qu, Z. Chai, Y. Zhao, Advanced nuclear analytical and related techniques for the growing challenges in nanotoxicology. Chem. Soc. Rev. **42**(21), 8266–8303 (2013). https://doi.org/10.1039/C3CS60111K
162. P.A. Schulte, F. Salamanca-Buentello, Ethical and scientific issues of nanotechnology in the workplace (0091-6765 (print)) (2007)
163. J.H. Shin, S.G. Han, J.K. Kim, B.W. Kim, J.H. Hwang, J.S. Lee, J.H. Lee, J.E. Baek, T.G. Kim, K.S. Kim, H.S. Lee, N.W. Song, K. Ahn, Yu IJ 5-day repeated inhalation and 28-day post-exposure study of graphene. Nanotoxicology **9**(8), 1023–1031 (2015)
164. S.T. Stern, S.E. McNeil, Nanotechnology safety concerns revisited. Toxicol. Sci. **101**(1), 4–21 (2008). https://doi.org/10.1093/toxsci/kfm169
165. R. Sequeira, A. Genaidy, R. Shell, W. Karwowski, G. Weckman, S. Salem, The nano enterprise: a survey of health and safety concerns, considerations, and proposed improvement strategies to reduce potential adverse effects. Hum. Factors Ergon. Manuf. Serv. Ind. **16**(4), 343–368 (2006). https://doi.org/10.1002/hfm.20057
166. M.Á. Gardetti, S.S. Muthu, *Handbook of Sustainable Luxury Textiles and Fashion*, vol. 1. Springer (2015)
167. H. Saleem, S.J. Zaidi, Sustainable use of nanomaterials in textiles and their environmental impact. Materials **13**(22), 5134 (2020)
168. S.S. Muthu, *Textiles and Clothing Sustainability: Nanotextiles and Sustainability* (Springer, Hong Kong, 2016)
169. S.S. Palaskar, A. Desai, S. Shukla, Plasma induced nano-finish for multifunctional properties on cotton fabric. Indian J. Fibre Text. Res. **41**(3), 325–330 (2016)

170. Z.A. Raza, S. Rehan, M.S. Naeem, K.J.F. Hassan, A sustainable approach in the development of nano-silver deposited conductive cellulosic fabric: a comparison of various bio-based reductants. Fibers Polym. **24**(8), 2731–2741 (2023)
171. P.S. Vankar, D. Shukla, Biosynthesis of silver nanoparticles using lemon leaves extract and its application for antimicrobial finish on fabric. Appl. Nanosci. **2**, 163–168 (2012)
172. X. Peng, M. Umer, M.N. Pervez, K.F. Hasan, M.A. Habib, M.S. Islam, L. Lin, X. Xiong, V. Naddeo, Y. Cai, Biopolymers-based microencapsulation technology for sustainable textiles development: a short review. Case Stud. Chem. Environ. Eng. **7**, 100349 (2023)
173. N. Singh, M. Yadav, S. Khanna, O. Sahu, Sustainable fragrance cum antimicrobial finishing on cotton: indigenous essential oil. Sustain. Chem. Pharm. **5**, 22–29 (2017)
174. S. Saif, A. Tahir, Y. Chen, Green synthesis of iron nanoparticles and their environmental applications and implications. Nanomaterials **6**(11), 209 (2016)
175. L. Huang, X. Weng, Z. Chen, M. Megharaj, R. Naidu, Green synthesis of iron nanoparticles by various tea extracts: comparative study of the reactivity. Spectrochim. Acta Part A Mol. Biomol. Spectrosc. **130**, 295–301 (2014)
176. F. Luo, D. Yang, Z. Chen, M. Megharaj, R. Naidu, One-step green synthesis of bimetallic Fe/Pd nanoparticles used to degrade orange II. J. Hazard. Mater. **303**, 145–153 (2016)
177. S. Machado, S. Pinto, J. Grosso, H. Nouws, J.T. Albergaria, C. Delerue-Matos, Green production of zero-valent iron nanoparticles using tree leaf extracts. Sci. Total. Environ. **445**, 1–8 (2013)
178. K. Elumalai, S. Velmurugan, Green synthesis, characterization and antimicrobial activities of zinc oxide nanoparticles from the leaf extract of azadirachta indica (l.). Appl. Surf. Sci. **345**, 329–336 (2015)
179. B. Ajitha, Y.A.K. Reddy, P.S. Reddy, Green synthesis and characterization of silver nanoparticles using lantana camara leaf extract. Mater. Sci. Eng. C **49**, 373–381 (2015)
180. G. Singaravelu, J. Arockiamary, V.G. Kumar, K. Govindaraju, A novel extracellular synthesis of monodisperse gold nanoparticles using marine alga, sargassum wightii greville. Colloids Surf. B **57**(1), 97–101 (2007)
181. K.N. Thakkar, S.S. Mhatre, R.Y. Parikh, Biological synthesis of metallic nanoparticles. Nanomedicine Nanotechnol. Biol. Med. **6**(2), 257–262 (2010)
182. G.E. Hoag, J.B. Collins, J.L. Holcomb, J.R. Hoag, M.N. Nadagouda, R.S. Varma, Degradation of bromothymol blue by 'greener' nano-scale zero-valent iron synthesized using tea polyphenols. J. Mater. Chem. **19**(45), 8671–8677 (2009)
183. Y. Kuang, Q. Wang, Z. Chen, M. Megharaj, R. Naidu, Heterogeneous fenton-like oxidation of monochlorobenzene using green synthesis of iron nanoparticles. J. Colloid Interface Sci. **410**, 67–73 (2013). https://doi.org/10.1016/j.jcis.2013.08.020
184. K. Mohan Kumar, B.K. Mandal, K. Siva Kumar, P. Sreedhara Reddy, B. Sreedhar, Biobased green method to synthesise palladium and iron nanoparticles using terminalia chebula aqueous extract. Spectrochim. Acta Part A Mol. Biomol. Spectrosc. **102**, 128–133 (2013). https://doi.org/10.1016/j.saa.2012.10.015
185. E.C. Njagi, H. Huang, L. Stafford, H. Genuino, H.M. Galindo, J.B. Collins, G.E. Hoag, S.L. Suib, Biosynthesis of iron and silver nanoparticles at room temperature using aqueous sorghum bran extracts. Langmuir **27**(1), 264–271 (2011). https://doi.org/10.1021/la103190n
186. M. Senthil, C. Ramesh, Biogenic synthesis of Fe_3O_4 nanoparticles using tridax procumbens leaf extract and its antibacterial activity on pseudomonas aeruginosa. Dig. J. Nanomater. Biostructures **7**(4), 1655 (2012)
187. T. Shahwan, S. Abu Sirriah, M. Nairat, E. Boyacı, A.E. Eroğlu, T.B. Scott, K.R. Hallam, Green synthesis of iron nanoparticles and their application as a fenton-like catalyst for the degradation of aqueous cationic and anionic dyes. Chem. Eng. J. **172**(1), 258–266 (2011). https://doi.org/10.1016/j.cej.2011.05.103
188. S. Thakur, N. Karak, One-step approach to prepare magnetic iron oxide/reduced graphene oxide nanohybrid for efficient organic and inorganic pollutants removal. Mater. Chem. Phys. **144**(3), 425–432 (2014). https://doi.org/10.1016/j.matchemphys.2014.01.015

189. T. Wang, J. Lin, Z. Chen, M. Megharaj, R. Naidu, Green synthesized iron nanoparticles by green tea and eucalyptus leaves extracts used for removal of nitrate in aqueous solution. J. Clean. Prod. **83**, 413–419 (2014). https://doi.org/10.1016/j.jclepro.2014.07.006
190. C.P. Devatha, A.K. Thalla, Green synthesis of nanomaterials, in *Synthesis of Inorganic Nanomaterials*. ed. by S. Mohan Bhagyaraj, O.S. Oluwafemi, N. Kalarikkal, S. Thomas (Woodhead Publishing, San Diego, 2018), pp.169–184
191. H. Bao, Y. Hong, T. Yan, X. Xie, X. Zeng, A systematic review of biodegradable materials in the textile and apparel industry. J. Text. Inst. **115**(7), 1173–1192 (2024)
192. M.A. Patwary, S. Surid, M. Gafur, Properties and applications of biodegradable polymers. J. Res. Updat. Polym. Sci. **9**, 32–41 (2020). https://doi.org/10.6000/1929-5995.2020.09.03
193. H. Kargarzadeh, M. Mariano, D. Gopakumar, I. Ahmad, S. Thomas, A. Dufresne, J. Huang, N. Lin, Advances in cellulose nanomaterials. Cellulose **25**, 2151–2189 (2018)
194. H.A. Khalil, A. Bhat, A.I. Yusra, Green composites from sustainable cellulose nanofibrils: a review. Carbohyd. Polym. **87**(2), 963–979 (2012)
195. Y. Chen, F. Wang, L. Dong, Z. Li, L. Chen, X. He, J. Gong, J. Zhang, Q. Li, Design and optimization of flexible polypyrrole/bacterial cellulose conductive nanocomposites using response surface methodology. Polymers **11**(6), 960 (2019)
196. S. Razzaq, S. Shahid, Y. Nawab, Applications and environmental impact of biodegradable polymers in textile industry: a review. Int. J. Biol. Macromol. **282**, 136791 (2024). https://doi.org/10.1016/j.ijbiomac.2024.136791
197. R. Dormanesh, S. Mohseni, Green synthesis of zinc oxide nanoparticles and their application in ring opening of epoxides. Prog. Chem. Biochem. Res. **4**(4), 403–413 (2021). https://doi.org/10.22034/pcbr.2021.293421.1193
198. S. Iravani, H. Korbekandi, S.V. Mirmohammadi, B. Zolfaghari, Synthesis of silver nanoparticles: chemical, physical and biological methods. Res. Pharm. Sci. **9**(6), 385–406 (2014)
199. J. Singh, T. Dutta, K.-H. Kim, M. Rawat, P. Samddar, P. Kumar, 'Green'synthesis of metals and their oxide nanoparticles: applications for environmental remediation. J. Nanobiotechnology **16**, 1–24 (2018)
200. B. Mahltig, Nanomaterials and textiles: Review on materials and applications. J. Eng. Fibers Fabr. **20**, 15589250251318010 (2025). https://doi.org/10.1177/15589250251318010
201. A. Liberman, N. Mendez, W.C. Trogler, A.C. Kummel, Synthesis and surface functionalization of silica nanoparticles for nanomedicine. Surf. Sci. Rep. **69**(2), 132–158 (2014). https://doi.org/10.1016/j.surfrep.2014.07.001
202. D.G. Papageorgiou, I.A. Kinloch, R.J. Young, Mechanical properties of graphene and graphene-based nanocomposites. Prog. Mater. Sci. **90**, 75–127 (2017)
203. T. Xu, Z. Zhang, L. Qu, Graphene-based fibers: recent advances in preparation and application. Adv. Mater. **32**(5), 1901979 (2020). https://doi.org/10.1002/adma.201901979
204. E. Pakdel, H. Zhao, J. Wang, B. Tang, R.J. Varley, X. Wang, Superhydrophobic and photocatalytic self-cleaning cotton fabric using flower-like n-doped TiO_2/PDMS coating. Cellulose **28**(13), 8807–8820 (2021). https://doi.org/10.1007/s10570-021-04075-3
205. N. Singh, Circular economy in textile and apparel supply chains, in *Innovation and Sustainability Through Circular Economy in Businesses*. ed. by S.A.R. Khan (IGI Global, Hershey, PA, 2025), pp.25–64
206. B. Cardoso, G. Nobrega, I.S. Afonso, J.E. Ribeiro, R.A. Lima, Sustainable green synthesis of metallic nanoparticle using plants and microorganisms: a review of biosynthesis methods, mechanisms, toxicity, and applications. J. Environ. Chem. Eng. **13**(3), 116921 (2025). https://doi.org/10.1016/j.jece.2025.116921
207. P.D. Sarvalkar, S.D. Barawkar, O.S. Karvekar, P.D. Patil, S.R. Prasad, K.K. Sharma, N.R. Prasad, R.S. Vhatkar, A review on multifunctional nanotechnological aspects in modern textile. J. Text. Inst. **114**(3), 470–487 (2022)
208. H. Yu, W. Feng, Carbon-based multifunctional nanomaterials. Nanomaterials **14**, 1600 (2024). https://doi.org/10.3390/nano14191600

209. C.M. Costa, V.F. Cardoso, P. Martins, D.M. Correia, R. Gonçalves, P. Costa, V. Correia, C. Ribeiro, M.M. Fernandes, P.M. Martins, S. Lanceros-Méndez, Smart and multifunctional materials based on electroactive poly(vinylidene fluoride): recent advances and opportunities in sensors, actuators, energy, environmental, and biomedical applications. Chem. Rev. **123**(19), 11392–11487 (2023). https://doi.org/10.1021/acs.chemrev.3c00196
210. J. Kim, S. Song, Y.-H. Kim, S.K. Park, Recent progress of quantum dot-based photonic devices and systems: a comprehensive review of materials, devices, and applications. Small Struct. **2**(3), 2000024 (2021). https://doi.org/10.1002/sstr.202000024
211. S. Bari, A. Chatterjee, S. Mishra, Biodegradable polymer nanocomposites—an overview. Polym. Rev. **56**, 00–00 (2016). https://doi.org/10.1080/15583724.2015.1118123
212. M. Sharma, N. Tellili, I. Kacem, T. Rouissi, Microbial biopolymers: from production to environmental applications—a review. Appl. Sci. **14**(12), 5081 (2024)

Chapter 9
Instrumental Analysis and Characterization Techniques in Textile Surface Modification and Functionalization

Sina Seipel and Mohammad Neaz Morshed

Abstract Textile surface modification and functionalization techniques are designed to induce controlled physicochemical changes at the fiber surface while maintaining the intrinsic bulk properties of the textile substrate. Precise characterization of these surface-specific changes is essential for understanding structure–property relationships. Surface-sensitive analytical methods are required to evaluate alterations in chemical composition, surface morphology, interfacial behavior, and nanoscale topography resulting from physical, chemical, and biological modification processes. Spectroscopic, microscopic, and interfacial measurement techniques play a critical role in quantifying functional group incorporation, surface energy variations, and microstructural evolution at the fiber interface. The chapter concisely describes commonly used instrumental analysis and characterization techniques with emphasis on wettability and surface energy evaluation, surface topology, surface chemistry, and functional property assessment.

Keywords Textile surface modification & functionalization · Instrumental characterization · Surface chemistry · Wettability and surface energy · Functional textile

S. Seipel · M. N. Morshed (✉)
Department of Textile Technology, Faculty of Textiles, Engineering and Business (Including the Swedish School of Textiles), University of Borås, Borås, Sweden
e-mail: mohammad_neaz.morshed@hb.se

S. Seipel
e-mail: sina.seipel@hb.se

M. N. Morshed and M. Shahid (eds.), *Advances in Surface Modification and Functionalization in Textiles*, Textile Science and Clothing Technology,
https://doi.org/10.1007/978-3-032-23377-6_9

9.1 Introduction

Textile surface modification and functionalization has gained significant importance in recent years due to the increasing demand for multifunctional and high-performance textile materials [1–3]. Unlike bulk modification, surface modification techniques are designed to selectively alter the outermost layers of fibers while preserving the intrinsic mechanical and structural properties of the textile substrate [4]. This approach enables the development of textiles with enhanced wettability, adhesion, dyeability, antimicrobial activity, UV protection, flame retardancy, electrical conductivity, and other advanced functionalities required for technical, medical, protective, and smart textile applications [1, 3, 5].

A wide range of surface modification and functionalization methods are employed in textile processing, including plasma treatment [6], chemical grafting [7], enzymatic modification [8], layer-by-layer deposition [9], sol–gel coatings [10], nanotechnology-based finishes [11], and beyond [12]. These processes induce changes at the molecular or nanoscale level, affecting surface chemistry, morphology, roughness, and interfacial properties [5]. Since the modified region is confined to a very thin surface layer, often only a few nanometers to micrometers in depth, conventional bulk characterization techniques are insufficient to fully capture these changes [4]. Consequently, advanced instrumental analysis and characterization techniques are used for accurate evaluation of textile surface modifications and functionalization. Instrumental characterization provides critical insights into the physicochemical nature of modified textile surfaces and helps establish correlations between surface structure and functional performance [5]. Surface-sensitive spectroscopic techniques such as Fourier transform infrared spectroscopy (FTIR), X-ray photoelectron spectroscopy (XPS), Raman spectroscopy, and secondary ion mass spectrometry (SIMS) are widely used to identify functional groups, elemental composition, and chemical bonding introduced during surface treatments [13–15]. In addition to chemical analysis, surface morphology and topography play a crucial role in determining textile performance, particularly in applications involving wettability, adhesion, friction, and coating durability. Microscopic techniques such as scanning electron microscopy (SEM), atomic force microscopy (AFM), and digital holographic microscopy enable detailed visualization of surface features, roughness, and nanoscale structures. Particle sizing and dispersion analysis using laser diffraction and dynamic light scattering further support the evaluation of nanomaterial-based surface treatments [1, 4, 14]. Wettability and surface energy analysis constitute another important aspect of textile surface characterization. Contact angle measurements, wicking tests, peel adhesion tests, dynamic surface tension analysis, and tensiometric methods are commonly employed to assess interfacial behavior and liquid–solid interactions [16]. These measurements are particularly relevant for applications involving moisture management, coating adhesion, and functional finishing processes [6, 7, 9]. Finally, functional property evaluation methods are essential for correlating surface characteristics with real-world performance and durability. Mechanical, thermal, electrical, optical, and biological testing methods are often used to assess the effectiveness and long-term stability of surface-modified textiles. In our previous studies on the surface modification and functionalization of polyester textiles using plasma, dendrimer, metal nanoparticles, as well as conductive coating, characterization techniques were essential to confirm successful surface modification and functionalization and to evaluate the resulting electrical conductivity and catalytic performance [14, 15, 17, 18]. They enabled correlation between nanomaterial deposition, surface morphology, and the textile's multifunctional properties, ensuring robust and durable performance. A similar approach can also be observed in various other literature, including Tudu et al. [19], Xu et al. [20], Meng et al. [21] and more.

A systematic understanding of instrumental analysis and characterization techniques is therefore indispensable for the rational design, optimization, and commercialization of advanced textile surface engineering technologies. Based on that, the following section of the chapter will discuss the common instrumental analysis and characterization techniques in textile surface modification and functionalization.

9.2 Common Instrumental Analysis and Characterization Techniques

A diverse range of characterization tools and techniques, including spectroscopic, microscopic, interfacial, thermal, and mechanical methods, are used in evaluating chemical, morphological, and functional changes in the textile surface induced by surface modification and functionalization. These analysis and characterization techniques will be described in this chapter by dividing them into four primary categories: (a) wettability and surface energy analysis; (b) surface topology analysis; (c) surface chemistry analysis; (d) functional properties analysis. The following sections will describe the common instrumental analysis and characterization techniques under each primary category.

9.2.1 Wettability and Surface Energy Analysis

Wettability describes how a liquid spreads on a textile surface and is most commonly quantified by the contact angle (θ). A low contact angle ($\theta < 90°$) indicates good wettability and strong solid–liquid interactions, while a high contact angle ($\theta > 90°$) reflects poor wettability and surface hydrophobicity. It is defined geometrically as the angle formed by a liquid at the three-phase boundary point where a liquid, gas, and solid intersect. The well-known Young Equation (Eq. 9.1) describes the balance at the three-phase contact of solid, liquid, and gas [22].

$$\gamma_{SV} = \gamma_{Sl} + \gamma_{lv} \cos \theta_Y \quad (9.1)$$

The equilibrium wetting behavior of a liquid on a solid surface is governed by the balance of interfacial tensions at the three-phase contact line. These include γ_{lv}, the surface tension of the liquid, γ_{sl}, the interfacial tension between the solid and the liquid, and γ_{sv}, the surface tension (or surface free energy) of the solid. The interplay among these three energies determines the equilibrium contact angle, commonly known as Young's contact angle (θ_Y), which represents the thermodynamic condition of wetting at equilibrium [16].

The common measurement approaches for wettability include the sessile droplet test, wicking test, T-peel test, and staining tests, which together assess droplet spreading, capillary transport, adhesion-related wetting, and visual penetration on textile surfaces [23, 24]. Additional methods include dynamic contact angle measurements (advancing and receding angles) to capture wetting kinetics and surface heterogeneity [25], capillary rise (Washburn) analysis to quantify liquid uptake in porous yarns and fabrics [26], drop penetration time tests to evaluate absorbency [27], and surface energy determination using multiple probe liquids to distinguish polar and dispersive interactions [28]. Practical wetting behavior is further characterized through liquid uptake and absorption capacity tests, spreading area analysis, standardized ink or dye wetting tests (AATCC/ISO), and microscopic or high-speed imaging techniques that visualize liquid spreading and penetration at fiber and fabric scales [29, 30].

On the other hand, surface energy (γ_s) is the thermodynamic driver behind wettability. High-energy surfaces promote spreading and adhesion, whereas low-energy surfaces resist wetting. One of the easiest ways to gather information about the surface energy of solids is to determine the contact angles that arise between the solid surface and liquids. Textile surface energy is typically resolved into dispersive (γ_s^d) and polar (γ_s^p) components [31]. Natural fibers like cotton exhibit higher polar contributions due to abundant hydroxyl groups, leading to intrinsically good wettability [32]. In contrast, synthetic fibers such as polyester and polypropylene possess low surface energy dominated by dispersive forces, resulting in poor wetting unless modified [33].

9.2.1.1 Sessile Droplet Test

The sessile drop method, often referred to as the static contact angle measurement, is the simplest and most widely used technique for evaluating surface wettability. It is typically performed using an optical tensiometer, also known as a contact angle goniometer or drop shape analyzer. In this method, a small droplet of liquid, most commonly water, is gently deposited onto the surface of the material, and the contact angle is determined while the droplet remains stationary (see Fig. 9.1) [16]. The calculation of the contact angle using Young's equation is based on the assumption of an ideal solid surface [16, 22]. Specifically, the model assumes that the surface is smooth, rigid, chemically uniform, insoluble, and non-reactive with the test liquid. In real materials, particularly textiles, one or more of these conditions are rarely satisfied. Surface roughness, chemical heterogeneity, and porosity all influence droplet

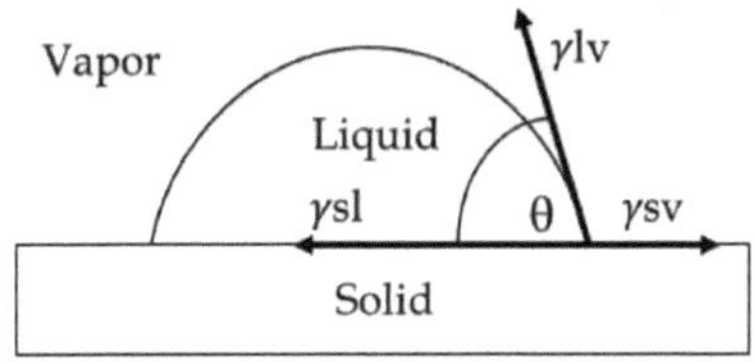

Fig. 9.1 Schematic illustration of Young's equation describing the contact angle in terms of interfacial forces

behavior, meaning that a single static value does not fully represent wetting behavior. As a result, the measured contact angle is better understood as part of a range of thermodynamically stable angles rather than a unique value. To characterize this range, dynamic contact angle measurements are employed, where the droplet volume is increased and decreased to determine the advancing and receding contact angles. The difference between these two values is known as contact angle hysteresis, which provides valuable insight into surface heterogeneity, roughness, and liquid–solid interactions [22].

9.2.1.2 Wicking Test

The wicking test is a simple and widely used method to evaluate the wetting and liquid transport behavior of textiles, particularly their ability to absorb and transfer liquids through capillary action. The test is based on the principle of capillarity, which arises from the balance between cohesive forces within the liquid (related to surface tension) and adhesive forces between the liquid and the solid textile surface (interfacial tension) [16, 34]. When a liquid wets the fibers, these capillary forces can overcome gravity in the narrow pores formed between fibers and yarns, allowing the liquid to rise through the textile structure [35, 36]. In a typical wicking test, one end of a textile specimen is brought into contact with a liquid, most commonly water, while the other end remains free. The height or mass of liquid transported through the fabric is recorded as a function of time, either vertically (capillary rise method) or horizontally (see Fig. 9.2). The rate and extent of wicking depend on several factors, including fiber surface chemistry and polarity, surface roughness, fiber spacing, yarn and fabric structure, and pore size distribution [37]. Any modification to the textile surface, such as plasma treatment, chemical functionalization, or thermal processing, can alter these parameters and thus change the wicking behavior. Although the wicking test does not directly identify the chemical nature of surface modifications, it provides a sensitive and practical indication of changes in wettability and surface-related properties [37]. For this reason, it is commonly used as a comparative and screening tool in textile research and finishing processes, complementing contact angle and surface energy measurements by reflecting liquid transport behavior in real, porous textile assemblies.

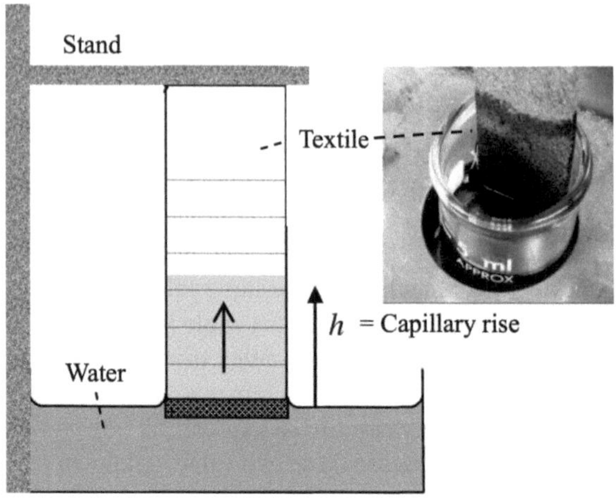

Fig. 9.2 Schematic illustration of the wicking test

9.2.1.3 Lap-Shear and T-Peel Test

The lap-shear strength and T-peel tests (ASTM D1876, DIN 55529) are standardized methods for assessing the adhesive strength between textiles or films and adhesive matrices [38]. Similar to water contact angle and wicking tests, these methods do not provide a direct distinction between chemical and morphological changes on the surface but offer a complementary perspective on surface–adhesive interactions. In both tests, differences in surface treatments are reflected as variations in the adhesive forces required to separate two bonded strips of fabric or film. The measured force not only indicates changes in chemical bonding and mechanical interlocking, but the observed mode of bond failure provides insight into the uniformity and quality of the applied surface treatment [39, 40]. The primary difference between the two tests lies in their setup and the direction of applied forces: lap-shear applies tensile force parallel to the bonded interface, whereas T-peel applies force to progressively peel the bonded layers apart. While the adhesive forces can be quantified as force per unit area using tensile testing instruments, a detailed analysis of bond failure typically requires microscopic examination to evaluate interfacial uniformity and the nature of adhesion.

9.2.1.4 Staining

Unlike the previously described methods, staining tests provide a direct indication of changes in chemical functionalities on textile or polymer surfaces rather than expressing them in terms of surface energy, polarity, or adhesive strength [41]. While surface energy still influences the wetting and interaction of liquids with the surface, the primary goal of staining is to specifically label functional groups with appropriate dyes and, where possible, determine their surface concentration. Various staining methods have been developed to detect and measure the density of functional groups such as carboxyl, carbonyl, hydroxyl, and amine groups on natural, synthetic, or bio-based polymers [42]. The mechanisms underlying these tests involve ion-exchange, electrostatic adsorption/desorption, or covalent chemical reactions with the target groups [40]. Some dyes are suitable only for qualitative detection, providing visual confirmation of the presence of specific functional groups, while others can be used for quantitative analysis. In quantitative staining, it is generally assumed that the dye binds to functional groups in an equimolar ratio, and a high affinity of the dye for the targeted functional group is required to ensure accurate measurement. Qualitative assessments can be conducted directly in solution or via microscopy after staining, whereas quantitative measurements are typically performed using spectrophotometric techniques, such as UV–Vis spectroscopy or colorimetry, to determine the extent of dye binding and, by inference, the concentration of functional groups on the surface.

9.2.2 Surface Topology Analysis

Surface topology analysis examines the micro- and nanoscale structure of textiles, which affects wetting, adhesion, mechanical properties, and functionalization efficiency. Fibers, yarns, and pores form complex hierarchical surfaces that govern liquid spreading, coating uniformity, and particle deposition. Techniques such as SEM, AFM, XPS, digital holographic microscopy, laser diffraction, and dynamic light scattering provide detailed qualitative and quantitative information on surface morphology, roughness, and composition.

9.2.2.1 Scanning Electron Microscopy (SEM)

SEM is a high-resolution imaging technique used to examine the surface morphology and microstructure of materials, including textiles. The principle of scanning electron microscopy is based on the interaction between a focused beam of high-energy electrons and the atoms at or near the surface of a specimen. When the electron beam scans the sample surface, various signals, including secondary electrons, backscattered electrons, and characteristic X-rays, are generated due to elastic and inelastic scattering. These signals are collected and processed to produce high-resolution images and compositional information [44]. SEM provides information on fiber surface topography, coating uniformity, and treatment effects, with resolutions down to a few nanometers. Often coupled with energy-dispersive X-ray spectroscopy (EDS/EDX), it can also identify elemental composition. Sample preparation may include sputter-coating with a conductive layer to prevent charging. SEM is widely used in textiles for evaluating fiber treatments, coatings, and surface damage. Figure 9.3 presents a sample SEM image of a textile treated with antibacterial silver nanoparticles [43].

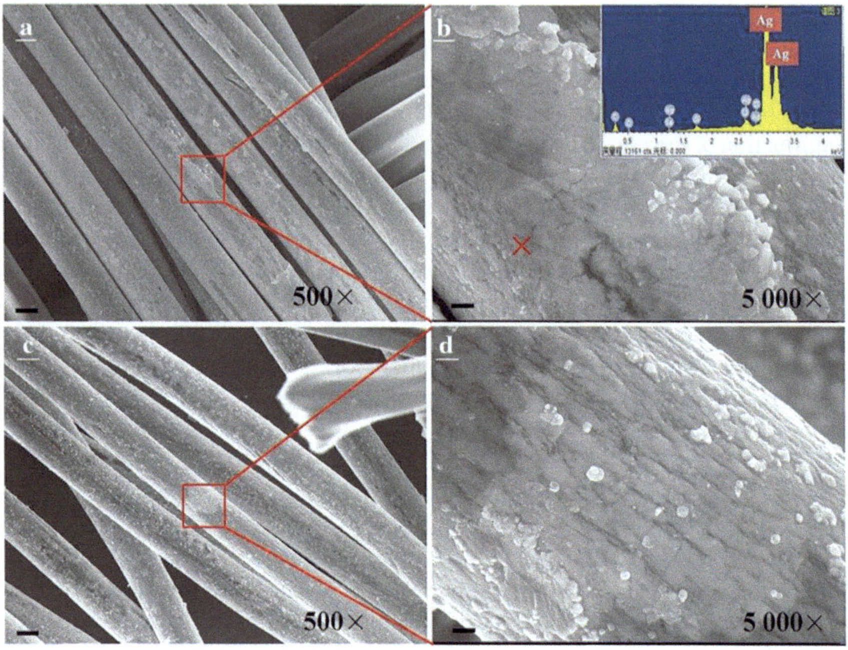

Fig. 9.3 Textile fibers untreated (**a**, **b**) and treated (**c**, **d**) with antibacterial silver nanoparticles; EDX in the inset (reused under CC4.0 licence from Ding et al. [43])

9.2.2.2 Atomic Force Microscopy (AFM)

Atomic force microscopy is a scanning probe microscopy technique that enables high-resolution surface imaging at the nanoscale [45, 46]. Unlike optical microscopy, AFM is not limited by light diffraction and can achieve resolutions down to fractions of a nanometer, depending on the probe tip geometry and scanner precision [47]. AFM operates by scanning a sharp nanoscopic tip, typically 2–20 nm in diameter, attached to a flexible cantilever across the sample surface [42]. Interactions between the tip and surface, such as van der Waals, electrostatic, chemical, and mechanical forces, cause cantilever deflection, which is detected by a laser–photodiode system and used to construct three-dimensional topographical images [48]. AFM can be operated in several imaging modes, including contact, tapping, and non-contact modes. In contact mode, the tip remains in continuous contact with the surface, making it suitable for rigid samples but potentially damaging to soft materials. Tapping and non-contact modes, also referred to as dynamic modes, minimize lateral forces and are therefore preferred for soft, flexible, or delicate materials such as textile fibers. In tapping mode, the cantilever oscillates near its resonance frequency, while changes in amplitude or phase are used for feedback and image generation. Non-contact mode relies on frequency shifts caused by long-range forces and is commonly performed under vacuum conditions [46, 49]. Beyond surface topography,

AFM enables nanoscale measurement of mechanical properties, including roughness, adhesion, stiffness, and elasticity. It is widely used to study fiber morphology, surface coatings, functionalization, and nano-modifications.

9.2.2.3 Digital Holographic Microscopy (DHM)

Digital holographic microscopy is an interferometry-based, optical, non-contact imaging technique that records interference patterns between a reference beam and light reflected from or transmitted through a sample. These digitally recorded holograms capture both amplitude and phase information and are computationally reconstructed to generate three-dimensional maps of surface morphology with high vertical sensitivity, typically at sub-micron resolution [50]. Because DHM does not require mechanical scanning, it enables real-time monitoring of surface dynamics, thickness variations, and coating uniformity over relatively large areas [51]. Through advanced phase analysis and complex parameterization of the reconstructed data, DHM can also provide contrast between different materials within the same field of view, although this capability may be influenced by material-specific factors such as manufacturing processes or surface pre-treatments [52]. Unlike SEM or AFM, DHM requires minimal sample preparation and can image soft, wet, or transparent samples under ambient conditions without causing surface damage. In textile applications, DHM is used to study fiber deformation, surface roughness, coating thickness, and uniformity of functional layers, offering quantitative height information across fiber assemblies and fabrics. Its high-speed, full-field imaging capability makes DHM particularly suitable for monitoring dynamic processes and surface changes on textile materials in real time [53].

9.2.2.4 Laser Diffraction (LD) and Dynamic Light Scattering (DLS)

LD and DLS are techniques used to measure particle or fiber size distributions. LD measures angular light scattering from suspended particles or fibers to calculate size distribution, covering a wide range from microns to millimeters. DLS analyzes temporal fluctuations of scattered light caused by Brownian motion to determine the hydrodynamic diameter of nanoscale particles [54, 55]. Both methods are useful for characterizing fiber fragments, nanoparticles, or coatings in textile applications. LD provides rapid analysis of bulk particle sizes, while DLS offers high-resolution measurement for nanoscale dispersions. Together, they are complementary tools for evaluating surface treatments and functional particle incorporation.

9.2.3 Surface Chemistry Analysis

Surface chemistry analysis focuses on the chemical composition, functional groups, and molecular interactions present at textile surfaces, which directly influence wettability, adhesion, and functionalization [56]. Techniques such as FTIR, Raman, XPS, EDX, and SIMS provide complementary information on elemental content, chemical states, and spatial distribution of surface modifications and functionalization. Understanding these chemical features is essential for optimizing coatings, grafting, plasma treatments, and other surface functionalization processes. By combining multiple analytical methods, researchers can correlate chemical changes with performance properties, ensuring effective and uniform textile surface modification and functionalization.

9.2.3.1 Fourier Transform Infrared Spectroscopy (FTIR)

Fourier transform infrared spectroscopy is a widely used analytical technique based on the interaction between infrared radiation and matter in solid, liquid, or gaseous states. The method relies on the absorption or transmission of infrared light, which induces vibrational transitions in molecular bonds [57]. Because these vibrations occur at characteristic frequencies, the resulting spectrum provides qualitative information about molecular structure and the presence of specific functional groups (such as –OH, $-NH_2$, –COOH, C=O, and C–O) within a material. FTIR spectra are obtained by recording the intensity of absorbed or transmitted radiation as a function of wavelength or frequency. Measurements can be performed in transmission mode for samples that are sufficiently transparent to infrared light, although this often requires extensive sample preparation, such as grinding and pellet formation. For textile materials, reflection-based techniques are more commonly employed due to their ease of use and minimal preparation requirements [58]. Among these, attenuated total reflection (ATR) is particularly popular, as it allows rapid analysis of fibers and fabrics. However, ATR is not strictly surface-sensitive, since the effective penetration depth extends to a fraction of the infrared wavelength, often reaching several micrometers below the surface. As a result, FTIR spectroscopy primarily provides information representative of the bulk material rather than changes confined to the outermost surface layers [59]. Functional groups present only at the surface may therefore produce weak signals that are difficult to distinguish from background noise. External reflection spectroscopy (ERS) offers improved surface sensitivity but is generally limited to thin films or monolayer systems. Spatial resolution depends on the measurement configuration and instrumentation, with advanced techniques such as micro-FTIR and nano-FTIR achieving resolutions down to the micrometer or nanometer scale. Overall, FTIR spectroscopy is valued for being non-destructive, fast, and capable of delivering molecular-level information, although its limited surface sensitivity restricts its application in detailed surface analysis [60].

9.2.3.2 Raman Spectroscopy

Raman spectroscopy is a vibrational spectroscopic technique that provides molecular and structural information on textile surfaces through inelastic scattering of monochromatic light, usually from a laser source. When photons interact with molecular vibrations, shifts in energy occur, producing Raman-active peaks characteristic of chemical bonds such as C=C, C–H, C–O, and N–H [61]. This makes Raman highly suitable for detecting surface modifications and functionalization, including chemical grafting, polymer coatings, or plasma-induced functionalization. Unlike FTIR, Raman is less sensitive to water and can probe carbon-based fibers, synthetic polymers, and composite coatings without interference. Confocal Raman microscopy enables spatially resolved chemical mapping, revealing the distribution of functional groups on fibers or coatings at micrometer resolution. Raman can also detect structural changes, such as crystallinity alterations in cellulose or polymer chains, after surface treatments [62]. In combination with complementary techniques like AFM or SEM, Raman spectroscopy provides detailed insight into chemical modifications, structural integrity, and functional group incorporation, allowing optimization of textile surface treatments for applications such as enhanced wettability, adhesion, dyeability, or catalytic activity.

9.2.3.3 Energy-Dispersive X-ray Spectroscopy (EDX/EDS)

Energy-dispersive X-ray spectroscopy is a widely used technique for elemental analysis and chemical characterization of materials, often coupled with scanning electron microscopy (SEM). The method is based on the interaction of a high-energy electron beam with the sample, which excites atoms and causes them to emit X-rays characteristic of the elements present. When an inner-shell electron is ejected, electrons from higher energy levels fill the vacancy, releasing energy in the form of X-rays. The energy of these X-rays is unique to each element and the specific electronic transition, enabling identification and mapping of elemental composition in the sample [63]. In EDX, the depth from which elemental information is obtained depends on the acceleration voltage of the electron beam, as well as the material's density and atomic number. For instance, in carbon-based polymers at 15 kV, the penetration depth can reach up to approximately 3 μm, with a lateral resolution of around 2 μm. Detection of light elements with atomic numbers below 11, such as sodium, can be challenging due to higher detection limits, and the technique is generally not surface-specific, limiting its ability to provide detailed information about ultra-thin surface modifications or functionalization [64, 65].

When combined with SEM, EDX provides valuable information about the elemental composition of textile surfaces and enables correlation between morphology and chemical distribution. It is extensively used to verify chemical or physical surface modifications, such as coatings, metal nanoparticle deposition, plasma treatment, or grafting. EDX allows semi-quantitative measurement of elements, including carbon, oxygen, nitrogen, sulfur, chlorine, and various metals, and maps their distribution across fibers, yarns, or fabric surfaces. This capability is particularly useful for confirming the presence of functional additives or reactive groups introduced by enzymatic functionalization, chemical finishing, or nanocoatings. Although EDX does not provide information on chemical bonding, it complements techniques such as XPS or FTIR by offering elemental and spatial insights, contributing to a more comprehensive understanding of surface modifications and functionalization. The combination of SEM and EDX enables researchers to evaluate the uniformity of treatments, the efficiency of coatings, and the relationship between fiber topography and elemental coverage, helping optimize textile functionality and performance.

9.2.3.4 X-ray Photoelectron Spectroscopy (XPS)

X-ray photoelectron spectroscopy is a well-established surface-sensitive analytical technique that provides information about the chemical composition of primarily solid materials. The method is based on the photoelectric effect, in which X-ray photons irradiate a sample and cause the ejection of photoelectrons from atomic orbitals near the nucleus. By measuring the kinetic energy of these emitted electrons, their binding energies can be calculated using the relation in Eq. 9.2 [66]:

$$E_B = E_{\text{photon}} - (E_{\text{kinetic}} + \Phi) \tag{9.2}$$

Here E_B is the binding energy of the electron, E_{photon} is the energy of the incident X-ray (e.g., 1486.7 eV for Al Kα radiation), E_{kinetic} is the measured kinetic energy of the electron, and Φ is the work function of the analyzer and sample surface. Both the binding and kinetic energies are characteristic of the element and the specific atomic orbital from which the electron originates. Variations in binding energy also reflect the chemical environment, including oxidation state, neighboring atoms, and bond hybridization, allowing XPS to provide chemical state information in addition to elemental identification and quantification [67].

XPS analysis is highly surface-sensitive, typically probing only a few nanometers into the material due to the limited inelastic mean free path of photoelectrons. Lateral resolution depends on the instrument configuration and can range from several micrometers to hundreds of micrometers. The technique can detect all elements except hydrogen and helium, which are not effectively ionized by X-rays, and generally achieves a detection limit of around 0.1 atomic percent [42]. Beyond elemental composition, XPS can also quantify different chemical species or functional groups on the surface, making it a valuable tool for detailed surface characterization. In textile research, XPS is extensively employed to evaluate functionalization efficiency, chemical durability of treatments, and changes in surface chemistry after processing or aging. When combined with other surface characterization techniques such as SEM and AFM, XPS provides complementary chemical information essential for understanding structure–property relationships in functional textiles and polymer surfaces.

9.3 Functional Property Analysis

Functional property analysis evaluates how textile materials perform in response to specific chemical, physical, or biological interactions resulting from their inherent composition or surface modifications. These properties are critical for determining the suitability of textiles for specialized applications, such as protective clothing, medical textiles, sportswear, or high-performance technical fabrics. Key functional properties include water repellency, moisture management, stain resistance, antimicrobial activity, flame retardancy, UV protection, mechanical strength, abrasion resistance, and chemical durability [3, 5, 10]. Each property reflects the textile's ability to interact with its environment, whether through liquid absorption, barrier performance, or reactive surface behavior. Wettability and water interaction are fundamental functional properties that govern liquid uptake, capillary transport, and drying behavior. These are influenced by fiber type, surface treatments, and finishing agents. Moisture management and wicking determine wearer comfort and drying efficiency, particularly in active or sports textiles. Similarly, water and oil repellency arise from surface energy modification, coatings, or hydrophobic finishes, ensuring barrier protection against liquids. Stain resistance combines repellency with chemical inertness, limiting the interaction of dyes, oils, or other contaminants with the fiber surface. Antimicrobial and biofunctional properties are increasingly critical, particularly for medical, hygiene, or sports applications. These properties depend on surface chemistry, functional coatings, or embedded biocidal agents, which inhibit microbial growth while maintaining textile integrity. Flame retardancy and UV protection provide safety and durability by reducing ignition potential or blocking harmful radiation, often through chemical additives or surface treatments. Mechanical properties such as tensile strength, tear resistance, abrasion resistance, and elasticity ensure that functional finishes do not compromise structural performance, while chemical durability confirms that properties remain stable after washing, exposure to chemicals, or environmental stress [3].

Standards and guidelines from organizations like ISO and ASTM provide validated and reproducible testing methods for most functional properties, enabling consistent evaluation and comparison across materials and processes. These methods include quantitative measurements for absorbency, repellency, mechanical performance, and chemical resistance, as well as qualitative or semi-quantitative assessments for surface coatings and functional finishes. Although specific methods may vary depending on the textile type, test conditions, and target application, standardized protocols ensure accuracy, reliability, and comparability in evaluating functional performance.

9.4 Conclusion

The characterization of textile surfaces is fundamental for understanding and optimizing their functional performance in diverse applications. Surface properties, including topography, chemistry, and energy, directly influence key textile behaviors such as wettability, adhesion, mechanical strength, and interaction with functional coatings or treatments. Techniques such as SEM, AFM, DHM, and DLS provide detailed insights into fiber morphology, roughness, and particle distribution, while spectroscopic and analytical methods like FTIR, Raman, XPS, and EDX reveal chemical composition, functional group distribution, and surface modifications and functionalization at micro- and nanoscale levels. Functional property analysis links these structural and chemical features to practical performance, encompassing wettability, moisture management, repellency, antimicrobial activity, flame retardancy, UV protection, and mechanical durability. Standardized testing protocols ensure reproducibility and comparability, enabling systematic evaluation of surface treatments and finishes. Overall, a combined approach integrating surface topology, chemistry, and functional performance analysis allows a comprehensive understanding of how textiles respond to environmental, chemical, and mechanical stimuli. Such insights are crucial for designing advanced, application-specific textiles with optimized properties, ensuring that surface modifications and functionalization translate effectively into durable, high-performing, and multifunctional materials.

AI-Usage Declaration

During the preparation of this chapter, the authors used artificial intelligence to improve the language of the manuscript. After using this tool/service, the authors reviewed and edited the content as needed and take full responsibility for the content of the publication.

References

1. A. Nadi et al., Evolution in the surface modification of textiles: a review. Text. Prog. **50**(2), 67–108 (2018)
2. G. Li et al., Surface modification and functionalization of silk fibroin fibers/fabric toward high performance applications. Mater. Sci. Eng. C **32**(4), 627–636 (2012)
3. Y. Zhang et al., Functional textiles with smart properties: their fabrications and sustainable applications. Adv. Func. Mater. **33**(33), 2301607 (2023)
4. M.P. Gashti, J. Willoughby, P. Agrawal, *Surface and Bulk Modification of Synthetic Textiles to Improve Dyeability*, vol. 13 (2011)
5. S. Periyasamy et al., Surface modification techniques in textiles: a review, in *International Conference on Functional Textiles & Clothing*. (Springer, 2023)
6. R.R. Mather, Surface modification of textiles by plasma treatments, in *Surface Modification of Textiles*. (Elsevier, 2009), pp.296–317
7. J. Ghosh et al., Grafting modification for textile functionalization: innovations and applications. Discov. Appl. Sci. **7**(1), 49 (2025)

8. V. Nierstrasz, Enzyme surface modification of textiles, in *Surface Modification of Textiles*. (Elsevier, 2009), pp.139–163
9. A.P. Gomes et al., Layer-by-layer deposition of antibacterial polyelectrolytes on cotton fibres. J. Polym. Environ. **20**(4), 1084–1094 (2012)
10. T. Textor, Modification of textile surfaces using the sol-gel technique, in *Surface Modification of Textiles*. (Elsevier, 2009), pp.185–213
11. N. Rawat et al., Textile surface modification and nano-finishing: a concise review of techniques and applications, in *Experimental and Computational Advances in Materials* (2026), pp. 311–335
12. Essa, D., *04. Recent techniques of surface modification for textile fabrics.* J. Meas. Sci. Appl., 2024. **4**(1): p. 66–87.
13. J. Hu et al., Surface-modified graphene oxide-based cotton fabric by ion implantation for enhancing antibacterial activity. ACS Sustain. Chem. Eng. **7**(8), 7686–7692 (2019)
14. M.N. Morshed et al., Surface modification of polyester fabric using plasma-dendrimer for robust immobilization of glucose oxidase enzyme. Sci. Rep. **9**(1), 15730 (2019)
15. M.N. Morshed et al., Development of a multifunctional graphene/Fe-loaded polyester textile: robust electrical and catalytic properties. Dalton Trans. **49**(47), 17281–17300 (2020)
16. T. Huhtamäki et al., Surface-wetting characterization using contact-angle measurements. Nat. Protoc. **13**(7), 1521–1538 (2018)
17. M.N. Morshed et al., Immobilizing redox enzyme on amino functional group-integrated tailor-made polyester textile: high loading, stability, and application in a bio-fenton system. ACS Sustain. Chem. Eng. **9**(26), 8879–8894 (2021)
18. M.N. Morshed et al., Modification of fibrous membrane for organic and pathogenic contaminants removal: from design to application. RSC Adv. **10**(22), 13155–13173 (2020)
19. B.K. Tudu, A. Sinhamahapatra, A. Kumar, Surface modification of cotton fabric using TiO_2 nanoparticles for self-cleaning, oil–water separation, antistain, anti-water absorption, and antibacterial properties. ACS Omega **5**(14), 7850–7860 (2020)
20. A.Y. Xu, D.J. McGillivray, A.J. Dingley, Active antibacterial coating of cotton fabrics with antimicrobial proteins. Cellulose **28**(12), 8077–8094 (2021)
21. X. Meng et al., Antibacterial and mothproofing wool fabrics modified by M-Arg/PHMG and ZnO. Appl. Surf. Sci. **680**, 161341 (2025)
22. A. Marmur, Thermodynamic aspects of contact angle hysteresis. Adv. Coll. Interface. Sci. **50**, 121–141 (1994)
23. G. Buyle, P. Heyse, I. Ferreira, Tuning the surface properties of textile materials, in *Plasma technology for hyperfunctional surfaces: food, biomedical, and textile applications* (2010), pp. 133–182
24. C. Gleissner et al., Surface activation of high performance polymer fibers: a review. Polym. Rev. **62**(4), 757–788 (2022)
25. J.-H. Kim, H.P. Kavehpour, J.P. Rothstein, Dynamic contact angle measurements on superhydrophobic surfaces. Phys. Fluids **27**(3), 032107 (2015)
26. M. O'Loughlin et al., Capillary rise dynamics of aqueous glycerol solutions in glass capillaries: a critical examination of the Washburn equation. J. Colloid Interface Sci. **411**, 257–264 (2013)
27. D. Leelamanie, J. Karube, A. Yoshida, Characterizing water repellency indices: contact angle and water drop penetration time of hydrophobized sand. Soil Sci. Plant Nutr. **54**(2), 179–187 (2008)
28. B.M. Oosterlaken, A. van den Bruinhorst, G. de With, On the use of probe liquids for surface energy measurements. Langmuir **39**(47), 16701–16711 (2023)
29. Y. Si et al., Wetting and spreading: fundamental theories to cutting-edge applications. Curr. Opin. Colloid Interface Sci. **36**, 10–19 (2018)
30. T. Bahners, J.S. Gutmann, Procedures for the characterization of wettability and surface free energy of textiles-use, abuse, misuse and proper use: a critical review. Prog. Adhes. Adhes. **5**, 259–293 (2020)
31. S. Zhu et al., Experimental determination of surface energy for high-energy surface: a review. Adv. Coll. Interface. Sci. **315**, 102905 (2023)

32. M.A. Hubbe, D.J. Gardner, Contact angles and wettability of cellulosic surfaces: a review of proposed mechanisms and test strategies. BioResources **10**(4), 1–93 (2015)
33. S.H. Zeronian, M.J. Collins, Surface modification of polyester by alkaline treatments. Text. Prog. **20**(2), 1–26 (1989)
34. D. Hettiarachchi et al., Liquid moisture wicking in textile and inter-fibre pore filling behaviour in yarns: a review. J. Text. Inst. **116**(11), 2500–2519 (2025)
35. M. Lei et al., Effect of weaving structures on the water wicking–evaporating behavior of woven fabrics. Polymers **12**(2), 422 (2020)
36. K.-P.M. Tang, C.-W. Kan, J.-T. Fan, Evaluation of water absorption and transport property of fabrics. Text. Prog. **46**(1), 1–132 (2014)
37. R. Fangueiro et al., Wicking behavior and drying capability of functional knitted fabrics. Text. Res. J. **80**(15), 1522–1530 (2010)
38. C. Heinzmann, C. Weder, L.M. de Espinosa, Supramolecular polymer adhesives: advanced materials inspired by nature. Chem. Soc. Rev. **45**(2), 342–358 (2016)
39. E. Botelho, M. Rezende, B. Lauke, Mechanical behavior of carbon fiber reinforced polyamide composites. Compos. Sci. Technol. **63**(13), 1843–1855 (2003)
40. L. Yang, J. Thomason, Development and application of micromechanical techniques for characterising interfacial shear strength in fibre-thermoplastic composites. Polym. Testing **31**(7), 895–903 (2012)
41. A. Guchait et al., Influence of nanofillers on adhesion properties of polymeric composites. ACS Omega **7**(5), 3844–3859 (2022)
42. J. Landsiedel, *Surface Activation, Modification and Functionalization of Natural and Synthetic Textiles* (Faculty of Chemistry and Pharmacy, Universitat Innsbruck, 2022)
43. D. Ding et al., Natural ageing process accelerates the release of Ag from functional textile in various exposure scenarios. Sci. Rep. **6**(1), 37314 (2016)
44. A. Mohammed, A. Abdullah, Scanning electron microscopy (SEM): a review, in *Proceedings of the 2018 International Conference on Hydraulics and Pneumatics* (HERVEX, Băile Govora, Romania, 2018)
45. S.M. Flores, J.L. Toca-Herrera, The new future of scanning probe microscopy: combining atomic force microscopy with other surface-sensitive techniques, optical microscopy and fluorescence techniques. Nanoscale **1**(1), 40–49 (2009)
46. B. Voigtländer, *Scanning Probe Microscopy: Atomic Force Microscopy and Scanning Tunneling Microscopy* (Springer, 2015)
47. S.V. Kontomaris et al., Overcoming challenges and limitations regarding the atomic force microscopy imaging and mechanical characterization of nanofibers. Fibers **11**(10), 83 (2023)
48. A. Trache, G.A. Meininger, Atomic force microscopy (AFM). Curr. Protoc. Microbiol. **8**(1), 2C. 2.1-2C. 2.17 (2008)
49. Y.F. Dufrêne et al., Imaging modes of atomic force microscopy for application in molecular and cell biology. Nat. Nanotechnol. **12**(4), 295–307 (2017)
50. M.K. Kim, Digital holographic microscopy, in *Digital Holographic Microscopy: Principles, Techniques, and Applications.* (Springer, 2011), pp.149–190
51. J. Behal et al., Toward an all-optical fingerprint of synthetic and natural microplastic fibers by polarization-sensitive holographic microscopy. ACS Photonics **9**(2), 694–705 (2022)
52. K.M. Yassien, M. Agour, M.A. El-Bakary, Determination of physical properties of irradiated PTFE fibers using digital holographic microscopy. Appl. Phys. B **125**(9), 180 (2019)
53. K.M. Yassien, M. Agour, M easuring opto-thermal parameters of basalt fibers using digital holographic microscopy. Microsc. Res. Tech. **80**(2), 211–219 (2017)
54. R. Sandhu et al., Dynamic light scattering (DLS) technique, principle, theoretical considerations and applications. Nanotechnol. Biochem. Tech. Assess. Qual. Saf. Milk Milk Prod. **8**, 135–137 (2018)
55. J.M. Schurr, K.S. Schmitz, Dynamic light scattering studies of biopolymers: effects of charge, shape, and flexibility. Annu. Rev. Phys. Chem. **37**(1), 271–305 (1986)
56. P. Fabbri, M. Messori, Surface modification of polymers: chemical, physical, and biological routes, in *Modification of Polymer Properties.* (Elsevier, 2017), pp.109–130

57. C. Berthomieu, R. Hienerwadel, Fourier transform infrared (FTIR) spectroscopy. Photosynth. Res. **101**(2), 157–170 (2009)
58. M. Hosseini Someah et al., FTIR-ATR spectroscopy: The method for identifying and separating the fibers of non-woody plants such as flax, hemp, kenaf and cotton. Iran. J. Wood Pap. Ind. **16**(1), 43–61 (2025)
59. S. Belfer et al., Surface characterization by FTIR-ATR spectroscopy of polyethersulfone membranes-unmodified, modified and protein fouled. J. Membr. Sci. **172**(1–2), 113–124 (2000)
60. L. Geminiani et al., Differentiating between natural and modified cellulosic fibres using ATR-FTIR spectroscopy. Heritage **5**(4), 4114–4139 (2022)
61. D. Puchowicz et al., Raman spectroscopy in the analysis of textile structures, in *Recent Developments in Atomic Force Microscopy and Raman Spectroscopy for Materials Characterization* (2022), pp. 1–21
62. M. Peterson, D. Kurouski, Non-destructive identification of dyes on fabric using near-infrared Raman spectroscopy. Molecules **28**(23), 7864 (2023)
63. A.J. Garratt-Reed, D.C. Bell, *Energy-Dispersive X-Ray Analysis in the Electron Microscope* (Garland Science, 2003)
64. J.C. Russ, *Fundamentals of Energy Dispersive X-ray Analysis: Butterworths Monographs in Materials* (Butterworth-Heinemann, 2013)
65. F. Hippert et al., *Neutron and X-Ray Spectroscopy* (Springer Science & Business Media, 2006)
66. D.N.G. Krishna, J. Philip, Review on surface-characterization applications of X-ray photoelectron spectroscopy (XPS): recent developments and challenges. Appl. Surf. Sci. Adv. **12**, 100332 (2022)
67. R. Ghobeira et al., Chemical characterization of plasma-activated polymeric surfaces via XPS analyses: a review. Surf. Interfaces **31**, 102087 (2022)

Chapter 10
Machine Learning (ML) and Artificial Intelligence (AI) in Textile Surface Modification

Badhon C. Mazumder, Nowrin Ahmed Urmi, Md. Himel Mahmud, Md. Shohave Sarkar, Sakil Mahmud, and Md. Reazuddin Repon

Abstract Machine learning (ML) and artificial intelligence (AI) have become popular in modern industries because they can transform traditional industrial practices through data-driven decision-making, pattern recognition, and predictive modeling. In textile surface modification, ML and AI are now widely used to optimize processes such as dyeing, printing, coating, and finishing. These essential tools can leverage an extensive database, optimize process parameters, and reduce human error. This chapter provides information on ML and AI models, including support vector machines (SVMs), artificial neural networks (ANNs), convolutional neural networks (CNNs), genetic algorithms (GAs), and fuzzy logic. They are incorporated with surface treatments to predict variables such as dye uptake, fabric strength, and moisture retention. Predicting critical parameters like temperature, time, and chemical concentration is a crucial part of surface-modification treatments. ML algorithms make the treatment process advanced by creating complex relationships between critical input parameters and consistent outputs. These technologies are used to reduce

B. C. Mazumder · N. A. Urmi · Md. Himel Mahmud · Md. Shohave Sarkar
Department of Textile Engineering, Northern University Bangladesh, Dhaka, Bangladesh
e-mail: md.sarkar@nub.ac.bd

S. Mahmud
Department of Chemistry and Physics, Ivory V. Nelson Center for the Sciences, Lincoln University, Oxford, PA, USA

Department of Textile Engineering, Faculty of Engineering, Daffodil International University, Dhaka, Bangladesh

S. Mahmud
e-mail: smahmud@lincoln.edu

Md. Reazuddin Repon (✉)
Department of Bioproducts and Biosystems, School of Chemical Engineering, Aalto University, Espoo, Finland
e-mail: reazmbstu.te@gmail.com

M. N. Morshed and M. Shahid (eds.), *Advances in Surface Modification and Functionalization in Textiles*, Textile Science and Clothing Technology,
https://doi.org/10.1007/978-3-032-23377-6_10

resource consumption and waste generation. Moreover, ML and AI models significantly improve the inspection system by identifying defects. Wrinkles, stains, discoloration, and uneven coating cannot be detected accurately by traditional human scanning systems. Despite its potential, many challenges and limitations exist, including the complexity of textile materials, limited data, and the computational intensity of deep learning approaches. However, the prospect of AI in textile surface modification is bound to continue revolutionizing manufacturing, bringing greener, more innovative, and more efficient processes.

Keywords Textile surface modification · Machine learning (ML) · Artificial intelligence (AI) · Sustainable textile manufacturing · Surface optimization

10.1 Introduction

Surface treatment of textiles is an integral part of producing high-performance, advanced fabrics that offer additional comfort, durability, and improved appearance. Traditional surface modification, e.g., dyeing, printing, and chemical processing, is a dated technology for textile engineers. These include fabric customization for specific applications through dyeing to improve colorfastness, as well as water- and UV-resistant coatings. However, these may not work in all cases despite their widespread usage. The industry faces problems of low-quality products, high-energy consumption, and the generation of large amounts of chemical waste. Furthermore, the manual manufacturing process only produces a batch of samples, and we continue to do so, which causes issues. ML and AI are also significant responses to the issues of textile surface treatment [1]. Deep learning and neural networks are two AI technologies that maximize the precision and efficiency required for predicting and estimating treatment outcomes. By shifting through massive data sets generated during textile production, they provide feedback to refine dyeing, finishing, and coating processes, enabling them to be performed more consistently and efficiently with less waste. Data-driven models, from raw materials and the environment to machines and the final performance of the product, increasingly use ML and AI algorithms to predict and optimize textile surface treatments such as dyeing, finishing, coating, and functionalization. To avoid non-informative trials and develop sustainable, cost-effective production systems, ML models, including support vector machines (SVMs) and artificial neural networks (ANNs), predict dye uptake as a function of fabric composition and dyeing conditions.

Particularly in situations such as hemicellulose hydrolysis, there is a noteworthy parallel between the use of ML in textile processing and that of global optimization models in biorefining. Similar techniques can be used to examine textile qualities after surface treatment. For example: ML-based forecasts of xylose yield in biorefining [2]. Manufacturers can choose specific pretreatment parameters using AI. AI technologies can accurately estimate various properties such as tensile strength, water absorbency, and color fastness after dyeing and coating. This helps make processes more efficient by reducing costs, waste, and energy use and ensuring consistent quality. Thus, enhanced sustainability in surface treatments is taking place in the textile sector. AI significantly improves the quality of clothing manufacturing by using computers to identify defects. Traditional inspection systems, dependent on human judgment, inherently limit the subjective component across all work domains (e.g., visual assessment). So, the final results become inconsistent. ML models have various categories, and convolutional neural networks (CNNs) are one of them. It can rapidly and precisely identify fabric defects. For example: wrinkles, absent coatings, and discolorations in high-resolution images. This is entirely a consequence of unpredictability. This automatic inspection method ensures quality and reduces the possibility of our customers receiving defective products. Water resistance, antimicrobial protection, and UV protection are fiber attributes that depend significantly on AI. Alongside showcasing these pairings, pharmaceutical researchers may devise effective therapies and minimize waste by conceptualizing complex interactions involving drugs, substances, and environmental factors. This chapter illustrates how AI may improve productivity, quality, and manufacturability using advanced shape-altering techniques, optimization, and predictive modeling. ML and AI might facilitate the textile industry's shift toward more sustainable, resource-efficient production practices, thereby enhancing environmental impact and product quality.

10.2 The Emergence of ML and AI in Textile

ML and AI are becoming notable in reshaping the existing textile industry. ML is based on extraordinary algorithms that can make their own decisions by learning from past data. On the other hand, AI is a set of diverse technologies used to reproduce human cognitive activities in problem-solving. Manufacturers can significantly improve their current processes by using data-driven advanced approaches. Therefore, it is possible to drive significant change within the textile industry by applying ML and AI, which may result in more efficient production and smarter, more forward-thinking decisions. Such alterations not only simplify work processes but also pave the way for more creative and advanced manufacturing methods.

ML and AI for advanced textile manufacturing involve supervised, unsupervised, and deep learning methods. They are built using other ML algorithms to predict the outcomes of fabric processes and to provide scientific support for optimizing dyeing and improving the properties of the final products. Similar models based on neural networks, decision trees, and SVMs have also been extensively used in fabric strength prediction after finishing, dyeing depth, etc. Supervised learning methods (e.g., SVMs, neural networks) have reached very high prediction accuracy in applications of intermediate textile processing [3]. These predictive models are used to develop additive manufacturing processes that could replace traditional, commonly manually controlled, and uncertain methods. The future seems very promising, with ML and AI likely to increase productivity and improve the quality of end products, while textiles manufacturing won't be left behind. In the textile industry, ML technology has enabled greater technical efficiency and predictive power, enabling integration of industrial control systems into surface finishing processes. This technical improvement puts sustainability first and reduces manufacturing costs. Particularly in fields such as pattern generation, defect detection, and predictive dyeing, AI is the key to building more intelligent, data-driven factory environments. To combine the image processing with these two classifiers as a front end, the ANN and SVM methods have been used, and achieved 91% path accuracy in color images, which is better than any previous methods for detecting different classes of fabric problems [3]. Fuzzy clustering and K-nearest neighbors (KNN) greatly enhance the accuracy of fabric classification in AI. These advances make it much easier to appraise textile quality. This method helps boost customer satisfaction and the return on resource allocation. The resulting prediction-based analytics and material use increase by 10 or more times. From there, they also support quick operations and recommend goods in line with individual tastes through interdisciplinary perspectives on marketing models. They also reduce costs associated with traditional textile machinery operations by up to 90%. Additionally, by reducing material waste during production, these technologies can lead to more eco-friendly practices in the textile industry. This not only helps protect our environment but also offers significant savings on natural resources [3].

10.3 Different Strategies for Textile Surface Modification

Surface-modification processes are vital for imparting or creating desired functionalities in materials, especially textiles. These techniques alter material surface properties to enhance its performance for specific applications, such as longevity, waterproofing, antimicrobial protection, and others. This section deals with essential strategies for surface modification of textiles, including mechanical, chemical, biological, plasma, and laser treatments, as well as other advanced approaches.

Physical methods of surface modification involve the mechanical processing of a material surface using mechanical energy. Mechanical textile finishing includes sanding and brushing processes. Sanding is the process of using an abrasive to ease or wear off the surface of a fabric. Excess fibers, irregularities in texture, or impurities may be eliminated by running the material through sandpaper rolls or abrasive rollers. This is necessary to modify the textile surface before further treatments, e.g., dyeing or finishing, to improve fabric hand (feel) and enhance dye wicking. Sanding can also be used to achieve a specific appearance, e.g., a soft or matte look on the fabric. Another mechanical process is brushing, which uses rotating brushes of varying stiffness to treat textile surfaces. This method is used to defuzz and clean the fabric, as well as to modify its surface texture. In textiles, brushing is used mainly for creating a soft feel or imparting a surface, such as velvet, especially on wool and cotton [4]. These mechanical methods often include regulating speed, pressure, and the abrasive used to achieve specific surface properties. Foolproof the texture and smoothness, or pre-treat to enhance them in some way.

Chemical surface modification refers to the use of chemicals to change the chemical composition and/or structure of a material's surface. Dyeing and printing are the two commonly employed chemical processes for achieving the appearance and functional properties of textiles. Dye immersion (dyeing) is the process of adding color to textile materials, in which the fabric is immersed in a dye bath under controlled conditions of time, temperature, and pH. The dye molecules are either attracted to the fiber molecules or react with them in a two-step process, depending on the type of fiber being dyed, forming a single molecule on the fiber surface. For instance, reactive dyes chemically bond to cotton (cellulose fibers), while disperse dyes are for synthetic fibers such as polyester. Also, during the dyeing operation, other desirable features such as ultraviolet resistance, antimicrobial treatment, and water repellency can be imparted to the fabric, depending on the dye's constitution. The success of dyeing depends on a variety of factors, including the type of dye used (e.g., reactive dyes for cotton or disperse dyes for polyester) and the textile material's composition. Dyeing has now joined ML and AI to create the best possible method. AI models can predict the effects of surface treatment based on past production data. They are also capable of predicting how a new formula for dye will perform over a series of many trials, which means lots of crash testing saved money. Using ANN eliminates dyeing problems, as techniques like support vector regression (SVR) and XGBoost can predict the exact color and dyeing conditions, thereby making the dyeing process more efficient [5]. Additionally, generalized adversarial networks (GANs) and deep learning algorithms are used to predict dye formulations based on desired color appearance and material characteristics, thereby reducing the time and resources required for conventional dyeing techniques. The reliance on many tests may diminish as AI-guided dye predictions conserve substantial amounts of dye and liquor, helping maintain color consistency between batches. The technique of using stencils, screens, or rollers to apply color or pattern to a specific area of fabric is known as printing. Moreover, the process can make specific patterns or drawings rather than solid-colored pieces. The three methods in this sector are digital printing, screen printing, and roller printing. To print, a paste containing pigments or dyes is applied to the foundation cloth, then heated or chemically fixed. This technique allows functional properties such as water repellency and antibacterial effects to be incorporated into the printed area of the cloth as well. Textile printing has been enhanced by applying an intelligent control system that combines a CNN and an ML-based pattern-detection algorithm. For example, in textile printing, defect checking and pattern recognition are performed using a CNN. These methods serve the objectives of quality control and for the delicate printing process, which means they result in the improvement of pattern formation on one hand and reducing the use of materials on the other [5]. AI models are also being developed to fine-tune the print process by modulating factors such as pressure, temperature, and ink formulation to achieve printable quality without resource waste. Generative AI processes can even generate entirely new fabric patterns, giving textile producers new design options. In addition, ML algorithms enable in-process printing monitoring and on-the-fly parameter control to maintain quality and consistency, thereby allowing quick identification and resolution of defects or artifacts at early stages of production. These two processes involve chemical reactions, where specific conditions and parameters, such as temperature, pressure, and chemical concentration, must be met to achieve a uniform look and feel on the textile. ML and AI technologies are being increasingly applied to optimize these chemical processes, reducing resource consumption, and enhancing the efficiency of textile production.

Biological surface-modification methods rely on natural processes, such as enzymes or microorganisms, for textile modifications in an eco-friendly way. Enzymes, e.g., cellulases, amylases, or proteases, are used to bring about selective changes in the surface properties of textiles. For example, cellulases are used to biopolish cotton and other cellulosic fibers, reducing excess fuzz and improving the fabric's softness and appearance. Such enzymatic treatments are environmentally friendly and more sustainable than conventional chemical processing. They may also improve the fabric's properties, such as hand, strength, and durability, without degrading the fibers themselves. Microbial processes involve using microorganisms, or microbial enzymes, to degrade or modify textile surfaces. In such applications, these methods can be instrumental because they offer environmentally friendly alternatives to traditional chemical treatments, reducing the use of harsh chemicals and water. Bacterial cellulose biosynthesis is an example of this biomimetic approach, in which microorganisms can be used to produce cellulose, which is subsequently incorporated into textiles to add new functional properties, including mechanical performance and biodegradability. Other microbial treatments could be used to impart functional characteristics, such as antimicrobial properties, or to assist with cleaning the textiles after the finishing solution is applied. They are increasing in popularity due to their green, eco-friendly process and reduced use of hazardous chemicals during surface modifications.

Plasma and laser techniques are high-energy methods for modifying surface properties at low cost and with minimal impact on stability. In composite fabrication, materials are treated in an ionized gas (plasma) generated under low or atmospheric pressure. The plasma is a surface treatment on the surface of the material, and it introduces functional groups (like hydroxyl group, carboxyl group, or amino) on the material to change its wettability [6]. Textile surface cleaning, activation, and functionalization by plasma treatments are well established. Plasma can, for example, increase the adhesion of coatings, inks, and or adhesives to textiles. It can also affect a fabric's wettability, converting hydrophilic fibers to hydrophobic or vice versa. Laser treatment uses intense laser beams to heat or melt a local spot on the surface, or to burn it. This technology now allows us to make precise adjustments on the micro- or nanoscale: we can print patterns and structures and even cut the fabric. In addition to the above niches, laser treatment is commonly employed for engraving, micro-patterning, and surface texturing, particularly in fashion and technical textiles. The technique is based on irradiating a material surface with intense light in which the material absorbs part of the light and is melted, ablated, or modified during nanosecond time scale exposure [7]. The advantage of both plasma and laser treatment is their intrinsically high precision, enabling modification of only local properties without losing the bulk material's characteristics.

Apart from standard techniques, some advanced surface-modification technologies have been developed to impart special functions to materials, such as textiles. Nanotechnology is the ability to move atoms and molecules around to produce new or enhanced properties. Nano-particles, such as Ag-nanoparticles or TiO_2, can be integrated into textiles to impart functionalities such as antimicrobial properties, self-cleaning, UV protection, and increased durability. These nanoparticles bind to textile surfaces, forming a thin functional layer, or integrating into the fiber structure [8]. Nanotechnology has enabled textiles with increased strength, better surface roughness, functionality, and an appealing appearance and/or feel. Electrospinning is applied to a high-voltage electric field for manufacturing nanofibers using a polymer solution. A positive electrostatic force, which elongates the polymer solution into fine fibers, can be applied to a substrate to form a nonwoven web. These nanofibers have high surface area-to-volume ratios and, therefore, are of interest for applications such as filtration, wound healing, or protective textiles. Furthermore, electrospun fibers can also be functionalized to yield other properties, such as antimicrobial activity, for producing special functionalities [9]. Microencapsulation is the encapsulation of actives (e.g., fragrances or pesticides) in small capsules and their attachment to a fabric. These microcapsules can release active agents, such as those listed above, over time, enabling the development of textile products with enhanced properties. For example, fabric could incorporate scent microcapsules that release aroma as the cloth is worn or rubbed [10]. This process also enables an antimicrobial finish or temperature-sensitive textile that releases actives according to demand. Sol–gel technology is a chemical process in which metal alkoxides are hydrolyzed and condensed to form a gel that can be applied to textiles as a thin film. Sol–gel coatings can provide different performance benefits (water repellency, flame retardancy, self-cleaning) for textiles. The technique is conducted at low temperatures, presenting it as an alternative for modifying textiles and other heat-sensitive materials [11]. These emerging-generation methods create a broad range of opportunities for surface functionalization of textiles. These approaches provide a significant performance boost while preserving the core traits and strengths of the materials.

10.4 Importance of Surface Modification in Textiles

Performance may be adjusted by changing the surface without replacing the fiber. While maintaining core mechanics, textiles that undergo grafting and cross-linking only in the outer layers exhibit improved moisture absorption, flame retardancy, abrasion and tearing resistance, wash-and-wear resistance, and water resistance. This is accomplished by either inter- or intramolecular cross-linking, which increases resistance to washing and usage, or grafting additional polymer chains onto the fiber (also known as "grafting to/from"). N-methylolacrylamide and vinyl systems are standard finishing chemistries used on cotton for strength and crease recovery; formaldehyde-based methods are avoided because of carcinogenic concerns.

Moisture management can be engineered by plasma-assisted grafting that creates graded wettability, like hydrophobic on the skin side, hydrophilic outside, so sweat is pumped outward while acrylic-acid grafts suppress odor formation [12]. In tests on one-sided treated cloth, this architecture effectively expelled sweat and controlled odor, making it well-suited for sportswear. Plasma steps also boost hydrophilicity and dye receptivity of synthetics like PET by introducing nitrogen-containing groups, delivering long-lasting super-hydrophilic behavior and improved coloration. UV protection is typically low in everyday cotton; UPF can be below 15, whereas 40–50 + is the benchmark for protective claims. Loading the surface with opacifying nanostructures, such as ZnO, TiO_2, Al_2O_3 multilayers, or Ag/TiO_2, raises UPF without sacrificing handle, and one-bath TiO_2–SiO_2 reactive dyeing has achieved high UV shielding with strong fastness [13]. Water repellency is available via nano-tiered finishes: fluoroalkyl-functional siloxanes, silica networks cross-linked with polycarboxylic acids, chitosan/fluoroanion systems, and ZnO nanoparticles have all produced durable beading on cotton and blends. Antimicrobial effects come from silver at controlled ppm levels on polyester and from in situ coatings on cotton, both of which suppress pathogens while supporting other finishes.

Functionalization for specialized applications is where surface work pays off. In sportswear, the hydrophobic-in/hydrophilic-out gradient created by atmospheric-pressure non-thermal plasma and acrylic-acid grafting improves sweat transport and odor control in real use. Sol–gel coatings add another layer to the toolkit: silica-based films deliver water repellency, antimicrobial and self-cleaning behavior, UV shielding, bioactivity, thermal effects, and reduced flammability, all laid down at low temperature and tunable via precursor, pH, and H_2O/Si ratio. For medical and hygiene textiles, plasma-grafted cationic polymers (e.g., DADMAC on polypropylene nonwovens) create antimicrobial hospital gowns, bedding, and drapes with strong adhesion and wear resistance [14]. PET surfaces can be bio-functionalized by building alkynyl "handles" via photo-initiated copolymer networks, enabling direct tethering of bioactive ligands for surgical drapes, vascular grafts, or ligament prostheses. Microencapsulation complements these routes: capsules on fiber surfaces enable controlled release (e.g., hygiene agents, drugs, fragrances) with improved stability against light and oxidation, and can also embed thermochromic or photochromic indicators for wearables and labeling. Surface modification enables precise control of moisture transport, UV shielding, antimicrobial action, repellency, and coloration through grafting, plasma activation, nanoparticle and sol–gel coatings, and microcapsules. The same toolbox scales from sweat-moving, odor-controlling sports apparel to bioactive, antimicrobial medical fabrics that must endure fluids, sterilization, and wear.

10.5 Role of ML and AI in Textile Surface Modification

ML and AI are revolutionizing industries, and the textile sector is no exception. AI is a branch of technology intended to simulate human intelligence. On the other hand, ML refers to AI that can independently learn from data to predict or make decisions without explicit programming. These technologies make production processes more innovative and more efficient by learning from data to optimize outcomes. There are several types of ML and AI algorithms, including deep learning, supervised learning, and unsupervised learning.

Supervised learning is a type of ML in which models are trained on labeled examples of the target. For example, in the case of fabric classification or defect detection, the algorithm is trained to identify specific patterns based on previous cases and act accordingly, meaning it is trained to "do something" with them. On the other hand, unsupervised learning is more about identifying patterns in the data itself, when we do not have our output. This method was formerly used in the textile industry for market and trend analysis or consumer prediction. In deep learning, a sub-genre of ML techniques that uses ANNs, signals are processed from one layer to the next at varying levels of abstraction. For many image-centric business tasks, such as fabric pattern identification or defect detection in fabrics, a hierarchy-based feature extraction algorithm often outperforms others due to its better understanding of complex patterns in woven fabrics [15]. For example, the application of a CNN for the detection of fabric defects, which consists of holes, stains, and color matching, thus automating a highly time-consuming process, required minimal human effort with a great potential of resulting in errors.

One such example is the evolution of CNNs, which have been widely employed to detect holes, stains, and color defects in fabric. It automates some tasks that were previously done manually and are often overlooked. These technologies are applied in finishing processes, such as dyeing, in textile manufacturing. The AI models can predict the effects of temperature, time, and chemical concentration on fabric performance characteristics, thereby enhancing industrial processes. This enables businesses to operate more efficiently, minimize their environmental footprint, and ultimately reduce costs. Not only can powered assembly lines predict, but they can also adjust process parameters autonomously. This ensures that each batch adheres to quality standards without excess resource wastage [5]. Decision tree, support vector machine, and neural network models have recently gained prominence in data science, enabling the prediction of the outcomes of surface finishing treatments with a high degree of accuracy and contributing to control over the quality and uniformity of textile production.

10.5.1 ML and AI in the Prediction of Textile Surface Modification Outcomes

ML can also predict how a surface treatment could influence textiles. That is, based on historical data, ML algorithms such as decision trees and random forests can be applied to predict the outcomes of surface treatments. By learning from various input quantities (e.g., chemical types and treatment times), these models can predict outputs such as color intensity, fabric tensile strength, and moisture absorption. These models can be trained on large-scale data collected across many textile processes to predict the effects on the final product from changes in input parameters. This leads to better control over production and quality, with less variation and greater consistency between batches. Recent advancements in ML and AI have significantly enhanced the ability to predict the outcomes of textile surface treatments. For example, there is an increasing trend of using AI models such as SVMs and ANNs to predict textile characteristics, such as color and strength. The latter ones forecast features that can make the testing process economically unviable. Their models are predictive of many variables, including fabric type, chemical type, temperature, and treatment time [5]. Moreover, predictive models could be developed using legacy data containing input features comparable to those used in textile surface treatment methods. Treatment variables can include temperature, chemical concentration, and fiber properties, and each textile property, such as color fastness, moisture absorption, or UV protection, can be an output variable.

A sufficiently large and diverse dataset is needed to train ML models for textile surface modification. Biorefining, training global-optimization decision tree models of xylose yield predictions on a dataset of 1955 data points from 71 experiments [2]. As textile production involves multiple surface treatments, the model could also be trained on extensive databases of previous production runs to predict the results of potential treatments. This data set may include information on fiber types, chemical formulations, and reaction conditions during textile processing. For instance, research in textile dyeing used datasets of colorimetric data (L, a, b* values in the CIELAB color space) to predict color variations on fashion fabrics. An AI model, such as feed-forward neural networks and support vector regression, was used to process these data sets with the aim of predicting changes in color properties due to various fabric treatments, and good prediction success (over 97%) was achieved. These models are really suitable for a mass production line with strict quality control requirements [5].

10.5.2 ML and AI in Predictive Modeling for Surface Modification

Predictive modeling is critical for predicting how the fabric's characteristics will change following surface treatments. For example, as a reliable model for forecasting xylose yield in biorefining, the global optimal decision tree (GODT) is employed for coating textile surfaces. The performance of ML algorithms, such as decision trees, random forests, and SVMs, for predicting the outcomes of different textile surface treatments has been successfully explored to optimize industrial production processes and reduce resource consumption. These models are trained using historical data and are used to make predictions on how various input parameters (now known as independent variables), such as the concentration of dye, temperature, and duration of treatment, will affect fabric properties (now called dependent variables) like color strength, fabric strength, and feel [5].

Predictive models, like the one under discussion, may be successfully applied in a variety of surface treatment procedures, such as coating and finishing, and go beyond dyeing applications. ML algorithms can leverage legacy data to provide manufacturers with a basic understanding of how these treatments affect fabric properties such as moisture management, UV protection, and stain resistance. It offers a defined path to ensure fabric is processed optimally, maximizing fabric efficiency while minimizing the margin for error and helping make better decisions. Decision trees and SVMs were used to design ML models for textile surface-modification conditions. For example, a decision tree-based model was shown to surpass others in accuracy for predicting xylose yield in biorefining paper. This can be applied in the textile context by determining the process parameters needed for dyeing, finishing, or coating that yield a suitable finish attribute, such as color intensity, to mask the disturbing SNPs. Another significant challenge probably hindering functional analysis is pseudogenization, a process that can convert a functional gene into a non-functional region (i.e., a pseudogene). Predicting the longevity of such treatments by observing their performance can also be assisted by AI-generated predictions, adding value to both quality control and performance. AI can help textile factories use fewer resources, optimize chemical use, and run their processes more effectively.

GANs and deep learning models are also proving useful for predictive modeling of textile surface finishes. These state-of-the-art AI techniques can generate synthetic data from real-world sources, enabling device makers to simulate and predict treatment outcomes across different scenarios. This capability of fabric simulation in the virtual world enables the production process to compensate before even producing anything, also limiting waste and product rejection [5].

10.6 Key Applications of ML and AI in Textile Surface Modification

AI is the advancement of computer systems that employ human capabilities, such as ML, deep learning, natural language processing, and image recognition, to process massive datasets and make predictions in accordance with this analysis [16]. ML and AI are also of great significance in the textile industry [17], especially with respect to surface modification. Textile surface modification with ML and AI is being assimilated to a greater extent for process parameters optimization, real-time monitoring, prediction of outcomes, and insertion in the design of new generation functional textiles [18].

10.6.1 Optimization of Parameters

The textile surface-modification process parameters, such as treatment time, power, chemical concentration, type, and temperature, have nonlinear, multidimensional impacts on the outputs. AI not only leads to improved product quality and decreased waste by optimizing these parameters but also enhances the overall sustainability of operations [19]. Input variables in those methods include gas type, power, exposure time, and pressure for plasma treatment and chemical grafting. Interaction between input variables is ensured to form predictive models, and optimal parameters are built using AI/ML models [20]. Textile surface treatment parameters can be optimized using AI-driven models, allowing for maintaining quality outcomes while minimizing material waste. In other use cases, it can help identify chemical concentration and reaction time during dyeing processes to ensure uniform color application. Additionally, AI models such as ANN, particle swarm optimization, and genetic algorithms (GA) provide strong platforms for the enhancement of quality and reduction of energy and environmental burden in industrial systems [18].

An ANN is a network of interlinked layers capable of processing input data via weighted connections. ANN is trained with input parameters such as treatment time, chemical amount, and substrate characterization, and predicts surface energy, tensile strength, and permeability [21]. An ANN prediction system can be used to optimize spinning processes and predict results from different finishing treatments. The employment of genetic algorithms (GA) and their combination with ANNs in hybrid models is a well-known technique to increase the efficiency of yarn engineering and to achieve a highly significant improvement in prediction capability [22]. The ANN approach involves multiple linear regressions and backpropagation with an adjustable learning rate. It can help evaluate fibers, predict dye recipes, predict yarn quality, and diagnose fabric problems [23].

A subset of ML and AI encompasses a wide range of modeling tools for data processing tasks. Over the past ten years, ML has become increasingly popular across most scientific disciplines. ML models are used to predict the toxicity and safety of nanomaterials. ML tools use exposure conditions against a range of organisms (bacteria and fungi) as inputs and growth inhibition measurements as outputs to predict nano-textiles with greater antimicrobial capacities [24]. ML tools are effectively used to predict properties based on polymer composition and structure, such as tensile strength, thermal stability, and glass transition temperature [25]. In the textile sector, ML algorithms enhance texture classification and ensure precise control over specific mass, weft, and warp parameters of fabrics, significantly outperforming manual measurement methods. Textile surface modification may alter bulk textile properties, such as thickness, porosity, and air permeability. ML helps to balance surface functionalization while preserving comfort [25].

10.6.2 Process Monitoring and Control

The integration of AI, internet of things (IoT), and smart sensors has led to significant advancements in the area of process monitoring and control in the textile surface modification [26]. Real-time monitoring of textile surface treatments can be achieved using AI-based systems. This implies that smart sensors embedded with ML algorithms can also be integrated to identify crucial treatment parameters, such as temperature and chemical concentration, in real time during the treatment process. Such systems can autonomously fine-tune processing conditions to ensure that the final product possesses the desired properties and functionality every time it is made. This real-time control enhances both product quality and production efficiency.

Smart sensors are beneficial for textile surface-modification treatments because they can provide real-time data. Due to the necessity for real-time feedback and prompt emergency responses, traditional laboratory-based monitoring and testing methods may soon become obsolete. More contemporary methods based on wireless sensor networks (WSNs) are being developed to address issues in energy management, coverage, and monitoring. [27]. It enables closed-loop control of critical variables such as temperature, pH, and chemical concentration. Different types of sensors, such as temperature sensors, control the temperature of the dye bath [28]. Potentiometric pH sensors maintain the pH of the dye bath [29]. Nano fibrous sensors are susceptible to pH, colorimetric textile gas sensors with excellent reversibility, reusability, and washability that can identify both gaseous strong alkalis and acids [30]. In the realm of smart textiles, piezo-resistive sensors are widely recognized. These are used in monitoring structures, such as those used in parachutes, canopies, or structural composites [31].

The textile manufacturing industry is being revolutionized by AI through real-time monitoring and control, thereby improving efficiency, quality, and adaptability. ML, neural networks, and computer vision are AI techniques used to optimize various stages of textile production, from predicting fabric properties to defect detection and quality assurance. The ANN approach uses multiple linear regressions and backpropagation with a configurable learning rate. Fiber grading, yarn quality forecasting, and fabric problem diagnosis can all benefit from it. Recently, yarn qualities have been evaluated using the adaptive neuro-fuzzy inference system (ANFIS) [32]. Today, most large yarn production companies use AI-based online monitoring and process management. It can identify flaws, evaluate failure rates scientifically, and modify control parameters to optimize spinning. Moreover, AI has been applied to examine yarn's programmable quality attributes. A high-resolution camera within the system takes a picture of a yarn flaw, which a computer system that has been preprogrammed to analyze it does [33].

The incorporation of smart sensors and IoT into the textile surface-modification treatment process affords much greater scope for automation and decision-making. Combining smart sensors and IoT enables more efficient, adaptive, and responsive textile manufacturing that may meet many of the critical challenges arising from traditional methods [34]. These networks continuously collect and analyze data related to [35]:

(a) Temperature, humidity, pressure.
(b) Plasma intensity and exposure time.
(c) Chemical concentration and pH.
(d) Fabric surface characteristics such as roughness, charge density.

The internet of things (IoT) is integrated with cloud platforms to enable real-time decision-making and control of process parameters during surface treatments. After receiving the sensor data, cloud platforms save a large amount of data for further processing [36]. They are situated at the edge of the controlling units so that immediate action can be taken [37].

10.6.3 Pattern Recognition for Quality Control

A variety of elements, including fabric, knitting, braiding, finishes, and prints, can be incorporated into fabric patterns. Manufacturers in the textile sector can prevent fatigue and human error in defect detection by replacing manual factory-level inspections with AI/ML-based inspection technologies. Camera-based inspection systems with large volumes of defect analysis images are typically implemented. The automatic surface inspection systems are a very effective method for continuous improvement when using image processing and analysis techniques [38]. These systems help identify defects such as holes, slubs, stains, misweaves, and uneven coating.

The image acquisition system (IAS) for visual inspection primarily comprises cameras, light sources, and imaging methods. Cameras typically capture surface images with high sensitivity and resolution. Proper lighting, such as LED, halogen, or fluorescent sources, is essential for precise imaging. Two primary imaging methods are used: intensity imaging (bright- and dark-field lighting) and range imaging, which captures 3D surface details using structured light. Before analysis, images undergo pre-processing to enhance quality for defect detection [39].

10.6.4 Automated Design of Surface Textures and Patterns

The application of AI in surface design illustrates its revolutionary impact, especially when integrating techniques such as ML, deep learning, and neural networks [40]. The term surface texture refers to the hand feel of fabric, including rough, shiny,

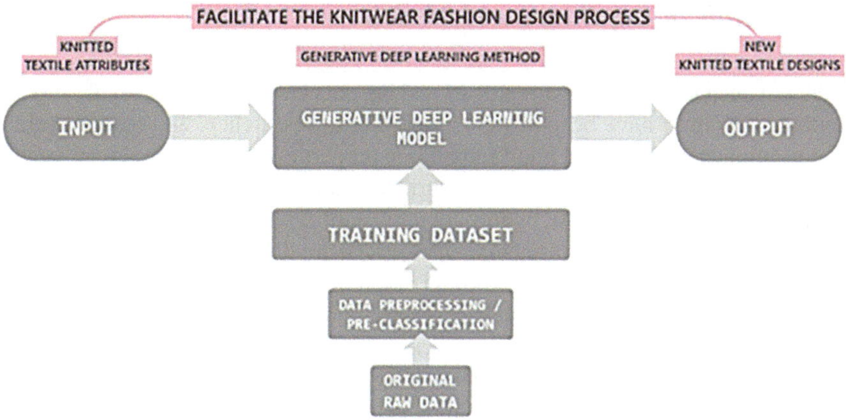

Fig. 10.1 Automated knitwear fashion design process [42] reused under CC

velvety, smooth, silky, soft, and so on. Fabric texture depends on the weaving techniques. An automated knitwear fashion design process is shown in Fig. 10.1. Any fabric, including linen, cotton, wool, leather, and silk, can have a texture added. In the textile industry, various types of fabric goods are made from textile materials. There are two main categories of textiles: natural and synthetic [38]. AI optimizes parameters for hydrophobic, UV-protective, and antimicrobial coatings to predict surface roughness and coating adhesion after plasma, laser, or enzyme treatments. Surface treatments affect tactile feel, color reflection, and mechanical properties; digital twins and virtual models help simulate the intensity of the effects [41]. Generative AI in fashion design has emphasized addressing general fashion design attributes (such as style, silhouette, pattern, print, and color) that are essential components of the fashion design process [42].

10.7 Data-Driven Approaches in Textile Surface Modification

Data-driven approaches enable the direct extraction of essential process information from large amounts of recorded data. Principal component (PC) analysis (PCA) and partial least squares (PLS) are two examples of process monitoring techniques that are widely used nowadays for process monitoring and fault diagnosis. Above all, the substantial advantage of multivariate statistical approaches is their capacity to handle a large number of strongly correlated variables. Data-driven approaches are frequently used for process monitoring in textile surface-modification treatments [43].

10.7.1 Role of Big Data in AI Models

In the last decade, the combination of AI and big data has transformed textile surface modification. Both have become very important to researchers and manufacturers for designing, analyzing, and improving fabric properties. These technologies enable the precise control and optimization of surface treatments, thereby improving functionalities such as dye uptake, hydrophobicity, antibacterial performance, and wear resistance. Big data has been used in textile surface modification to enable significant developments in the prediction and optimization of process parameters. Large-scale data analysis is used to improve the accuracy of predictive models of surface treatments by integrating process parameters, material composition data, and treatment results to achieve optimal outcomes in textile applications. AI techniques like ML and deep learning may be used to predict fabric behavior, optimize finishing parameters, and enhance product quality and functionality [44].

Big data is an enormous amount of structured, semi-structured, and unstructured data of many different kinds that is produced quickly. Four processes are often involved in big data: data generation, capture, storage, and analysis. Big data requires more real-time analysis than traditional data since it contains enormous volumes of unstructured data of all kinds. To improve big data collection in an IoT system, wireless sensor networks (WSN) using mobile collectors to reduce energy consumption [45]. In color management and surface treatments, data are collected using sensors, spectrophotometers, machine logs, and vision cameras that record various parameters and defect information in real time [46]. Feature extraction and selection techniques have emerged as a vital approach to mitigate two major issues related to big data [47]:

(a) The growth in computational resources required for processing data
(b) Irrelevant information linked to the curse of dimensionality effect.

To improve visualization and facilitate interpretation, feature extraction involves transforming the original data by identifying its inherent patterns. On the other hand, feature selection identifies and selects a subset of the most instructive aspects to improve simplicity and efficiency, leading to better performance of AI models.

10.7.2 Integration of ML and AI with Textile Manufacturing Systems

ML and AI can improve manufacturing efficiency, productivity, and sustainability. The fourth industrial revolution (Industry 4.0) is enabled mainly by the development of ML and AI from computer science theory into practical technologies, as shown in Fig. 10.2, to the extent that it incorporates AI/ML with other cutting-edge technologies to change the industry [48]. AI encompasses a broad spectrum of ML, where the secret to success is [49]:

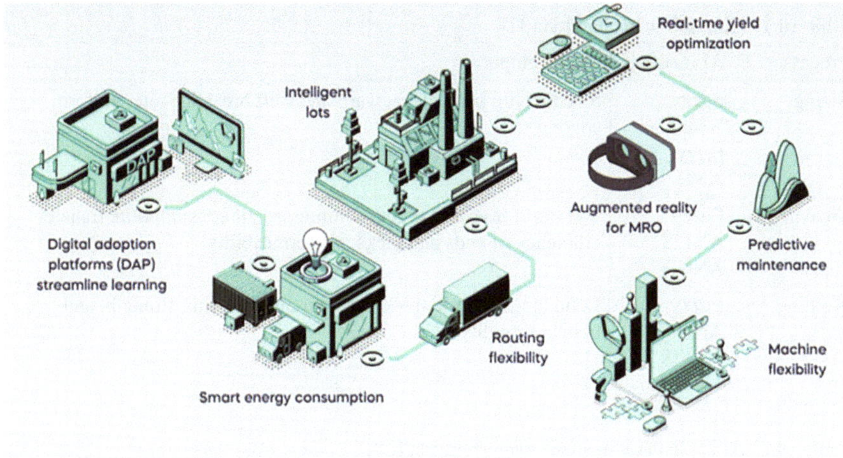

Fig. 10.2 Application of AI in advanced manufacturing [50] reused under CC

(a) Pattern recognition for highly nonlinear data
(b) Unstructured data analysis
(c) Resilience to repetitive work
(d) Fast computation speed
(e) High interpretability.

The desired product quality in yarn manufacturing is ensured by evaluating various physical and mechanical properties of fibers, slivers, roving, and yarns. Typically, this process is intricate, time-consuming, and requires skilled operatives and costly machinery. ANNs seek to uncover the complex functional relationships between fiber and yarn structure. By identifying yarn characteristics, it becomes possible to optimize qualities for diverse applications while also reducing costs [51]. An ANFIS has been developed to predict yarn tenacity and unevenness using cotton fiber parameters like strength, elongation, upper half mean length, uniformity index, fineness, and short fiber content [32]. The application of AI in fabric production is listed in Table 10.1.

AI is integrated to formulate reliable dyes, match colors, and detect defects without human intervention during dyeing and finishing [52]. ANN-based system for printing selects the optimal pigment mixtures. Neural networks can effectively recognize color or pigment combinations in textile printing. An automatic color matching prediction model enhances the capacity to extract high-dimensional characteristics from spectral data by combining a multilayer perceptron (MLP) and a CNN model [23]. The application of AI in smart manufacturing in textiles is listed in Table 10.2.

Table 10.1 AI in fabric production [32]

Process	AI methods	Model output
Sizing	Fuzzy Neuro fuzzy ANN	Size take-up percentage, warp and weft breakage rate, uniform coating
Weaving	Fuzzy ANFIS ANN	Weft yarn insertion speed, compressed air consumption, transfer efficiency of ends and picks, air permeability
Knitting	Fuzzy ANN ANFIS	Loop length, fabric thickness, spirality, bursting strength, and bending rigidity

Table 10.2 AI application in smart manufacturing [50]

AI application area	Description of usage	Outcomes
Predictive maintenance	Monitors machinery vibrations, temperature, and operational data to identify potential failure	Reduce downtime
Quality control	Computer vision algorithms to identify real-time fabric defects with over 90% accuracy	Minimized production waste
Generative and clever pattern design	Analyze millions of layouts to maximize material efficiency	Promoted circular fashion
Intelligent textile and wearable technology	IoT-enabled sensors are embedded into fabrics to allow interactive and responsive functionality	Health monitoring, integrating comfort with reactive solutions
Intelligent material simulation	AI-driven material science platforms	Innovative sustainable textile materials to ensure durability and performance
Automated robotics and intelligent manufacturing	AI-integrated robotic systems to perform precision cutting, sewing	Increased productivity

10.8 Industry Examples

Surface-modification methods are essential processes for improving the properties of textiles to meet specific needs. One of the advanced processes for preparing nanofiber coatings on textile surfaces is electrospinning. Electrospinning is the production of nanofibers from a polymer solution using a high-voltage electric field. The high surface area, tunable porosity, and mechanical flexibility of these nanofibers enable specialized surface interactions for potential applications such as filtration, wound healing, and protective textiles. Electrospinning is a method that can alter the textile surface but not the core material. Process parameters, such as polymer concentration and applied voltage, can be controlled by manufacturers in such a way that there is tight control over the properties of the nanofibers, including diameter and surface energy. Such functionalities make electrospinning an essential method for functional textiles; examples include the properties of water repellency, antimicrobial, and UV protection [53]. Integration of stacked ML and AI algorithms is increasingly embedded in the surface-modification process during electrospinning to achieve optimal results in the minimum time with a minimal number of experiments. AI methods, such as ANNs and SVM-based ML algorithms, have been applied to predict and optimize a variety of textile treatments, thereby reducing or eliminating the need for the conventional trial-and-error approach. AI models, mostly deep learning methods such as CNNs, are applied for predicting the color changes in fabrics in the dyeing process [54]. These models forecast the effects of factors such as dye concentration, fabric type, and temperature on the final color and ensure consistency while minimizing dye waste. In addition, AI streamlines finishing processes by enhancing fabric characteristics such as moisture retention, UV protection, and stain resistance.

Predicting the results of surface treatment like chemical finishing, coating, and surface functionalization using AI. For example, AI algorithms forecast the impact of treatments such as plasma or enzymatic finishing on fabric properties, including dirt barrier, water repellence, and antimicrobial properties. By employing historical data to fine-tune parameters, these models help manufacturers optimize toward their targets while reducing chemical use and environmental impact. ML algorithms are trained on electrospinning experimental data to improve the prediction of optimal process parameters and resulting fiber characteristics. Therefore, this integration helps optimize electrospinning parameters for nanofibers with specialized characteristics, such as diameter, porosity, and surface energy. This electrospinning process, augmented by ML at the outset, affords economic benefits in material, energy, and production costs.

Many of the top textile-producing companies have successfully implemented AI-based Models that improve the surface-modification process. One such case is a European textile manufacturer that incorporated AI color prediction models into its dyeing process, enhancing color consistency and minimizing dye waste by 30% [5]. In the same vein, an ML algorithm tracks how and to what extent a fabric withstands treatment, so that, after passing through a textile finishing company, it can predict the fabric's strength and durability. The company produced analyses of fiber types, chemical treatments, and environmental conditions, which not only helped eliminate defects but also enhanced customer satisfaction.

10.9 Challenges and Limitations

The use of ML and AI in textile surface modification has enabled new ways to optimize process parameters and develop innovative products. In various textile surface-modification treatments, such as plasma treatment, nanocoating, enzyme finishing, and chemical grafting, ML and AI are widely used to build models that predict complex relationships between inputs and outputs. These models help to reduce experimental effort and waste. Though AI/ML models are now widely used in textile surface-modification treatments, they face significant limitations and challenges. These challenges are related:

(a) Complex nature of textile materials
(b) Huge variations in processing conditions
(c) Limited availability of high-quality datasets.

Sometimes, the data that is used to train the ML models is biased, unbalanced, and insufficient to predict the final result [55]. ML and AI models require a large amount of high-quality data. ML algorithms require accurate, noise-free data. In most cases, sufficient labeled data is unavailable. Moreover, data collection is a costly and time-consuming process. Strict data privacy regulations in some treatments limit data availability. Algorithmic bias is a critical issue in AI/ML systems. Biased data from sampling errors, inaccurate labeling, and systemic inequities are part of the data collection process [56]. AI in surface-modification treatments has great potential but also presents many challenges. In capturing the mechanical properties of textiles using AI/ML, the approach faces challenges because the data is often inaccurate. Research shows that while AI-based simulations can potentially streamline manufacturing processes, there exist inconsistencies in accuracy compared to physical measurements [57]. Sensitive data from various textile processes needs to be protected and requires robust cybersecurity measures [32]. The application of ML algorithms to textile surface-modification poses several challenges. These are mainly based on the complexity and variability of the textile materials as they interface with the algorithms [58].

Because textile materials exhibit complex relationships, it is tough to predict the final output using ML algorithms [59]. The different treatment processes might interact in unpredictable ways. Models trained on specific datasets may fail to perform perfectly when exposed to other types or batches of textiles with different properties or compositions. High computational demand is the drawback of applying deep learning to textile-related applications. This includes high costs of data annotation and computational resources for model training and deployment [60].

10.10 Future Trends and Innovations

The future scope of ML and AI in textile surface modification is closely related to the digitalization of the textile industry, with data-based intelligence embedded at every step from recipe design to finishing. Continuous development of new surface treatment processes, such as dyeing, printing, and nano-finishing, will rely on data-mining pipelines that compile vast datasets spanning chemical compositions and environmental conditions, and on self-learning systems that further improve predictive models and decision-making. Traditional AI techniques (ANNs, SVMs, and decision trees) can help predict several material properties of textile materials, while hybrid and advanced learning approaches will emerge in future methods. Deep learning approaches using surface image data, such as convolutional and attention-based models, will advance surface inspection and defect recognition. For processing complex data, ensemble learning methods are expected to be used, and for improving data relatedness, unsupervised methods, such as K-means^{++}, are used. [3]. In addition, sequential patterns will be mined to understand the temporal dependencies in surface quality behavior, while generative models will be used to suggest new production combinations to meet performance goals.

As the need for sustainability grows in textile manufacturing, AI/ML tools will soon be introduced to handle surface alteration, helping reduce water, energy, and chemical consumption during the process without compromising quality. Subsequent studies will couple process optimization with life-cycle assessments to forecast environmental impact in parallel with quality measures. It will also address the problem of "*small textile data*" in these datasets by applying dimensionality reduction and transfer learning to improve model accuracy with small datasets. AI will enable circular-economy initiatives to assess the recyclability of surface treatments. AI will play a role in the development of smart textiles that enable surfaces to act as interactive interfaces capable of sensing and communicating. So, they will help develop intelligent improvements using conductive materials and sensors to measure health and performance in a predictive way. Ontologies developed for textiles will help extract knowledge to improve the design and integration of bright surfaces.

The evolution of AI into surface-modification systems will lead to increasingly sophisticated, sensor-network-based, and analytics-enabled services for real-time quality monitoring and automation, assisting automated decision-making to improve defect size detection and minimize waste. While text mining extracts valuable information from unstructured text data for effective decision-making, process mining reveals hidden inefficiencies in production processes. These would probably lead to the final digital twin platforms providing a virtual model of production environments, in which new strategies can be simulated before implementation.

Ultimately, the impact of ML and AI in textile alteration will depend not only on the development of more algorithms but also on their integration into human-centered workflows. Engineers need model recommendations to understand them, and to do that, they need a trusted environment for data sharing and governance frameworks. Instead, face those challenges to drive an underlying paradigm shift in surface modification, enabling next-generation textiles for sustainable, customized functional performance.

10.11 Conclusion

ML and AI presented new solutions to centuries, if not millennia, of manufacturing problems, bringing excellent efficiency to many aspects of the process from production to application, as well as ensuring the field becomes increasingly sustainable while still enabling the full potential in surface customization in the future. From dyeing and printing to finishing and coating, ML and AI are redefining the future of manufacturing by optimizing surface treatments. These technologies are increasingly used in the textile industry to enhance product quality, reduce costs, and minimize environmental impact. Analyzing data from previous manufacturing cycles, ML and AI enable accurate control of textile treatments by predicting based on the chemical composition and processing conditions of the desired color intensity, fabric strength, and durability. The trial-and-error methods waste resources, prolong lead times, and increase the risk of failure. This capability removes them.

ML and AI enable real-time process monitoring and optimization through smart sensors and IoT. Using this, manufacturers can track key variables during treatments to minimize variability and improve productivity and efficiency. When errors, inefficiencies, or deviations from the expected process are detected in real time, production systems respond more quickly. Developing new bounds for textile modification through ML and AI, as well as for process optimization. Output from generative adversarial networks (GANs), which simulate treatments or textile patterns and have thus also been used to produce advanced textiles with antibacterial and self-cleaning properties. With the arrival of innovative materials that offer functionality comes the opportunity for AI to flex its muscle in fields such as sensor-based textiles that can pave the way for wearable intelligence and health-monitoring textiles.

However, these developments cannot be directly transposed to textile surface modification, as the extreme heterogeneity of textile materials hampers the generalization of models across fiber and treatment types. Adaptive AI algorithms, data availability, including privacy, and computing resources are barriers to practical applications. These issues must be addressed before the textile sector can successfully adopt it.

The future of AI in surface modification, enabled by big data analytics and smart sensors, will enhance the accuracy and efficiency of industrial processes. AI-driven resource optimization and waste reduction solutions will promote recycling and material reuse as the sector prioritizes sustainability. ML and AI will transform textile production, creating new potential for innovation and environmental stewardship.

AI-usage Declaration

During the preparation of this chapter, the authors used AI tools to improve the language of the manuscript. After using this tool/service, the authors reviewed and edited the content as needed and take full responsibility for the content of the publication.

References

1. J. Qin et al., AI-assisted dyeing optimization of polyimide fibers via backpropagation neural networks. Prog. Org. Coat. **209**, 109611 (2025). https://doi.org/10.1016/j.porgcoat.2025.109611
2. J. Ren, K. Hua, H. Trajano, Y. Cao, Global optimal explainable models for biorefining, in *Computer Aided Chemical Engineering*, vol. 52, ed. by A.C. Kokossis, M.C. Georgiadis, E. Pistikopoulos (Elsevier, 2023), pp.1339–1346
3. P. Yildirim, D. Birant, T. Alpyildiz, Data mining and machine learning in the textile industry. WIREs Data Min. Knowl. Discov. **8**(1), e1228 (2018). https://doi.org/10.1002/widm.1228
4. M.N. Morshed, Understanding mechanical finishing techniques in textile processing, in *Advancements in Textile Finishing: Techniques, Technologies, and Trends*. ed. by M. Shahid, S. Biranje, M. Yusuf, R. Adivarekar (Singapore, Springer Nature Singapore, 2025), pp.1–26
5. N. Ingle, W.J. Jasper, A review of deep learning and artificial intelligence in dyeing, printing and finishing. Text. Res. J. **95**(5–6), 625–657 (2024). https://doi.org/10.1177/00405175241268619
6. A. Zille, F.R. Oliveira, A.P. Souto, Plasma treatment in textile industry. Plasma Process. Polym. **12**(2), 98–131 (2015). https://doi.org/10.1002/ppap.201400052
7. M. Ayesh, A.R. Horrocks, B.K. Kandola, The impact of atmospheric plasma/UV laser treatment on the chemical and physical properties of cotton and polyester fabrics. Fibers **10**(8), 66 (2022). https://doi.org/10.3390/fib10080066
8. A.Z.M. Mofasser, B.C. Mazumder, I.K. Alam, N. Ibn Sayed, M.S.N. Mazumder, R. Mia, Application of nanoparticles in textile coloration, in *Nanoparticles Integrated Functional Textiles*. ed. by M.R. Repon, D. Mikučioniene, A. Haji (Springer Nature Switzerland, Cham, 2025), pp.387–416
9. M.T. Raihan, M.H. Mahmud, B.C. Mazumder, M.N.H. Chowdhury, M.T. Islam, A review of PEO (polyethylene oxide) assisted electrospinning of chitosan: innovation, production, and application. J. Polym. Mater. **42**(3), 677–711 (2025). https://doi.org/10.32604/jpm.2025.068356
10. E. Abdel-Aziz, M. Bakr, M. Zayed, H. Othman, A.G. Hassabo, Microencapsulation and its application in textile wet processing: a review. J. Text. Color. Polym. Sci. **19**(2), 189–202 (2022). https://doi.org/10.21608/jtcps.2022.140954.1121
11. W.N.W. Ismail, Sol–gel technology for innovative fabric finishing—a review. J. Sol Gel Sci. Technol. **78**(3), 698–707 (2016). https://doi.org/10.1007/s10971-016-4027-y
12. C. Makhlouf, S. Marais, S. Roudesli, Graft copolymerization of acrylic acid onto polyamide fibers. Appl. Surf. Sci. **253**(12), 5521–5528 (2007). https://doi.org/10.1016/j.apsusc.2006.12.086
13. A. Nadi, A. Boukhriss, A. Bentis, E. Jabrane, S. Gmouh, Evolution in the surface modification of textiles: a review. Text. Prog. **50**(2), 67–108 (2018). https://doi.org/10.1080/00405167.2018.1533659
14. M. Mazloumpour, P. Malshe, A. El-Shafei, P. Hauser, Conferring durable antimicrobial properties on nonwoven polypropylene via plasma-assisted graft polymerization of DADMAC. Surf. Coat. Technol. **224**, 1–7 (2013). https://doi.org/10.1016/j.surfcoat.2013.02.022
15. N. Ingle, W.J. Jasper, A review of the evolution and concepts of deep learning and AI in the textile industry. Text. Res. J. **95**(13–14), 1709–1737 (2025). https://doi.org/10.1177/00405175241310632
16. S. Malik, K. Muhammad, Y. Waheed, Artificial intelligence and industrial applications-a revolution in modern industries. Ain Shams Eng. J. **15**(9), 102886 (2024). https://doi.org/10.1016/j.asej.2024.102886

17. N. El Messaoudi et al., The role of artificial intelligence in optimizing photocatalytic degradation technologies of dyes in textile wastewater: recent advances, challenges, and prospects. J. Water Process. Eng. **77**, 108457 (2025). https://doi.org/10.1016/j.jwpe.2025.108457
18. S. Bbumba et al., Prediction and optimization of process parameters using artificial intelligence and machine learning models. Asian J. Appl. Chem. Res **16**, 11–33 (2025). https://doi.org/10.9734/ajacr/2025/v16i1317
19. M. Seçkin, A.Ç. Seçkin, P. Demircioglu, I. Bogrekci, FabricNET: a microscopic image dataset of woven fabrics for predicting texture and weaving parameters through machine learning. Sustainability **15**(21), 15197 (2023). https://doi.org/10.3390/su152115197
20. D. Essa, 04. Recent techniques of surface modification for textile fabrics. J. Meas. Sci. Appl. (JMSA) **4**(1), 66–87 (2024). https://doi.org/10.21608/jmsa.2024.360350
21. S. Bhambure, A. Dhavale, P. Kadole, D. Kodavade, Artificial neural network & its applications in textiles. J. Text. Assoc. **74**(1), 31–37 (2013)
22. S. Arora, A. Majumdar, Machine learning and soft computing applications in textile and clothing supply chain: bibliometric and network analyses to delineate future research agenda. Expert Syst. Appl. **200**, 117000 (2022). https://doi.org/10.1016/j.eswa.2022.117000
23. İC. Şahin, C. Eyüpoğlu, Textile dyeing process and dyeing recipe prediction using artifical intelligence. İstanbul Ticaret Üniversitesi Teknoloji ve Uygulamalı Bilimler Dergisi **6**(2), 1–20 (2024)
24. M. Mirzaei, I. Furxhi, F. Murphy, M. Mullins, A supervised machine-learning prediction of textile's antimicrobial capacity coated with nanomaterials. Coatings **11**(12), 1532 (2021). https://doi.org/10.3390/coatings11121532
25. I. Malashin et al., Machine learning in polymeric technical textiles: a review. Polymers **17**(9), 1172 (2025). https://doi.org/10.3390/polym17091172
26. G. Ferreira et al., Energy harvesting and movement tracking by polypyrrole functionalized textile for wearable IoT applications. J. Energy Chem. **102**, 230–242 (2025). https://doi.org/10.1016/j.jechem.2024.10.028
27. K. Yılmaz, İÖ. Aksu, M. Göçken, T. Demirdelen, Sustainable textile manufacturing with revolutionizing textile dyeing: deep learning-based, for energy efficiency and environmental-impact reduction, pioneering green practices for a sustainable future. Sustainability **16**(18), 8152 (2024). https://doi.org/10.3390/su16188152
28. C.-C. Wang, Y.-H. Li, Machine-learning-based system for the detection of entanglement in dyeing and finishing processes. Sustainability **14**(14), 8575 (2022). https://doi.org/10.3390/su14148575
29. L. Manjakkal, W. Dang, N. Yogeswaran, R. Dahiya, Textile-based potentiometric electrochemical pH sensor for wearable applications. Biosensors **9**(1), 14 (2019). https://doi.org/10.3390/bios9010014
30. Y.K. Park et al., Colorimetric textile sensor for the simultaneous detection of NH_3 and HCl gases. Polymers **12**(11), 2595 (2020). https://doi.org/10.3390/polym12112595
31. N. Trifigny, F.M. Kelly, C. Cochrane, F. Boussu, V. Koncar, D. Soulat, PEDOT: PSS-based piezo-resistive sensors applied to reinforcement glass fibres for in situ measurement during the composite material weaving process. Sensors **13**(8), 10749–10764 (2013). https://doi.org/10.3390/s130810749
32. M.P. Sikka, A. Sarkar, S. Garg, Artificial intelligence (AI) in textile industry operational modernization. Res. J. Text. Appar. **28**(1), 67–83 (2022). https://doi.org/10.1108/RJTA-04-2021-0046
33. Reenu, S. Singh, S. Vig, S. Dwivedi, Impact of uses of artificial intelligence (AI) in textile industry in India, in *2024 15th International Conference on Computing Communication and Networking Technologies (ICCCNT)*, 24–28 June 2024, pp. 1–9. https://doi.org/10.1109/ICCCNT61001.2024.10725995
34. A.K. Singh, A. Sharma, P.K. Singh, S. Kesarwani, A.P. Singh, Internet of things and sensor networks in industry 5.0: connecting devices and machines, in *Emerging Technologies in Digital Manufacturing and Smart Factories*. ed. by A. Hassan, P.K. Dutta, S. Gupta, E. Mattar, S. Singh (IGI Global Scientific Publishing, Hershey, PA, USA, 2024), pp.67–78

35. R. Kumar, S. Lalitha, R. Bharathy, G. Mariappan, IOT (Internet of things) smart applications in textiles. Asian Text. J. **5**, 51–55 (2021)
36. G.B. Ramaiah, Theoretical analysis on applications aspects of smart materials and Internet of Things (IoT) in textile technology. Mater. Today Proc. **45**, 4633–4638 (2021). https://doi.org/10.1016/j.matpr.2021.01.023
37. W. Shi, J. Cao, Q. Zhang, Y. Li, L. Xu, Edge computing: vision and challenges. IEEE Internet Things J. **3**(5), 637–646 (2016). https://doi.org/10.1109/JIOT.2016.2579198
38. F. Pereira, V. Carvalho, R. Vasconcelos, F. Soares, A review in the use of artificial intelligence in textile industry, in *Innovations in Mechatronics Engineering*. ed. by J. Machado, F. Soares, J. Trojanowska, S. Yildirim (Springer International Publishing, Cham, 2022), pp.377–392
39. B. Tang, L. Chen, W. Sun, Z.-K. Lin, Review of surface defect detection of steel products based on machine vision. IET Image Process. **17**(2), 303–322 (2023). https://doi.org/10.1049/ipr2.12647
40. F. Al-Akayleh, A.S.A. Ali-Agha, R.A. Abdel-Rahem, M. Al-Remawi, A mini review on the applications of artificial intelligence (AI) in surface chemistry and catalysis. Tenside Surfactants Deterg. **61**(4), 285–296 (2024). https://doi.org/10.1515/tsd-2024-2580
41. M.M. Rathore, S.A. Shah, D. Shukla, E. Bentafat, S. Bakiras, The role of AI, machine learning, and big data in digital twinning: a systematic literature review, challenges, and opportunities. IEEE Access **9**, 32030–32052 (2021). https://doi.org/10.1109/ACCESS.2021.3060863
42. X. Wu, L. Li, An application of generative AI for knitted textile design in fashion. Des. J. **27**(2), 270–290 (2024). https://doi.org/10.1080/14606925.2024.2303236
43. S. Yin, S.X. Ding, X. Xie, H. Luo, A review on basic data-driven approaches for industrial process monitoring. IEEE Trans. Industr. Electron. **61**(11), 6418–6428 (2014). https://doi.org/10.1109/TIE.2014.2301773
44. A. Paramasivan, Big data to better care: the role of AI in predictive modelling for healthcare management. Int. J. Innov. Res. Creat. Technol. **6**(3), 1–9 (2020). https://doi.org/10.5281/zenodo.14551652
45. A.C. Djedouboum, A.A. Abba Ari, A.M. Gueroui, A. Mohamadou, Z. Aliouat, Big data collection in large-scale wireless sensor networks. Sensors **18**(12), 4474 (2018). https://doi.org/10.3390/s18124474
46. S. Liu, Y.K. Liu, K.-Y.C. Lo, C.-W. Kan, Intelligent techniques and optimization algorithms in textile color management: a systematic review of applications and prediction accuracy. Fash. Text. **11**(1), 13 (2024). https://doi.org/10.1186/s40691-024-00375-x
47. D. Ruano-Ordás, Machine learning-based feature extraction and selection. Appl. Sci. **14**(15), 6567 (2024). https://doi.org/10.3390/app14156567
48. S.J. Plathottam, A. Rzonca, R. Lakhnori, C.O. Iloeje, A review of artificial intelligence applications in manufacturing operations. J. Adv. Manuf. Process. **5**(3), e10159 (2023). https://doi.org/10.1002/amp2.10159
49. S.W. Kim, J.H. Kong, S.W. Lee, S. Lee, Recent advances of artificial intelligence in manufacturing industrial sectors: a review. Int. J. Precis. Eng. Manuf. **23**(1), 111–129 (2022). https://doi.org/10.1007/s12541-021-00600-3
50. A.B. Anik Biswas, AI-driven industrial innovation: transforming fashion, textile, and advanced manufacturing for the digital economy. AI Driven Ind. Innov. Transform. Fash. Text. Adv. Manuf. Digit. Econ. **2**(11), 36–46 (2025). https://doi.org/10.31149/ajta.v2i11.2858
51. J. Hossen et al., Application of artificial neural networks in yarn manufacturing industry: a critical review. J. Eng. Fibers Fabr. **20**, 15589250251329416 (2025). https://doi.org/10.1177/15589250251329415
52. M. Elkhaoudi, M. El Bakkali, R. Messnaoui, O. Cherkaoui, A. Soulhi, Literature review on artificial intelligence in dyeing and finishing processes. Data Metadata **3**, 360–360 (2024). https://doi.org/10.56294/dm2024360
53. Y. Al-Saif, R. Cselkó, Revolutionizing electrospinning: a review of alternating current and pulsed voltage techniques for nanofiber production. Processes **13**(7), 2048 (2025). https://doi.org/10.3390/pr13072048

54. J. Zhang, X. Zhang, J. Wu, C. Xiao, Dyeing recipe prediction of cotton fabric based on hyperspectral color measurement and an improved recurrent neural network. Color. Technol. **137**(2), 166–180 (2021). https://doi.org/10.1111/cote.12516
55. E. Barbierato, A. Gatti, The challenges of machine learning: a critical review. Electronics **13**(2), 416 (2024). https://doi.org/10.3390/electronics13020416
56. N. Zhou, Z. Zhang, V.N. Nair, H. Singhal, J. Chen, A. Sudjianto, Bias, fairness, and accountability with ML and AI Algorithms. arXiv preprint arXiv:2105.06558 (2021)
57. S. Youn, A. West, K. Mathur, Evaluation of a new artificial intelligence-based textile digitization using fabric drape. Text. Res. J. **94**(17–18), 2001–2018 (2024). https://doi.org/10.1177/00405175241236881
58. C. Duan, A. Nandy, H.J. Kulik, Machine learning for the discovery, design, and engineering of materials. Annu. Rev. Chem. Biomol. Eng. **13**, 405–429 (2022). https://doi.org/10.1146/annurev-chembioeng-092320-120230
59. Y.-F. Tu, M.-Y. Kwan, K.-L. Yick, A systematic review of AI-driven prediction of fabric properties and handfeel. Materials **17**(20), 5009 (2024). https://doi.org/10.3390/ma17205009
60. Z.-Y. Yang, W.-K. Xia, H.-Q. Chu, W.-H. Su, R.-F. Wang, H. Wang, A comprehensive review of deep learning applications in cotton industry: from field monitoring to smart processing. Plants **14**(10), 1481 (2025). https://doi.org/10.3390/plants14101481

MIX
Papier aus verantwortungsvollen Quellen
Paper from responsible sources
FSC® C105338

If you have any concerns about our products,
you can contact us on
ProductSafety@springernature.com

In case Publisher is established outside the EU,
the EU authorized representative is:
Springer Nature Customer Service Center GmbH
Europaplatz 3, 69115 Heidelberg, Germany

Printed by Libri Plureos GmbH
in Hamburg, Germany